Progress in Colloid and Polymer Science · Volume 127 · 2004

Progress in Colloid and Polymer Science

Recently Published and Forthcoming Volumes

Surface and Colloid Science
Volume Editor: Galembeck, F.
Vol. 128, 2004

Analytical Ultracentrifugation VII
Volume Editors: Lechner, M. D., Börger, L.
Vol. 127, 2004

Trends in Colloid and Interface Science XVII
Volume Editors: Cabuil, V., Levitz, P., Treiner, C.
Vol. 126, 2004

From Colloids to Nanotechnology
Volume Editors: Zrinyi, M., Hórvölgyi, Z. D.
Vol. 125, 2004

Aqueous Polymer Dispersions
Volume Editor: Tauer, K.
Vol. 124, 2004

Trends in Colloid and Interface Science XVI
Volume Editors: Miguel, M., Burrows, H. D.
Vol. 123, 2004

Aqueous Polymer - Cosolute Systems
Volume Editor: Anghel, D. F.
Vol. 122, 2002

Molecular Organisation on Interfaces
Volume Editor: Lagaly, G.
Vol. 121, 2002

Lipid and Polymer-Lipid Systems
Volume Editor: Nylander, T.
Vol. 120, 2002

Progress in Colloid and Polymer Science
Editors: F. Kremer, Leipzig and W. Richtering, Aachen

Volume 127 · 2004

Analytical Ultracentrifugation VII

Volume Editors:
M.D. Lechner
L. Börger

Springer

The series Progress in Colloid and Polymer Science is also available electronically (ISSN 1437-8027)

- Access to tables of contents and abstracts is *free* for everybody.
- Scientists affiliated with departments/institutes subscribing to Progress in Colloid and Polymer Science as a whole also have full access to all papers in PDF form. Point your librarian to the LINK access registration form at http://www.springerlink.com

ISSN 0340-255X
ISBN 978-3-662-14525-8
ISBN 978-3-540-45197-6 (eBook)
DOI 10.1007/978-3-540-45197-6

http://www.springeronline.com

Originally published Springer-Verlag Berlin Heidelberg New York in 2004
Softcover reprint of the hardcover 1st edition 2004

Typesetting: SPS, Chennai, India

Cover: Estudio Calamar,
F. Steinen-Broo, Pau/Girona, Spain

Printed on acid-free paper

Progr Colloid Polym Sci (2004) 127: V

PREFACE

Symposia on Analytical Ultracentrifugation take place every two years at different places in Germany. Since their initiation in 1978 these meetings have developed to be an inherent part of the scientific community concerning Ultracentrifugation. Starting as mainly national meetings not just the number of nationalities of participants, but also the number and range of topics covered by the contributions has increased since then. This fact once more shows the general power of an often underestimated analytical technique resident in the exciting intersection between chemistry, biology and physics.

Following this tradition the 13th meeting took place at the university of Osnabrück from March 6 to 7, 2003. The meeting was joined by more than 50 participants from groups from different countries from all over the world. 20 oral and several poster presentations were given. Of course the interesting discussions as well as establishing new contacts between scientists were essential parts of the meeting. The presentations, posters and discussions were in the frame

- Technical Methods, Data Analysis, Innovations
- Polymers, Colloids, Supramolecular Systems
- Biological and Interaction Systems
- Hydrodynamics and Modelling

19 contributions were selected for publication in this special volume of "Progress in Colloid and Polymer Science". F. Kremer and Dr. M. Hertel from Springer-Verlag are gratefully acknowledged.

The Volume editors as well as most contributors are already looking forward to the 14th meeting that will take place in the beginning of 2005 in Lausanne, Switzerland.

The 13th symposium was kindly sponsored by Beckman Coulter GmbH (Unterschleißheim), BASF AG (Ludwigshafen), Bayer AG (Leverkusen) and Universitätsgesellschaft (Osnabrück). The symposium took place under the patronage of BASF AG, Ludwigshafen.

L. Börger (Ludwigshafen), M. D. Lechner (Osnabrück)

Progr Colloid Polym Sci (2004) 127: VI–VIII

CONTENTS

Progr Colloid Polym Sci (2004) 127: 1–8
DOI 10.1007/b94237

A.J. Rowe

Determination of the absolute fringe number in Rayleigh interference patterns in systems at sedimentation equilibrium via a newly defined function ($k\Delta$)

A.J. Rowe
NCMH Business Centre,
School of Biosciences,
University of Nottingham,
Sutton Bonington,
Leicestershire LE12 5RD, UK
e-mail: arthur.rowe@nottingham.ac.uk
Tel.: +44-115-9516156
Fax: +44-115-9516157

Abstract Analysis of data from the analytical ultracentrifuge used in sedimentation equilibrium modes calls for knowledge of the absolute (as distinct from relative) final solute mass distribution as a function of radial distance. This poses particular problems when Rayleigh interference optics are used, as the resultant fringe patterns provide estimates only for the relative concentration values. Methods for circumventing this problem are discussed, and their limitations described. Computer simulation is used to demonstrate the limitations of the most popular approach ("floating" the offset value in nonlinear fitting). A new approach to tackling this problem is defined, which is not limited by the number of components present, nor by their interactions if any. A function $k\Delta(E)$ is shown to have the value zero only when an assumed offset (E) in the values acquired is equal to the "genuine" offset. A "proof of concept numerical algorithm" has been constructed, which shows that the new approach is capable in principle of yielding offset values of a precision similar to that of the data set.

Keywords Interference optics · Analytical ultracentrifuge · Sedimentation equilibrium · Baseline offset

Introduction

The analysis of the mass distribution of solute with respect to radial distance forms the basis of the technique of sedimentation equilibrium analysis. Rayleigh interference optics provide the most precise (although not necessarily the most sensitive) approach to the practical measurement of that solute distribution. This optical system suffers, however, from one serious limitation: the signal recorded is the fringe shift (relative to some selected radial position), whereas for computational purposes it is the absolute fringe number which is required – at least for sedimentation equilibrium (in contrast to sedimentation velocity) experiments. Thus, at every radial position, r, there is an offset or "baseline" quantity, E, which has to be added on to the relative fringe increment, j_r, at that radius in order to yield the true fringe number, J_r. The same problem can arise when absorption optics are employed, especially in the presence of optically absorbing cosolutes of low mass.

The problem has been tackled over the years in a number of ways. Clearly, if there is a region in the solution column where there is total depletion of macromolecular solutes, then $J_r = 0$ at that point by definition and the problem vanishes. This "meniscus depletion" method [1] has serious limitations. In particular, the evaluation of data places the greatest emphasis on the region where errors are highest, i.e. in the upper part of the cell, where fringe increments are lowest, and there is strong selection of components in favour of those of lower-than-average mass, with the higher-mass component being either heavily depleted or even pelleted. Attempts have thus been made to devise specific procedures applicable to low or intermediate speed equilibrium [2, 3, 4]. These procedures are reviewed here, and their merits and limitations are appraised. Alternatively, E has been simply "floated" as a parameter in nonlinear least-squares analysis [5]. This latter approach has been by far the most popular. Unfortunately, as we now confirm by computer simulation experiments, it is a hazardous approach, which can in many circumstances lead to incorrect conclusions being drawn.

The question arises as to whether it is actually possible even in theory and with "perfect" data to determine an accurate value for E, the "offset" value which must be added on to all relative fringe shift readings in order to yield a data set comprising true fringe numbers as a function of radial distance, r. In the format in which data are usually analysed [6], the equilibrium distribution is given as the sum of a potentially infinite set of exponential terms, from which the deconvolution of E might be expected to be complex. This question actually received a positive answer in principle some time ago [3], albeit with an approach which when implemented as a practical procedure makes poor use of a data set. We now derive a formulation which shows that E can be estimated via the use of a function of $E - k\Delta(E)$ – which makes use of the full range of a data set. We further describe a preliminary approach to the practical use of this function via a "proof of concept numerical algorithm" (PCNA).

Methods

Computer simulation was carried out using pro Fit software (Quantum Soft, Zurich) running on an Apple G4 computer. Appropriate modules for sedimentation equilibrium analysis were written using the pro Fit inbuilt language, which is similar to Pascal. The equations used followed the generally accepted format [6]. These modules also serve to generate simulated data. A module ADDERROR was used to add simulated "noise" with a Gaussian distribution to initially generated "perfect" data sets. It has been demonstrated that the results obtained from the use of these modules to fit data sets are identical to those obtained when software such as NONLIN [5] is employed. The fact that files stored by the Beckman XL-A software can be opened directly within pro Fit has made such comparison simple.

Theory and simulations

We use the following conventions. The baseline offset (as defined later) is denoted by the symbol E. The difference between an estimated baseline and the true value of E is denoted by E'. Following an earlier convention, and to avoid confusion with a common statistical parameter (σ), we use the symbol k to denote the reduced buoyant molecular weight, as in the basic equation for sedimentation equilibrium derived from the equivalence at all r of the chemical and the sedimentation potential:

$$dc/dr = kcr, \tag{1}$$

where c is the concentration at radial position r (subscripts in Eq. 1 are omitted for generality but are added to c and k as and when necessary).

Currently available methods for the estimation of a value for E

E is the numerical factor which must be added (algebraically) to all values of the relative "concentration" values logged by the ultracentrifuge software. In this context, concentration is used to denote any quantity (fringe increment, optical absorbance) which can be treated as being – at least to within acceptable error – a linear function of concentration. It is convenient to discuss firstly the range of specific methods which have been developed for the estimation of E. Then, secondly, we consider the extent to which it is valid to circumvent the problem of "unknown E" by simply treating it as a parameter to be floated in nonlinear fitting algorithms. It should be noted that we do not use the term nonlinear least squares as we make regular use of a Robust fitting algorithm [7].

Specific procedures for the estimation of E

There are three main approaches which have been developed over the years:

Fringe tracking during overspeed

Overspeeding after attainment of the equilibrium distribution has long been used in conjunction with absorbance optics to assess the location of the baseline after all (macromolecular) solutes have been pelleted. Whilst its use is only feasible when the solute under investigation has an s value sufficiently high for pelleting to occur, this approach copes well with those cases where a much lower mass cosolute is present and contributes a background absorbance. It should be pointed out that the estimate for E should be made at a radial position in the centre of the solution column, to allow for any redistribution of such a cosolute. An obvious limitation of this method is that the optical conditions (e.g. exact window alignment) may not be identical at high as opposed to low rotor speed, but this consideration is usually ignored.

The method has not been used to any extent with Rayleigh interferometric optics. Here, the effect of window distortion is perceived to be significant, and in any case a simple overspeed would lose the integral part of the fringe number values. This latter problem is unique to interference optics. However, it was pointed out by Rowe [8] that providing one was willing to "track fringes" during both the acceleration phase of overspeed and subsequent solute depletion, then the integral part of the fringe number could be retained. The effect of window distortion was allowed for by a back extrapolation to zero post-acceleration reduced time ($\omega^2 t$, where ω is the angular rotor velocity and t is the elapsed time). Providing that a radial position in the central region was selected for the fringe tracking, a very simple but adequate approximation to the Lamm equation showed that this back extrapolation would be linear in time, using data from the time domain following the completion of acceleration [9]. Practical application of this method to 16S rRNA [9] showed that this latter expectation was met, and good results could be obtained.

Obviously this method, however sound in theory, was extremely tedious to use in the days when data were logged to film and subsequently scanned. There would be no problem in acquiring time sequence data in a contemporary analytical ultracentrifuge (Beckman XL-I). Furthermore, the much rapider acceleration time of the modern instrument would mean that the necessary back extrapolation would be a very short one indeed, and hence errors should be minimal. This approach deserves contemporary use. However, one cannot circumvent the major limitation noted earlier for absorption optics: if the solute under investigation cannot be pelleted, the method cannot be applied.

Function extrapolation

Creeth and Harding [3] showed that a novel average molecular mass, M^*, could be defined, which had useful properties in evaluating weight-averaged molecular weights for polydisperse systems. Whilst these properties are not relevant to the present discussion, these authors also showed that in practical implementation of their M^* approach (k^* in our terms), an extrapolation could be performed which enabled E to be estimated at the meniscus position. From the following plot [3]

$$j/(r^2 - a^2) \text{ versus} \int rj \cdot \mathrm{d}r/(r^2 - a^2), \qquad (2)$$

where j is the fringe increment at r, and a the radial value at the meniscus, it could be shown that the following relationships held at the meniscus:

$$\text{limiting slope} = 2k_a^*, \quad \text{intercept} = k_a^* J_a, \qquad (3)$$

and hence J_a, the true fringe increment at the meniscus, is simply given by the ratio of the limiting slope to the intercept of the plot described by Eq. (2).

This method has limitations in practical application. As the authors point out, and as was discussed later [4], there is no simple and mathematically sound way in which the form of the extrapolation can be defined. This requires that the extrapolation should be a "short" one, and this was achieved by the use of "intermediate" rotor speeds to give extensive (although not complete) solute depletion at the meniscus. One is thus again faced with the problem of reliance upon only a small part of the data set, that part being the most error-prone. Precision is thereby restricted to around $\pm 15\%$ [3].

However, a very important – although generally unremarked – feature of this work [3] is that it clearly demonstrates that a solution to the problem of the determination of E is actually possible in mathematical terms for any system no matter how polydisperse. This result is neither obvious nor trivial, and is in no way vitiated by the limitations in precision of the practical method suggested.

Use of an accessory synthetic boundary experiment

Providing that the solution under study has been rigorously dialysed against solvent, then a single synthetic boundary run establishes the total fringe count associated with the loading concentration of the solution. This is a long-known historical approach. In recent work its use has been revived [4]. These authors followed a classical approach in assigning the fringe count thus established to the "hinge point" in the solution column, i.e. the point at which the total solute concentration remains unchanged with time and equals the cell loading concentration. The hinge point is identified by accessory use of the absorption optics to identify the radial position at which the absorption is equal to the original plateau level.

There are problems which limit the applicability of this approach. Interference optics are often used because of the lack of a suitable solute chromophore; hence, the method given for identifying the hinge point must fail in all such cases. The assumption of conservation of mass may not be valid, and even if it is, then in a multicomponent system where there are variations in extinction coefficient, the hinge point will not be correctly located. Finally, it is often not feasible to dialyse solutes, because of their limited availability or low mass (leading to membrane permeability).

Is it valid simply to float E in nonlinear fitting methods?

This is by far the simplest approach to the "baseline problem". All one has to do is to select E as a parameter to be floated, in a fitting procedure which one would be doing in any case [5]. Assuming that plausible initial guesses have been selected for a (sensibly limited range of) parameters floated, the fit is normally stable. If one has used a fitting algorithm such as Levenberg–Marquardt (nonlinear least squares) which returns error estimates for the parameters, these often appear plausible.

There is just one basic problem with this approach: the estimates for the parameters can often be "wrong". By this we mean that an apparently plausible estimate is given for (say) the reduced buoyant molecular weight, but that estimate differs from the "true" value by more than can reasonably be attributed to experimental error. How can this happen? It is a phenomenon which we have explored using computer simulation, in which standard fitting algorithms are applied to sets of "perfect" data, normally perturbed by the addition of "noise". The latter is fixed at a level consonant with experimental experience; it is constant in an average level with respect to radius, and is normally (Gaussian) distributed.

It is a basic requirement of all nonlinear (or indeed other) fitting procedures that the model used (i.e. the equation employed) should be correct, i.e. it must accurately represent the system under study. Fitting a data set with an incorrect model may well yield numerical values for the (incorrectly specified) parameter set: but those numerical values will not accurately represent the state of the system.

So far, the argument may appear trivial. Surely any serious departure from the assumed model will be detected by a simple look at the distribution of the residuals between the fit and the data set [6] Does polydispersity not give rise to a "tick"-shaped distribution, whilst nonideality results in a distribution showing downward curvature at both ends [6]? The answer to these question is positive – but only if E has been correctly specified. Consider two simulated examples (Figs. 1, 2). In Fig. 1 we see the results of a nonlinear fit to a data set generated on the basis of a single solute component being present, with a thermodynamic nonideality term (2BM) equal to 40 ml/g (the ordinate values being in grams per millilitre with a root-mean-square error of ± 0.00005 g/ml), and with $E=0$. The fit used assumed the presence of a single ideal component, and E was floated. An apparently excellent fit results – but the estimates for the parameters are nowhere near the correct values (Fig. 1). Thus even though the error simulated was at least as low as could be attained in actual experimental work, the outcome is without value. Only if one forces E to have its true value does the inapplicability of the model used become apparent (Fig. 1). But this assumes that one knows in advance a value for E.

In Fig. 2 we see the results of applying the same simple model (one ideal component) to a highly polydisperse system (Fig. 2), again with E floated. Here the buoyant mass values cover no less than a fourfold range of size.

Fig. 1 Fit to a set of simulated data. The data set parameters used were $c_i=0.002$, $k=1$, $2BM=40$, $E=0$, radial distances from 6.8 to 7.0 cm. Simulated noise of ± 0.00005 was added. The fit used was for a single species, ideal model, with c_i, k and the baseline floated. The residuals between the fit and the points used are plotted above. Values estimated for $c_i(=a[1])$, $k(=a[2])$ and the baseline ($=a[5]$) are shown on the *right-hand side* of the plot of the residuals. The fit is good, but the parameter estimates are very poor (k_{app} at the hinge point is expected to be around 0.87). When the baseline ($=a[5]$) is fixed at its true value (plot not shown, residuals and fit values displayed at the *top*), the parameter values are very much better, but the plot of the residuals (inverted in accord with customary convention) now shows the downward curvature typical of nonideality

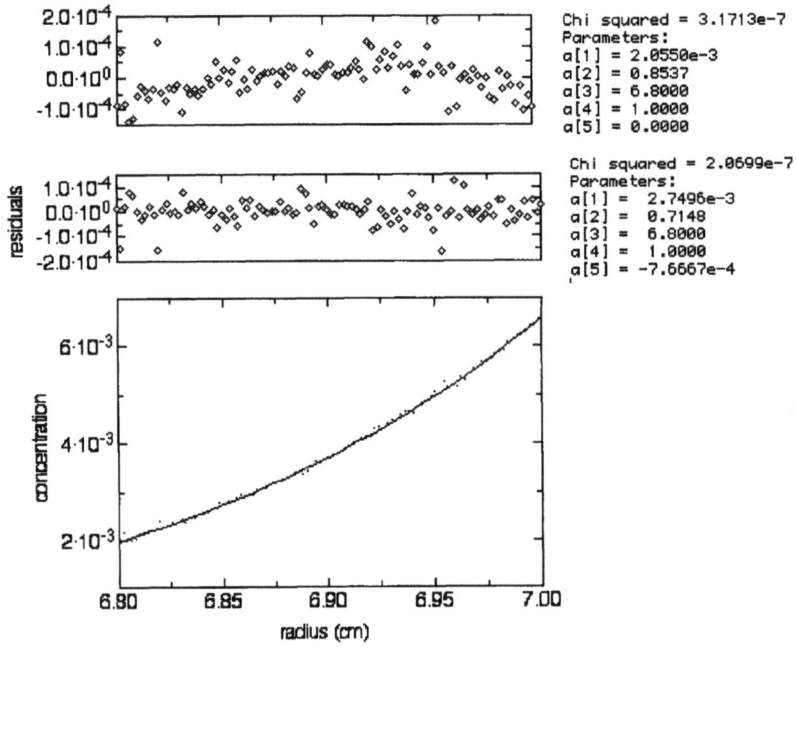

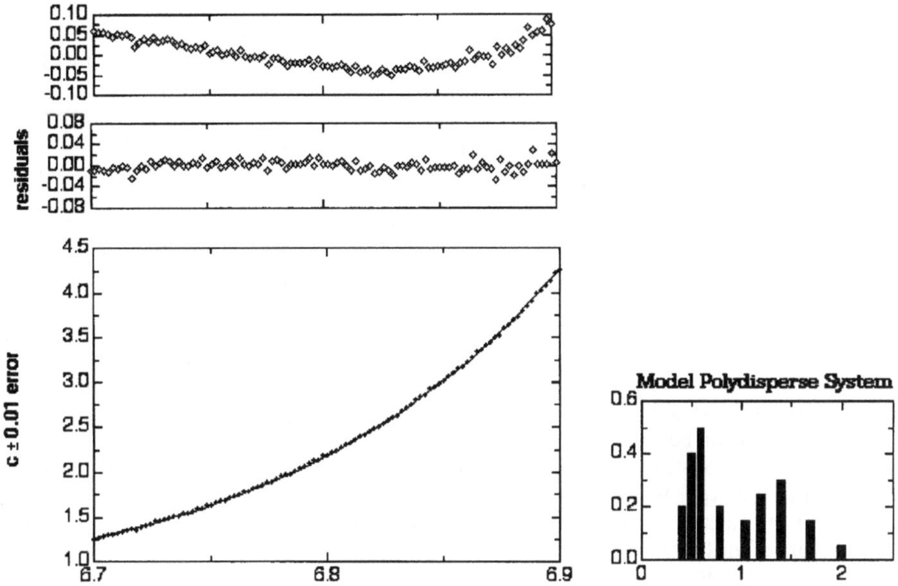

Fig. 2 A single species ideal fit to a set of simulated data for five species ranging from $k = 0.4$ to 2.0 (as shown in the histogram, *right*). The true average k_w at the hinge point is computed independently to be 0.8966. The baseline was floated. Residuals are displayed immediately above the plot, and appear to be random or close to random in their distribution. The estimate yielded for k_w was 1.1287, in error by some 29%. In contrast, a fit (plot not shown) in which the baseline was fixed at its true value gave an estimate for $k_w = 0.922$ (error of 2.9%), but the residuals (*top*) show a very pronounced trend

Yet the "single ideal component" fit gives a distribution of residuals which if not perfect is plainly no worse than many which have been published in the literature. Again, it is only by fixing E at its true value that the polydispersity present becomes instantly apparent. And again, the estimate for the reduced buoyant mass is of no worth when E is floated – it gives a very poor estimate for the weight average of the defined mass distribution.

Examples of this effect could be multiplied many times over. There is in fact a simple explanation for why floating E when using an incorrect model can give fits which seem to be good or even excellent. A standard consideration in all nonlinear fitting is the need to avoid correlation between parameters (or between parameters and the data set). E is to a significant extent correlated with the reduced mass parameter, and perhaps even more so with the nonideality term. This ensures that within the space allowed by experimental error, a "best" fit can be achieved with a set of parameters which have no simple relation to the set used to generate the initial data.

Our conclusion is that floating E in a nonlinear fit based upon an assumed model can generate a fortuitously good fit. Thus the all-too-common practice of "testing a range of models and seeing which gives the best fit" is simply invalid. We must emphasise that we are not saying that "one cannot ever float E". Clearly, if one has sufficient prior knowledge of one's solute system based upon independent evidence that it really is a single ideal component system (or whatever), then floating E is valid, although one still has to be aware that in all nonlinear fitting, more parameters floated means less precision in estimates. The worry is this: the need for quite detailed prior knowledge of one's solute system seriously limits the potential use of sedimentation equilibrium as an investigative tool.

Development of a new approach to estimation of E via the function $k\Delta(E)$

It is clear that an independent estimate of E would be of great utility in practical use of the sedimentation equilibrium procedures. We have seen that such a method can theoretically exist [3], albeit no high precision practical implementation is available as yet. Ideally, a procedure for the estimate of E would

- Be universally applicable, i.e. to all speed regimes, to both optical systems.
- Be free of any need for predialysis.
- Involve no extrapolations – all data would be used with similar weighting.
- Involve no assumption as to conservation of mass.
- Call for no additional experimentation.
- Yield a precision in the estimated E comparable to the precision in the data set.

Our approach to this problem has been as follows. We have defined two novel operational averages for k, evaluating their definition relative to the more conventional weight average, k_w, by the usual method of summation over all species. We start with Eq. (1), which can be employed in either its stated (derivative) form or in logarithmic form. We denote the k average derived from the derivative form as $k_{DF'}$ and from the logarithmic form as $k_{LN'}$, it being obvious that

$$k_{DF'} = k_{LN'} = k_w \tag{4}$$

since

$$k_w = \sum_1^n w_i \frac{dc_i/dr}{cr} = \sum_1^n \frac{c_i}{c} \frac{dc_i/dr}{cr} = \frac{dc/dr}{cr}, \tag{5}$$

and an identical result is obtained via

$$k_w = \sum_1^n d\ln c_i/dr^2 = \sum_1^n \frac{c_i}{c} d\ln c_i/dr^2$$

$$= \sum_1^n \frac{c_i}{c} \frac{dc_i/dr}{c_i r} = \frac{dc/dr}{cr}. \tag{6}$$

What happens, however, if we accept that in these novel averages the c values in the operational (first) term of the equations can be replaced by $(c+E')$ values, where E' is a numerical constant (the "baseline error")? Clearly neither $k_{DF'}$ nor $k_{LN'}$ will be identical to the k_w value, other than in the case where only a single component is present (proof not presented, but trivial). But for the general case, will either or both have the same numerical relationship to k_w? To find out, we repeat the previous summations with the c values duly replaced by $(c+E')$, and the c_i values by (c_i+w_iE'). Note we are thereby assigning a mass-weighted fraction of the baseline to each individual component. This satisfies the obvious requirement that $\Sigma_1^n w_i E = E'$. But what is w_i? It cannot be simply c_i/c, as this would be appropriate only for the simple, weight-average case. Logically, it should be a weight specified by the form of average being defined – but that assumes we know the answer. For the moment we proceed without defining this quantity. First, defining k_{DF}, we can write

$$k_{DF} = \Sigma_1^n w_i \frac{d(c_i + w_iE')/dr}{(c+E')r}$$

$$= (dc/dr)\Sigma_1^n \frac{(c_i + w_iE')}{(c+E')} \frac{1}{(c_i + w_iE')}$$

$$= \frac{dc/dr}{cr} \frac{c}{(c+E')}. \tag{7}$$

We have piece of good fortune: w_i cancels out. The last term, $c/(c+E')$, gives in simple form the relation between the novel average k_{DF} and k_w (the penultimate term) when $E' \neq 0$.

But for the case of k_{LN}, where

$$k_{LN} = 2\Sigma_1^n w_i \frac{d\ln(c_i + w_iE')}{dr^2}$$

$$= 2\Sigma_1^n \frac{(c_i + w_iE')}{(c+E')} \frac{d\ln(c_i + w_iE')}{dr^2} = ? \tag{8}$$

we have problems. We can only proceed, in a manner analogous to Eq. (6) if we can assume the identity

$$d\ln(c_i + w_iE') = dc/(c_i + w_iE'), \tag{9}$$

which as w_i is an unknown function of r and hence in general of c is untenable. We have to date not found a useful rearrangement of Eq. (8). All that one can say is that where E' has a finite value, then $k_{DF} \neq k_{LN}$, and $k_{DF}-k_{LN} \neq 0$. The latter difference is a function of E', and this function can be written as $k\Delta(E)$, the superscript being dropped in recognition that it can be defined for any value of E.

$k\Delta(E)$ has been computed over r for sets of simulated data, the necessary derivative being obtained either analytically or via the pro Fit inbuilt DERIVATIVE function. The results for a simple two-component system are shown in Fig. 3: mass ratio 50, k values of 0.7 and 1.0 – with a range of assumed E values. We have carried out many such simulations. The greater the polydispersity, the greater the difference between the $k\Delta(E)$ values at given r.

It is now clear that the "true" value of E can be determined simply by assuming a range of E values and finding – graphically or otherwise – the assumed E value for which $k\Delta(E)=0$. At first sight, this approach suffers from a major drawback, inasmuch as it cannot be applied successfully to a monodisperse (or in practice nearly monodisperse) system. This is easily circumvented, however. One merely adds to the actual data set a table of simulated perfect data, so producing an "artificial" polydispersity. After estimation of E for this system, the data added are stripped away, including whatever baseline was added, to leave the true baseline.

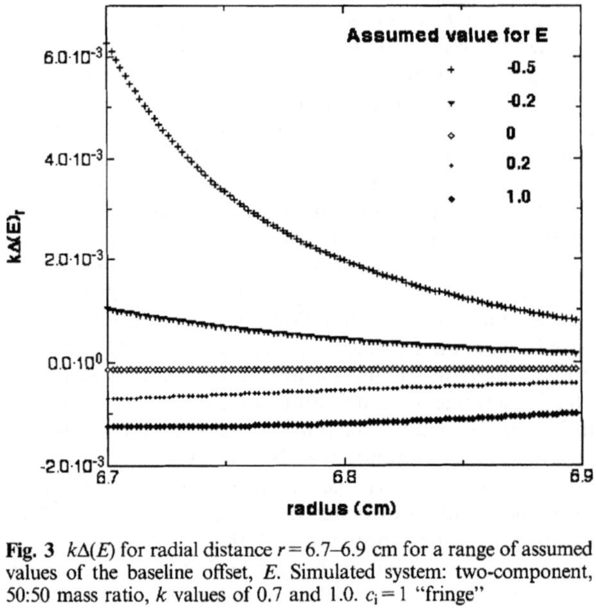

Fig. 3 $k\Delta(E)$ for radial distance $r=6.7$–6.9 cm for a range of assumed values of the baseline offset, E. Simulated system: two-component, 50:50 mass ratio, k values of 0.7 and 1.0. $c_i=1$ "fringe"

A PCNA

How can the use of the function $k\Delta(E)$ be put into routine practice? If an explicit expression for $k\Delta(E)$ was available, one could tabulate for a range in E, and then globally fit over r. Unfortunately, as already noted, we do not have a useful formulation available to us. Moreover, all approaches starting with real data involve the use of numerical differentiation procedures, which are notoriously hazardous. In order to assess whether the use of the $k\Delta(E)$ function could be of genuine utility in finding (unknown) E values, we have devised a PCNA. This is an algorithm which is capable of being implemented to the level where one can at least judge the feasibility of finding E, starting from a simulated data set with realistic levels of error.

As noted, numerical differentiation is the key step in determining estimates for $k\Delta(E)$. Routine approaches, such as the use of orthogonal polynomials or sliding strip polynomials, have – at least in our hands – proved of limited utility. Our own approach has involved the development of a two-stage fitting procedure, in which the novel first stage is suggested to us by the fact that all equilibrium data sets are sums of multiple exponential terms. So, our first stage is to fit the data set to the function $Fcve(r)$, where

$$Fcve(r) = a[3] + \sum_{i=1}^{i=3} \exp\{0.01ia[1](r - a[2])\}. \qquad (10)$$

This gives, in a single stage, a remarkably good fit, of which the derivative is computed as the analytical derivative of $Fcve(r)$ using coefficients $a[1]$ and $a[2]$. An improved fit is then generated by the use of a second stage, in which the residuals between $Fcve(r)$ and the data set are fitted to a simple (order 3) polynomial. The final derivative values used are then the algebraic sum of the derivatives of the $Fcve(r)$ and of the polynomial. Use of this two-stage approach has been implemented in pro Fit, and Fig. 4 shows how, having computed $k\Delta(E)$ as a function of r for assumed E, one can use a graphical procedure to estimate the "true" E. It should be borne in mind however that in this simulation work we have used error values (± 0.01 fringe) which are only attainable in practical work under optimal conditions. Further optimisation of the whole procedure will thus be called for.

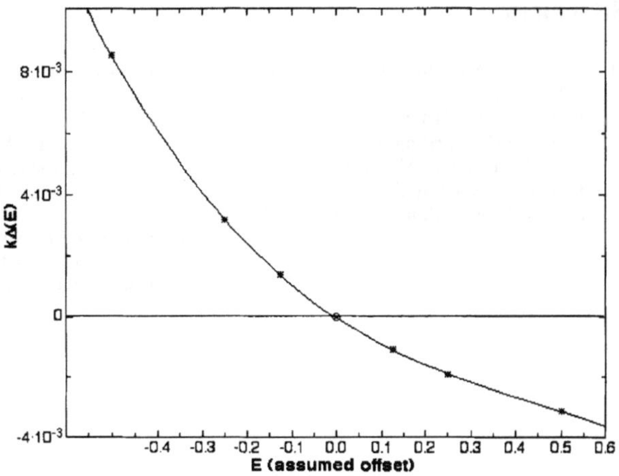

Fig. 4 $k\Delta(E)$ against assumed E assessed via the proof of concept numerical algorithm, for a simulated two-component system (as in Fig. 3). The values of $k\Delta(E)$ for the central region (hinge point $\pm 25\%$ of column length) were averaged to minimise noise. Noise of ± 0.01 fringes was added to the simulated data values. A simple exponential fit was used to enable an estimate to be made of the E value (–0.007) for which $k\Delta(E) = 0$. This estimate would give a value of 0.847 (as against the true value of 0.85) for k_w

Discussion

So far as we are aware, and the literature on analytical ultracentrifugation is vast, the fact that the equivalence of the average reduced mass values when evaluated via the derivative and via the logarithmic plot is only applicable when $E = 0$ has not previously been reported. This gives rise to a method of potential importance in sedimentation equilibrium analysis. We have specified six conditions which should be met in any "ideal" approach to the "baseline problem", and the use of the $k\Delta(E)$ function satisfies all of these. We still do need for choice an analytical solution in numerically usable form of Eq. (8). This would enable a definitive working algorithm for evaluation of E to be written. In the meantime, our PCNA is slow and rather cumbersome to use, but is adequate for demonstrating that this approach is a feasible one, capable of yielding estimates for E of a precision probably no worse than around twice the root-mean-square error on individual c_i values. This level of precision is very adequate for almost all practical work. However, when real (as compared with simulated) data are being used, then the prior removal of noise which is not simply Gaussian in form becomes necessary. The methods of Schuck [10] would be particularly applicable. On the positive side, it should also be borne in mind that we have carried out our simulations using a narrow spread of mass values, and under "low speed equilibrium" conditions. Additional simulations (not reported here) confirm that the precision is markedly increased with higher levels of either solute polydispersity (which as noted earlier can be generated artificially) or by the use of higher ("intermediate") rotor speeds. Further work is proceeding with the intention of producing a full, working algorithm, capable – hopefully – of "single keystroke" implementation by the user.

References

1. Yphantis DA (1964) Biochemistry 3:297–317
2. Rowe AJ (1975) J Mol Biol 97:202–205
3. Creeth JM, Harding SE (1982) J Biochim Biophys Methods 7:25–34
4. Hall DA, Harding SE, Winzor DJ (1999) Prog Colloid Polym Sci 113: 62–68
5. Johnson ML, Correia JJ, Yphantis DA, Halvorson HR (1981) Biophys J 36:575–588
6. McRorie DK, Voelker PJ (1993) Self-associating systems in the analytical ultracentrifuge. Beckman Instruments, CA
7. Simanca SR, Sutherland S (2002) http://www.math.sunysb.edu/~scott/Book331/Robust_fitting.html
8. Rowe AJ (1975) J Mol Biol 97:202–205
9. Turnock G, Pearce TC, Rowe AJ (1975) J Mol Biol 97:193–201
10. (a) Schuck P (1999) Anal Biochem 272:199–208; (b) http://www.analyticalultracentrifugation.com/systematic_noise_analysis.htm

Progr Colloid Polym Sci (2004) 127: 9–13
DOI 10.1007/b94243

H. G. Müller

Determination of very broad particle size distributions via interference optics in the analytical ultracentrifuge

Abstract Dispersions play a big role in industry. Important characteristics such as viscosity and gloss depend on the particle size and the particle size distribution of these systems. Industrial dispersions often have a broad particle size distribution. If the particle size is in the submicron range, particle size distributions of this type can best be analysed by analytical ultracentrifugation as shown in a big round-robin test which was carried out by Bayer. When particles become smaller and their transmission increases, turbidity optics have to be substituted by interference optics. This leads to a more tedious way of transforming measured data into particle size distributions. This has to do with the enormous broad range of *s* values which often belongs to such a particle size distribution, for example from 1 to 300 Sv. As a consequence not only several interference pictures have to be taken at different times but also pictures at different rotor speeds. For this example the rotor speeds are 3,000, 10,000, 25,000 and 40,000 rpm. The new computer program for data evaluation converts the $i=f(r)$ picture into a $\delta\, i=f(d)$ function using the well-known Stokes law. Then it collects the number of fringes for a given diameter interval. This concentration is computed for each of the 500–1,500 interference pictures and an average value is taken for the final result. By the application of our new program we can do our interference work in a semiautomated way and avoid the arbitrary man-made selection of pictures, which saves time and increases precision.

Keywords Analytical ultracentrifugation · Interference optics · Broad particle size distributions · Nanoparticles · Computer program

H.G. Müller
Q-18, Bayer Industry Services,
51368 Leverkusen, Germany

Dispersions play a big role in industry. Important characteristics such as viscosity and gloss depend on the particle size and the particle size distribution of these systems. An example for the influence of the particle size distribution on viscosity is given in Fig. 1.

At a solid content of 70% the product which possesses a narrow distribution of fine particles is a solid; the same product with a broad distribution is a liquid.

Industrial dispersions often have a broad particle size distribution. If the particle size is in the submicron range, particle size distributions of this type can best be analysed by analytical ultracentrifugation as shown in a big round-robin test which was carried out by Bayer [1].

The high precision of the ultracentrifugal method to determine particle size distributions is demonstrated in Fig. 2, which shows the result for a mixture from nine different calibration latices.

As long as the dispersions were turbid they were analysed by a specifically modified preparative ultracentrifuge [2, 3, 4, 5] which was equipped with turbidity optics. An automated computer program evaluated the turbidity data from a light transmittance $I=f(t)$ function, which was measured at the middle of the analytical cell.

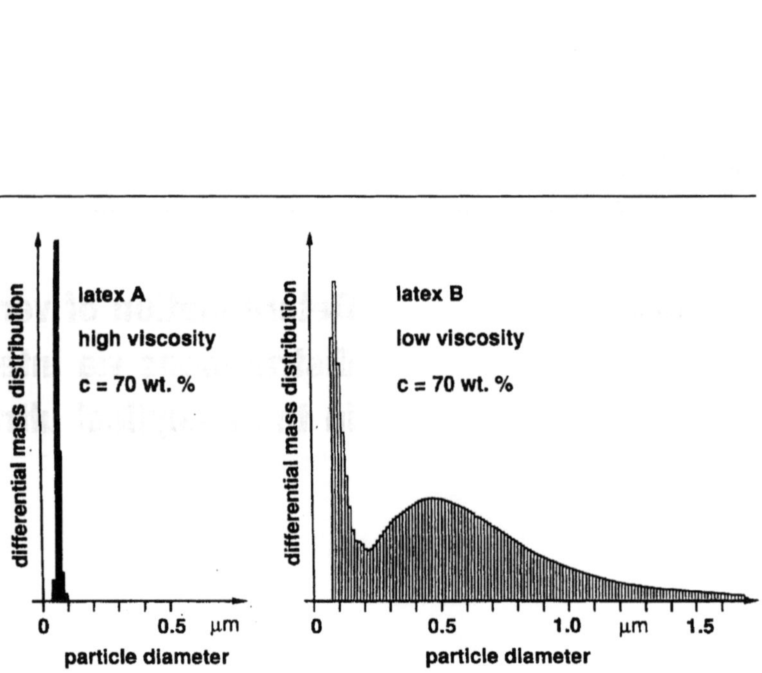

Fig. 1 Particle size distributions of different widths for two latices with identical solid content resulting in different viscosities

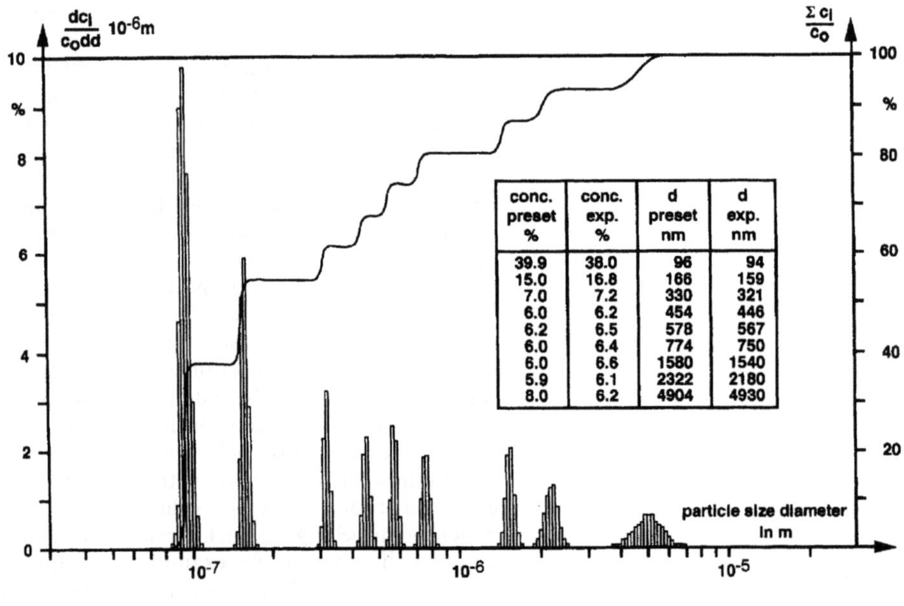

conc. preset %	conc. exp. %	d preset nm	d exp. nm
39.9	38.0	96	94
15.0	16.8	166	159
7.0	7.2	330	321
6.0	6.2	454	446
6.2	6.5	578	567
6.0	6.4	774	750
6.0	6.6	1580	1540
5.9	6.1	2322	2180
8.0	6.2	4904	4930

Fig. 2 Resolving power of the ultracentrifuge method for the determination of particle size distributions

For details see Ref. [2]. This procedure can easily be automated but as Mie theory [2, 6] has to be applied to transform these data into the desired concentration – diameter relation – extreme baseline stability is necessary. Often this means that for an exact determination of small particles in the sample a second experiment at a higher concentration has to be carried out, and the result is combined with the result of the first experiment as described in the literature [5, 7].

The situation changes when particles become smaller and their turbidity decreases. In this case turbidity optics can no longer be applied and have to be substituted by interference optics. So we can change from our self-made turbidity analytical ultracentrifuge to the much more sophisticated Beckman XLI analytical ultracentrifuge.

Changing from turbidity optics to interference optics means that Mie theory no longer has to be used because the number of interference fringes is proportional the particle concentration. As a consequence only one experiment is necessary to determine a particle size distribution instead of at least two experiments with turbidity optics. This advantage, however, is in contrast to the more tedious way of transforming interference data into particle size distributions. This has to do with the enormous broad range of *s* values which belongs to broad particle size distributions. An example for such a finely dispersed sample with a broad distribution is given

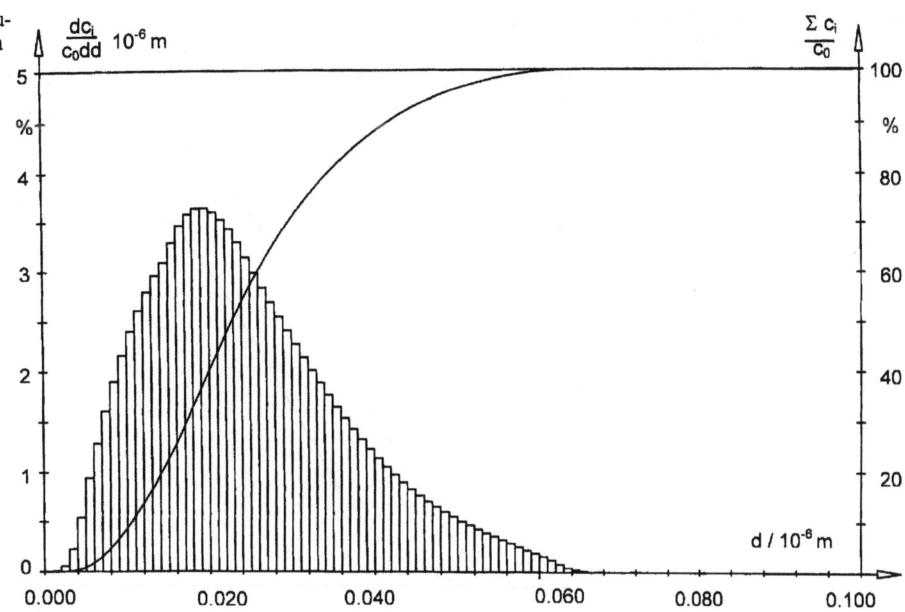

Fig. 3 Broad particle size distribution of a finely dispersed sample in the nanometer range

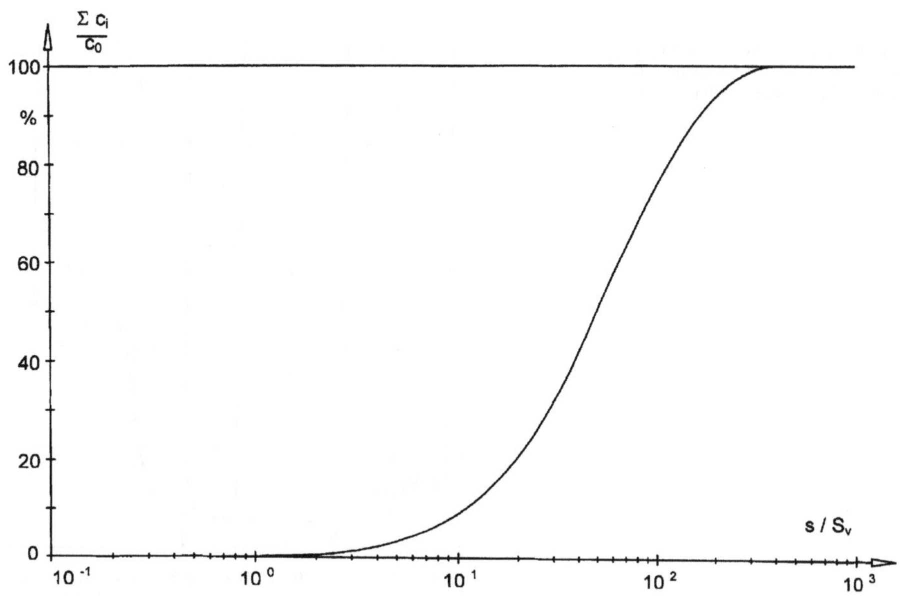

Fig. 4 Sedimentation coefficient distribution of the sample in Fig. 3

in Fig. 3. Its particle size distribution ranges from 4 to 68 nm, the U_{90} value of this distribution as defined by

$$U_{90} = (d_{90} - d_{10})/d_{50}$$

is 1.4. In this equation d_{90} is the 90% value, d_{10} the 10% value and d_{50} the median value of the mass particle size distribution. The corresponding range of s values is between 1 and 290 Sv (Fig. 4). This means the biggest s value is about 300 times greater than the smallest s value of this distribution. As a consequence not only several interference pictures have to be taken at different times but also pictures at different rotor speeds. For this example the rotor speeds were 3,000, 10,000, 25,000 and 40,000 rpm.

Existing computer programs, for example, "ultrascan" from Demeler [8], make it possible to analyse a lot

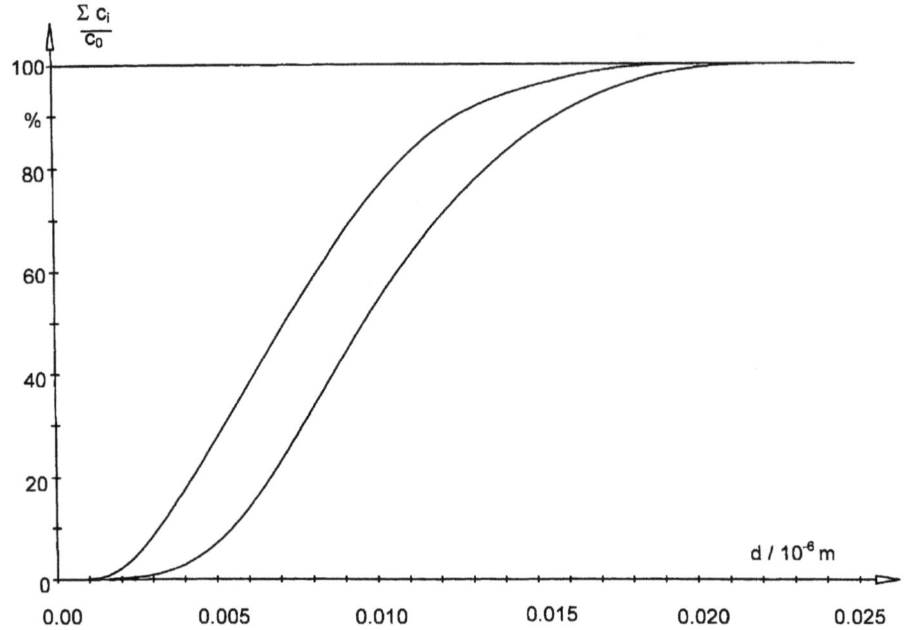

Fig. 5 Two results of evaluation of the same experiment by different persons on a graphical basis

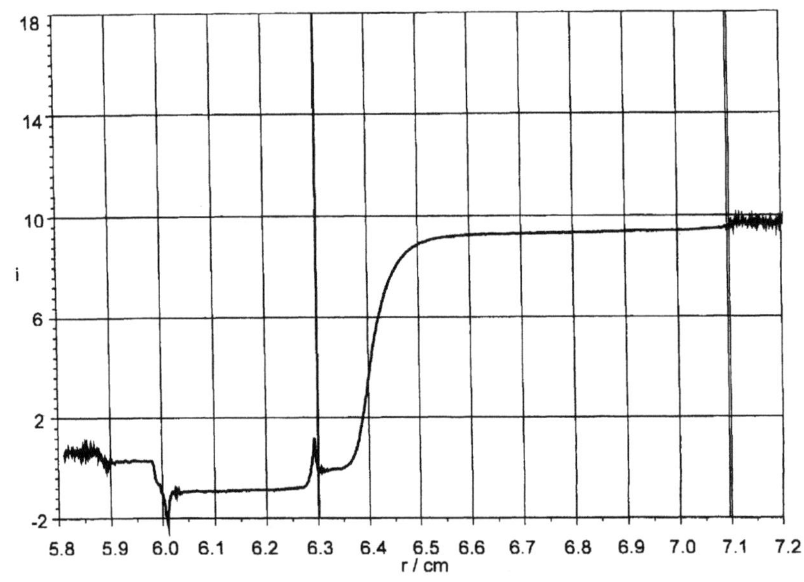

Fig. 6 Interference picture from a Beckman XLI for a dispersed sample in the nanometer range

of pictures in a nearly automated way, but at a constant rotor speed only. So in the past we had to evaluate interference pictures at different times and rotor speeds. Then we added the resulting concentration diameter plots to the whole distribution on a graphical basis. This is a time consuming and in some respects arbitrary method as shown in Fig. 5, where two results of evaluation of the same experiment are given.

So we had to develop a new program of our own. In the first step in our new program the position of meniscus and the bottom in the interference picture are fixed at the starting speed of 3,000 rpm. To increase the accuracy of the velocity data, artificial borders for evaluation are introduced. In the case of sedimentation, the evaluation meniscus is the physical meniscus plus 0.05 cm, the evaluation bottom is the physical minus 0.1 cm; the

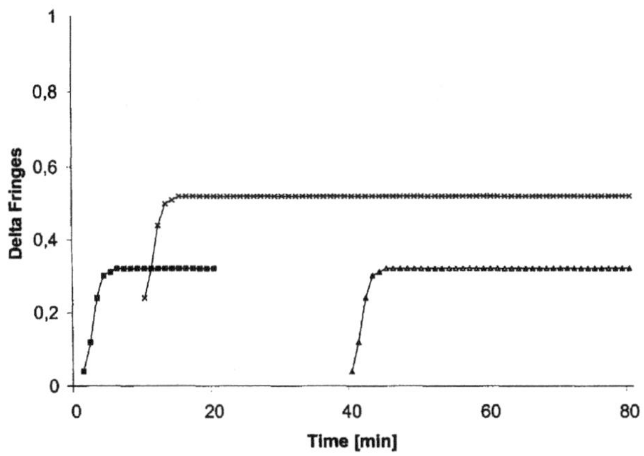

Fig. 7 Concentration of different diameter intervals as a function of time: big (*squares*), medium-sized (*crosses*) and small (*triangles*) particles

greater distance is taken to avoid disturbances by accumulated material. In the case of flotation, the evaluation bottom is the physical bottom minus 0.05 cm, the evaluation meniscus is the physical meniscus plus 0.1 cm.

After the dilution effect due to the sectorial shape of the cell has been taken into account the interference picture $i=f(r)$ (Fig. 6) is converted into the $i=f(d)$ function using the well-known Stokes law.

As the number of interference fringes, *i*, is proportional to the particle concentration we have a concentration–diameter function and we get information about the concentration of a given diameter species (diameter interval).

At the beginning of the experiment at time t_1 with a low rotor speed from this $i=f(d)$ function, the concentration of the biggest particles is determined, which at this short time have covered the longest distance in the cell. Additional information about smaller particles comes from the number of fringes for shorter path lengths.

At time t_2 the biggest particles have reached the bottom of the cell, small particles have just entered the described range of evaluation and medium-sized particles are to be found in the middle. Again an $i=f(d)$ function is plotted and doing the same for t_3, t_4 and so on, a lot of $i=f(d)$ functions with increasing times result.

In this way "two dimensional" $i=f(d)$ plots are created and are evaluated at constant diameters, resulting in a number of plots for the dependence of concentration as a function of time; see Fig. 7 for a big, a medium-sized and a small diameter of a broad distribution. These plots differ in width, because big particles settle faster, whereas small particles enter the range of evaluation later in the given example after about 40 min.

The number of data points for each curve depends on the rotor speed and varies, for example, between 20 and 1,500. From these plots an average value of the concentration for a given diameter interval is taken. When we plot these average values as a function of diameter we find the desired mass particle size distribution. It should be added that the collection of data points is restricted to constant rotor speeds because of offset problems with change of speeds. So gaps in the plots of Fig. 7 come into existence, and are filled by interpolation.

This concept works well for the radial derivative, which can be successfully applied for samples with a sufficiently high number of fringes; problems, however, come up when the total number of fringes for a whole sample is smaller than, for example, 2 or 3. In this case the radial derivative has to be replaced by the time derivative [9] to get rid of the time-independent baseline noise. So even samples with lower number of fringes (concentrations) can be analysed successfully.

To test the accuracy of this new method of data evaluation a 1:1:1 mixture of 3 calibration latices with diameters of 20, 50 and 100 nm were successfully analysed with a precision referring to diameters and mass fractions of better than plus/minus 3%.

By the application of our new program for data evaluation we can do our interference work in a semi-automated way and avoid the arbitrary man-made selection of pictures, which saves time and increases precision.

References

1. Lange H (1995) Part Part Syst Charact 12:148–157
2. Scholtan W, Lange H (1972) Kolloid Z Z Polym 250:782–796
3. Mächtle W (1984) Makromol Chem 185:1025–1039
4. Müller HG (1989) Colloid Polym Sci 267:1113–1116
5. Mächtle W (1988) Angew Makromol Chem 162:35–52
6. Mie G (1908) Ann Phys 25:377–395
7. Müller HG (1997) Prog Colloid Polym Sci 107:180–188
8. Demeler B http://www.ultrascan.uthscsa.edu
9. Stafford WF (1992) In: Harding SE, Rowe AJ, Horton JC (eds) Analytical ultracentrifugation in biochemistry. Royal Society of Chemistry, Cambridge, p 35

Progr Colloid Polym Sci (2004) 127: 14–18
DOI 10.1007/b94245

W. Borchard
A. Straatmann

Sedimentation and diffusion coefficients in polymer solutions and gels

Abstract Recently Lamm's differential equation, which frequently was applied in ultracentrifugal sedimentation experiments of polymer solutions, was extended to cross-linked swollen polymers. If a polymer solution is chemically cross-linked to a swollen polymer network (gel), the transport of a solvent in a gel in a centrifugal field consists of velocities due to sedimentation, diffusion and deformation. It was shown that the elastically active coefficient describing the deformation velocity has the same units as the diffusion coefficient. The sum of both coefficients in a gel is proportional to the mobility of the polymer and the thermodynamic factor. The latter can be found in the literature in papers of G. Rehage and coworkers. As diffusion coefficients are also known for some systems, the sedimentation coefficients for solutions and gels can be calculated and compared at the same concentrations and temperatures. The transport coefficients in thermodynamically good and bad solvents are discussed.

Keywords Coefficients of diffusion and sedimentation · Thermodynamic factors · Polystyrene/organic solvents

Dedicated to Prof. Dr. G. Rehage (deceased)

W. Borchard (✉) · A. Straatmann
Fakultät für Naturwissenschaften,
Universität Duisburg–Essen,
47048 Duisburg, Germany
e-mail: borchard@uni-duisburg.de;
Fax: +49-203-3792110

Introduction

Recently Lamm's differential equation for polymer solutions suited to describe the time-dependent change of the polymer concentration as a function of diffusion and sedimentation of the polymer component in a centrifugal field has been extended to gels [1]. Bearing in mind that the phenomenological definition of the flux is valid for polymer solutions as well as for gels a generalised differential equation was derived which contains the coefficients of diffusion, D, the sedimentation of the polymer component, s_2, and an elastically active coefficient, E. All changes between a dissolved polymer in solution and a cross-linked swollen polymer have been attributed to this coefficient. D and E have the same units.

Ultracentrifugal measurements of a gelatin/water gel, which is physically cross-linked, did not allow a comparison between the transport coefficients of linear and gelled polymer, because at the same temperature and concentration both states, solution and gel, cannot be realised; therefore, literature data of Rehage and coworkers of the polystyrene (PS)/chlorobenzene system were used to illustrate the predicted difference in the transport coefficients. In this contribution further results of Rehage and coworkers have been prepared to demonstrate why diffusion and sedimentation are different in solution and gels, especially at relatively low polymer concentrations.

Some relations

Based on the theory of thermodynamics of irreversible processes the mutual diffusion coefficient in a binary solution (index s) is known to be the product of the mobility u^{s}_{k} and the thermodynamic factor

$$F_{\text{th}}^{\text{s}} = \left[\rho_k(\partial\tilde{\mu}_k/\partial\rho_k)_{T,P}\right]^{\text{s}}$$

$$D^{\text{s}} = u_{\text{k}}^{\text{s}}\left[\rho_k(\partial\tilde{\mu}_k/\partial\rho_k)_{T,P}\right]^{\text{s}}, \text{ with } k=1,2, \tag{1}$$

where ρ_{k} is the partial density or mass concentration of component k and $\tilde{\mu}_k$ the partial specific free enthalpy of mixing or the specific chemical potential of component k [2, 3]. The same relation holds for a gel (index g) where the corresponding quantities have the same meaning as in Eq. (1) [1]:

$$(D+E)^{\text{g}} = u_{\text{k}}^{\text{g}}\left[\rho_k(\partial\tilde{\mu}_k/\partial\rho_k)_{T,P}\right]^{\text{g}}, \text{ with } k=1,2, \tag{2}$$

We use k = 1 for the solvent and $k=2$ for the polymer. Thus, we have a single mutual diffusion coefficient which is related to the two mobilities and thermodynamic factors. As the thermodynamic factors and mobilities depend on concentration at constant temperature and pressure, the concentration dependence of the diffusion coefficient may be rather complicated [4, 5, 6]. According to classical thermodynamics, the thermodynamic factors in Eqs. (1) and (2) have to vanish at stability limits. It is known that Fick's law, in which the flux density is proportional to the concentration gradient, does not hold at the coexistence curve of a demixed system. If the flux density is zero, then D has to vanish, because the concentration gradient at the phase boundary of two demixed phases is very high.

The mobilities $u^{\text{s}}{}_2$ and $u^{\text{g}}{}_2$ are related to the sedimentation coefficients by

$$s_2^{\alpha} = u_2^{\alpha}(1-\bar{V}_2\rho)^{\alpha}, \tag{3}$$

where $\bar{V}_2$ is the partial specific volume of the polymer and ρ the density of the system for $\alpha = \text{s, g}$.

From Eqs. (1), (2) and (3) we obtain

$$\frac{(D+E)^{\text{g}}}{D^{\text{s}}} = \frac{s_2^{\text{g}}\left[\rho_2(\partial\tilde{\mu}_2/\partial\rho_2)_{T,P}\right]^{\text{g}}(1-\bar{V}_2\rho)^{\text{s}}}{s_2^{\text{s}}\left[\rho_2(\partial\tilde{\mu}_2/\partial\rho_2)_{T,P}\right]^{\text{s}}(1-\bar{V}_2\rho)^{\text{g}}} = \frac{s_2^{\text{g}}F_{\text{th}}^{\text{g}}(1-\bar{V}_2\rho)^{\text{s}}}{s_2^{\text{s}}F_{\text{th}}^{\text{s}}(1-\bar{V}_2\rho)^{\text{g}}}. \tag{4}$$

For the same composition of gel and solution the ratio of the buoyancy terms in Eq. (4) is close to 1, which reduces Eq. (4) to

$$\frac{(D+E)^{\text{g}}}{D^{\text{s}}} = \frac{s_2^{\text{g}}}{s_2^{\text{s}}}F_I, \text{ with } F_I = \frac{F_{\text{th}}^{\text{g}}}{F_{\text{th}}^{\text{s}}}. \tag{5}$$

By the use of the Gibbs–Duhem equation (Eq. 6), at constant temperature and pressure we have a relation between the derivatives of the chemical potential of components 1 and 2 with respect to mass fraction w_k of component k:

$$w_1\left(\frac{\partial\tilde{\mu}_1}{\partial w_2}\right)_{T,P} + w_2\left(\frac{\partial\tilde{\mu}_2}{\partial w_2}\right)_{T,P} = 0 \tag{6}$$

and

$$\left(\frac{\partial\tilde{\mu}_2}{\partial w_2}\right)_{T,P} = -\frac{w_1}{w_2}\left(\frac{\partial\tilde{\mu}_1}{\partial w_2}\right)_{T,P} = \frac{w_1}{w_2}\left(\frac{\partial\tilde{\mu}_1}{\partial w_1}\right)_{T,P}. \tag{7}$$

The derivative of the specific chemical potential $\tilde{\mu}_1$ is related to the derivative of the molar chemical potential of the solvent:

$$M_1\left(\frac{\partial\tilde{\mu}_1}{\partial w_2}\right)_{T,P} = \left(\frac{\partial\mu_1}{\partial w_2}\right)_{T,P} \tag{8}$$

In the present contribution the transport coefficients and the thermodynamic factors have been gathered from different sources. In different papers various concentration scales have been used and did not allow direct comparison. All useful data were converted into the base molar fraction scale to enable the calculation of the ratio of, for example, $(D+E)^{\text{g}}/D^{\text{s}}$.

For solutions the base molar fraction of the solvent $x^*{}_1$ is given by

$$x_1^* = \frac{n_1}{n_1 + rn_2}, \text{ with } x_1^* + x_2^* = 1, \tag{9}$$

where n_1 is the amount of matter of the solvent, n_2 the amount of matter of the polymer, r is the number-average degree of polymerisation and $x^*{}_2$ is the base molar fraction of the polymer. In a similar way, the base molar fraction of the solvent, $x^*{}_1$, and the cross-linked component in a gel are defined by

$$x_1^* = \frac{n_1}{n_1 + zv^*}, \text{ with } x_1^* + x_2^* = 1. \tag{10}$$

Here v^* is the amount of matter of the network chains and z is their number-average degree of polymerisation. The total mass of a polymer in solution, m_2, is given by $m_2 = n_2rM_0$, M_0 being the molar mass of a polymer repeating unit. In gels, the total polymer mass is represented by $m_2 = v^*zM_0$. From these relations the base molar fractions may be easily calculated. The advantage of the use of base molar quantities instead of molar functions is to achieve a better symmetry in the representation of thermodynamic quantities with respect to $x^*{}_1 = x^*{}_2 = 0.5$ in systems with molecules differing in size, like macromolecules and low molecular weight solvent molecules in solutions.

The thermodynamic factor is obtained by use of the Flory–Huggins–Staverman–van Santen theory [7, 8, 9] for real polymer solutions in which the chemical potential of the solvent, μ_1, in a solution reads as a function of the base molar concentration:

$$\mu_1^{\text{s}} = \mu_{01} + RT\left[\ln x_1^* + \left(1-\frac{1}{r}\right)x_2^* + \chi_{\text{x}}^{\text{s}}(x_2^*)^2\right], \tag{11}$$

where μ_{01} is the chemical potential of the pure solvent, r is the degree of polymerisation and χ^s_x is the interaction parameter. The enthalpy part of χ^s_x is proportional to the excess interaction energy, Δw, given by $\Delta w = 2w_{12} - (w_{11} + w_{22})$. w_{ij} are the interaction energies of the pairs ij, all of which are negative. In a good solvent $|w_{12}| > |w_{11}|$ and $|w_{12}| > |w_{22}|$. The system is athermal if $\Delta w = 0$. In this case, w_{12} is the arithmetic means of $(w_{11} + w_{22})$. If the solvent is bad, we have $|w_{12}| < |w_{11}|$ and $|w_{12}| < |w_{22}|$.

The thermodynamic factor of the swelling agent in a gel may be obtained by use of the result of the statistical theory of polymer solutions combined with the theory of rubber elasticity due to Flory [10]. The chemical potential of the solvent in the gel, μ^g_1, is given by

$$\mu^g_1 = \mu_{01} + RT\left[\ln x^*_1 + x^*_2 + \chi^g_x (x^*_2)^2 + \frac{(x^*_2)^{1/3}}{z}\right] , \qquad (12)$$

where μ_{01} is the chemical potential of the pure solvent, χ^g_x is the interaction parameter of the gel and z is the degree of polymerisation of the network chains between cross-links. By comparison of Eqs. (11) and (12) the interaction parameters χ^s_x and χ^g_x are very similar if r and z tend to high values.

In some systems the interaction parameters are not constant and may be empirically described by a series:

$$\chi^\alpha_x = \chi^\alpha_{x,1} + \chi^\alpha_{x,2} x^*_2 + \cdots , \qquad (13)$$

where α is related to the solution (s) or the gel (g).

By use of Eqs. (7) and (8) the thermodynamic factors in phase α are calculated:

$$F^\alpha_{th} = \rho^\alpha_2 \left(\frac{\partial \tilde{\mu}_2}{\partial \rho_2}\right)^\alpha_{T,P} = \rho^\alpha_2 \left[\left(\frac{\partial \tilde{\mu}_2}{\partial w_2}\right)\left(\frac{\partial w_2}{\partial \rho_2}\right)\right]^\alpha_{T,P}$$

$$= \left(\rho_2 \frac{w_1}{w_2}\right)^\alpha \left[\left(\frac{\partial \tilde{\mu}_1}{\partial x^*_1}\right)\left(\frac{\partial x^*_1}{\partial w_1}\right)\left(\frac{\partial w_1}{\partial \rho_2}\right)\right]^\alpha_{T,P} . \qquad (14)$$

The result for the ratio F_I from Eq. (5) is given at the same composition and temperature by use of Eqs. (11) and (12) for a single interaction parameter in solution and in a gel:

$$F_I = \frac{(1/x^*_1) - 1 - 2\chi^g_x x^*_2 - (1/3z)(x^*_2)^{-2/3}}{(1/x^*_1) - (1 - 1/r) - 2\chi^s_x x^*_2} . \qquad (15)$$

With known values χ^g_x, χ^s_x, z and r the ratio F_I could be calculated. Because the first term dominates in the nominator and denominator the ratio tends to $F_I \to 1$ for $x^*_1 \to 0$. χ^s_x and r are taken from osmotic results [11, 12, 13]. χ^g_x and z are gathered from swelling and deswelling results [14]. $(D+E)^g$ and D^s have been determined experimentally [4, 5, 6, 15, 16]. The ratio s^g_2/s^s_2 was obtained by using Eq. (5).

Results and discussion

The diffusion coefficients of the PS/toluene system have been measured in solutions [4, 6] and chemically cross-linked gels [6]. The diffusion coefficients of both systems increase strongly with concentration of the polymer, pass through a maximum and reach very low values close to the glassy region of the polymer. At the freeze-in concentration the diffusion coefficient is lower than 10^{-9} cm^2 s^{-1}. There is an intermediate range of concentrations $0.2 < (\rho_2)/(\mathrm{g\ ml^{-1}}) < 0.35$ where D^s is larger than the diffusion coefficient of the swollen cross-linked polymer. At $x^*_2 \approx (\rho_2)/(\mathrm{g\ ml^{-1}}) \approx 0.65$ both curves merge, which means that at this high concentration of the polymer cross-links have no influence on the mutual diffusion of solvent and polymer. For concentrations below the saturation concentration $c_{2s} \approx 0.2$ g ml^{-1} the gel is metastable. In classical thermodynamics the diffusion coefficient $(D+E)^g$ has to vanish at the extrapolated stability limit of the gel, because the thermodynamic factor becomes zero.

By use of Eq. (15) F_I has been calculated for 303.15 K with the values $\chi^s_x = 0.442$, $r = 2095$ [8] and $\chi^g_x = 0.468$ and $z = 2090$ [11]. F_I is represented in Fig. 1 together with the ratio $(D + E)^g/D^s$ versus x^*_2.

It can be seen that in the overlapping range of concentrations $0.2 < x^*_2 < 0.35$ F_I and $(D+E)^g/D^s$ are smaller than 1, which is in agreement with the findings in the PS/chlorobenzene system [1]. In both cases this is due to the elastically active coefficient E which is part of the diffusion coefficient of gels.

From the molecular point of view, diffusion in macromolecular solution is the process of mutual penetration of macromolecules and low molecular weight

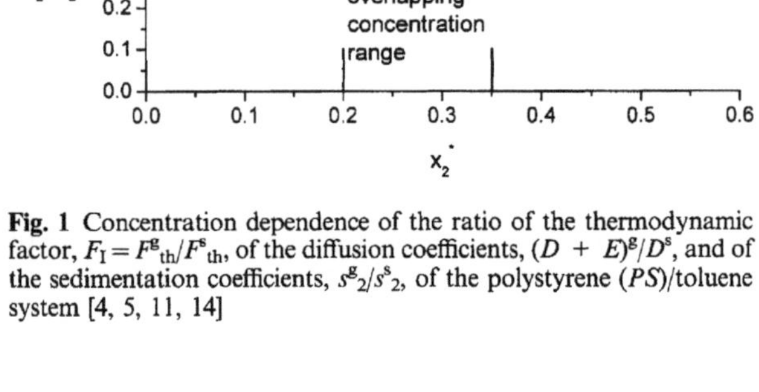

Fig. 1 Concentration dependence of the ratio of the thermodynamic factor, $F_I = F^g_{th}/F^s_{th}$, of the diffusion coefficients, $(D + E)^g/D^s$, and of the sedimentation coefficients, s^g_2/s^s_2, of the polystyrene (*PS*)/toluene system [4, 5, 11, 14]

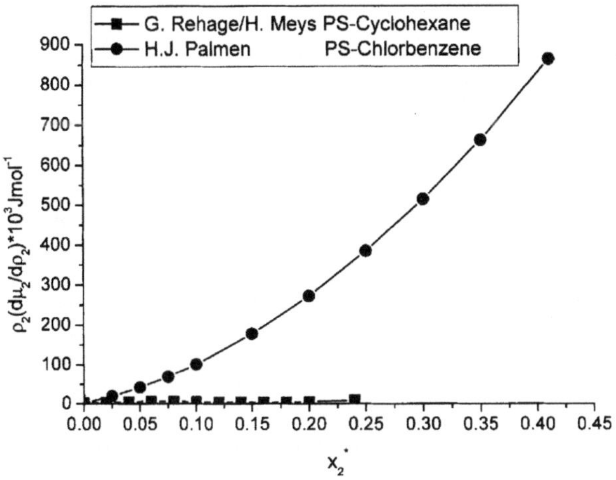

Fig. 2 Thermodynamic factor versus x^*_2 in the PS/chlorobenzene and PS/cyclohexane systems at 303.15 K [11, 12, 13]

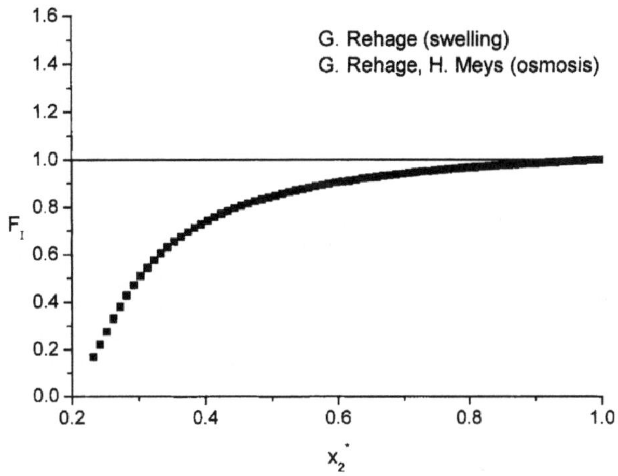

Fig. 3 Concentration dependence of $F_I = F^g_{th}/F^s_{th}$ of the diffusion coefficients of gels and solutions in the PS/cyclohexane system at 303.15 K [11, 14]

solvent. In gels the diffusion process is superimposed by the deformation of the network chains. Therefore, $(D + E)^g$ is smaller than D^s in the overlapping concentration range.

Additionally the ratio of sedimentation coefficients s^g_2/s^s_2 versus x^*_2 is presented in Fig. 1 and was calculated by use of Eq. (5). It can be seen that the sedimentation coefficients of the gels are smaller than the corresponding values in solutions at the same polymer/solvent composition, which seems mainly to be a consequence of the elastically active coefficient.

Until now we have compared the ratios of the transport quantities and also the ratios of thermodynamic factors of gels and solutions in the same solvents. From the literature it is known that the thermodynamic factors, which have been determined by osmotic and differential osmotic measurements [11, 12, 13], depend on the quality of the solvent.

The thermodynamic factors of PS/chlorobenzene and PS/cyclohexane solutions are presented in Fig. 2. There is a strong increase of $\rho_2\ (\partial\mu_2/\partial\rho_2)$ in the PS/chlorobenzene system, whereas the values of the PS/cyclohexane system remain nearly independent of concentration up to $x^*_2 \approx 0.25$. Chlorobenzene is a good solvent for PS; cyclohexane is a bad one, because in the PS/cyclohexane system a miscibility gap with an upper critical point exists. At a critical point the thermodynamic factor has to vanish. Measurements of the diffusion coefficients of PS/cyclohexane solutions versus x^*_2 at 28 °C, which is 1 °C above the critical point, reveal a minimum close to 9 base molar percent [4, 5]. This minimum although a little smoother is also formed at 50 °C. The total course of the diffusion coefficient versus x^*_2 is rather complicated, because the minimum is followed by a maximum. This was already mentioned for other solvents of PS. For

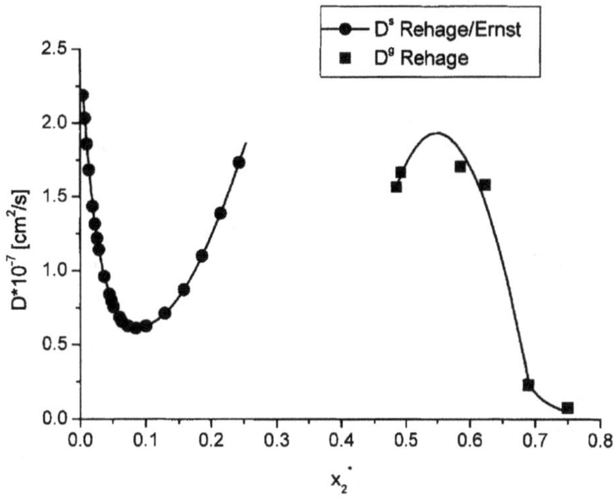

Fig. 4 Dependence of the diffusion coefficients of solutions and gels upon concentration (x^*_2) in the PS/cyclohexane system [4, 6]

the same composition, a minimum of the thermodynamic factor of the PS/cyclohexane system has been found by Scholte [17] using equilibrium ultracentrifugation for solutions at 30 °C. At high polymer concentrations the thermodynamic factor increases strongly. From these results it can be determined that the concentration dependence of the thermodynamic factor dominates in concentration regions up to $x^*_2 \approx 0.3$. At higher concentrations, the mobility and with it the sedimentation coefficient of the polymer decrease [5, 15]. This behaviour explains the maximum of the diffusion coefficient and finally its decrease at higher concentrations, before the system becomes glassy.

Table 1 Thermodynamic data of cross-linked polystyrene in various swelling agents

Solvent	x_1* (50 °C)	χ^g_x (50 °C)	α_x	β_x (grad)
Chlorbenzene	0.9650–0.9000	0.441	0.497	–18
Toluene	0.9500–0.8250	0.470	0.513	–14
Cyclohexane	0.7500–0.5000	0.590–0.704	0.480–0.296	36–132

The ratio of the thermodynamic factors of gels and solutions versus x^*_2 is presented for the PS/cyclohexane system in Fig. 3, where two parameters for the χ_x-function following Eq. (13) are used [11, 14]. The data set for 30 °C is $x^s_{x1} = 0.4776$, $x^s_{x2} = 0.4828$, $r = 1714$ and $\chi^g_{x1} = 0.463$, $\chi^g_{x2} = 0.534$, $z = 1081$. F_I tends to very low values close to $x^*_2 \approx 0.2$ owing to the nearness of the stability limit of the gel. F_I is 0.7 at $x^*_2 = 0.4$, whereas F_I of the PS/toluene system is about 0.9 for the same concentration, which can be taken from Fig. 1. This difference is mainly due to the values of z and r. The F_I curve has been presented only for positive values which correspond to stable or metastable phase behaviour.

The experimentally determined diffusion coefficients, D^s, of Rehage and Ernst [5, 15] are plotted in Fig. 4 versus x^*_2 for the PS/cyclohexane system together with the $(D+E)^g$ value of Rehage for the gel [6]. It is seen that D^g for the lowest concentration at $x^*_2 = 0.48$ corresponds to the value at swelling equilibrium. The function$(D+E)^g(x^*_2)$ goes through a maximum and reaches very low values at $x^*_2 = 0.75$. There is no range of overlapping concentrations for $D^s(x^*_2)$ and $(D+E)^g(x^*_2)$. An extrapolation of $D^s(x^*_2)$ to higher polymer concentrations was not performed in the range $0.24 < x^*_2 < 0.5$, because the χ_x functions are valid only in the range measured.

The thermodynamic data of cross-linked PS in the various swelling agents considered in this paper are gathered in Table 1 [14]. The quality of the swelling agents decreases in the sequence chlorobenzene, toluene, cyclohexane for the gels. The χ_x-parameters increase in the same direction. With the relation $\chi^g_x = \alpha_x + \beta_x/T$ the entropy term (α_x) and the enthalpy term (β_x) of the χ^g_x-parameters are listed. There is a decrease of α_x and an increase of β_x. The temperature dependence of χ^g_x means that mixing entropy and mixing enthalpy are independent of temperature. A negative β_x-value is characteristic for an exothermal system. From the x^*_1 values it is seen that the maximum degree of swelling is observed in the best solvent. The quality of the solvents for gels and solutions reveals the same sequence [16].

Conclusion

The diffusion coefficient in solution and gels is composed of a mobility and a thermodynamic factor.

The diffusion in gels is slower than in solutions if the degree of polymerisation of the linear polymer and the degree of polymerisation of the network chains are comparable at identical compositions. This is explained by the elastically active coefficient which enters into the diffusion coefficient of gels.

A comparison of different solvents clearly reveals the importance of the thermodynamic factor, which may vary by orders of magnitude depending on the quality of the solvent.

If a solvent is a good solvent for a polymer in solution it is also a good swelling agent for a swollen cross-linked system.

The ratios of the sedimentation coefficients of gels and solutions are slightly smaller than 1.

References

1. Kisters D, Straatmann A, Borchard W (2002) Prog Colloid Polym Sci 119:92–100
2. Fujita H (1962) Mathematical theory of sedimentation analysis. Academic, New York
3. Haase R (1963) Thermodynamik der irreversiblen Prozesse. Steinkopf, Darmstadt
4. Rehage G, Ernst O (1964) Kolloid Z Z Polym 197:64–70
5. Rehage G, Ernst O (1964) DECHEMA Monogr 49:157–179
6. Rehage G (1959) Symposium on macromolecules, Wiesbaden, IIA 15
7. Flory PJ (1942) J Chem Phys 10:51
8. Huggins ML (1942) Ann NY Acad Sci 43:1
9. Staverman AJ, van Santen JM (1941) Rec Trav Chim Pays-Bas 60:76
10. Flory PJ (1950) J Chem Phys 18:108
11. Rehage G, Meys H (1958) J Polym Sci 30:271–284
12. Palmen HJ(1960) Diploma thesis. Aachen
13. Palmen HJ (1965) Doctoral thesis. Aachen
14. (a) Rehage G (1964) Kolloid ZZ Polym 194:97; (b) Rehage G (1964) Kolloid Z Z Polym 196:16
15. Ernst O (1962) Doctoral thesis. Aachen
16. Ernst O (1959) Diploma thesis. Aachen
17. Scholte THG (1970) J Polym Sci A2 8:841–868

Progr Colloid Polym Sci (2004) 127: 19–25
DOI 10.1007/b94246

L. Börger
M. D. Lechner
M. Stadler

Development of a new digital camera setup for the online recording of Schlieren optical pictures in a modified Beckman Optima XL analytical ultracentrifuge

L. Börger (✉) · M. Stadler
Polymer Research,
BASF Aktiengesellschaft,
67056 Ludwigshafen,
Germany
e-mail: lars.boerger@basf-ag.de
Tel: +49-621-6095339
Fax: +49-621-60-92281

M. D. Lechner
Physical Chemistry, University,
Barbarastrasse 7, 49069 Osnabrück,
Germany

Abstract Because Schlieren optical systems (e.g. essential for density gradients with steep refractive index gradients) are commercially not available a homemade system has been developed by transferring the existing Schlieren optics of a Beckman model E analytical ultracentrifuge to a preparative Beckman Optima XL analytical ultracentrifuge with an eight-cell multiplexer. In this paper we describe the expansion and further development of this system by introducing a new optical alignment that allows the online recording of Schlieren optical pictures using a digital camera. The digital black and white charge-coupled-device camera allows the control of the whole recording process, including, for example, control of picture brightness and contrast. Going along with the development of the physical setup, a new modular software concept has been worked out that clearly distinguishes three steps of picture recording, editing of the pictures and transforming them into analyzable data and finally analyzing the data with regard to the desired physical information. In this contribution details of the optical setup are described as well as the attendant software developments. Examples from industrial practice demonstrate the abilities and power of this new development in research and development applications as well as in routine industrial analytical measurements.

Keywords Analytical ultracentrifugation · Schlieren optics · Online recording · Polymer characterization · Colloid characterization

Introduction

The analytical ultracentrifuge (AUC) is a powerful tool in characterization of polymers and colloidal systems [1, 2]. Its fractionating ability allows the analysis of complicated multicomponent systems that are important especially in industrial research and development. Many industrial products, like paints, coatings and adhesives, are based on such colloidal systems, mainly polymeric dispersions. The detailed characterization of these products or preproducts leads to a better understanding of structure–property relationships. That is the basis for both the improvement of existing and the development of new products. The underlying phenomena are very often of basic colloidal nature. For studying this, advantageously different methods of AUC can be applied, including sedimentation velocity runs, synthetic boundary and density gradient experiments [3]. These different types of experiments give access to information on various properties, like the molar mass distribution, the particle size distribution and chemical heterogeneity. In the AUC laboratory at BASF we study in the order of 3,000 samples a year. Mostly particle size distributions measurements and density gradient runs are performed on these samples using turbidity and Schlieren optics. The principle measurement for any of these experiments is the solute concentration as a function of the radial position.

Up to now basically five different types of detectors can be built in ultracentrifuges: turbidity optics [4], interference optics [5], UV/vis absorption optics [6], fluorescence optics [7] and Schlieren optics. Both fluorescence and turbidity optics have been realized in only very few laboratories throughout the world. In our laboratory homemade turbidity optics are integrated and used to measure particle size distributions in four ultracentrifuges.

As far as the AUC community is concerned, the most commonly used detectors are UV/vis absorption optics (limited to absorbing samples) and universal interference optics. Schlieren optics were a widely used detector system as the precursor model E AUC comprised a Schlieren optical system. With the launch of the Optima XL-I/A system this changed, because the Schlieren optical system is not integrated in the Beckman XL-I/A instrument anymore, whereas UV/vis absorption and interference optics are. In an earlier work it was described how the model E Schlieren optics were transferred to a Beckman Optima XL [8]. We expanded and further developed this existing system. This is justified because Schlieren optics are indispensable for our daily work owing to certain unique features. Just two are named here:

- It allows the use of monosector cells that are easy to handle and assure highest reproducibility.
- Even steepest refractive index gradients (as they often result from steep density gradients) can be followed. As already mentioned density gradient measurements are among the most frequently performed experiments in our laboratory; therefore, it is essential for everyday work to be able to analyze this kind of experiment under manifold conditions.

The expansion and further development of the existing Schlieren optical system as described in the following creates new advantages compared with the preceding system; only a few advantages are mentioned here:

- The time- and money-consuming step of photographic film development is skipped.
- Optimal picture quality is achievable as the operator can customize the photographic parameters for each cell and each rotor speed individually.
- With the new computer-based triggering of the flash lamp the centrifugation processes can be followed at all rotor speeds, even below 10,000 rpm.
- The moment of overlaying in a fast dynamic density gradient as well as in overlaying experiments can be observed directly.
- Long run experiments, like static density gradients, can be performed automatically; therefore, the average utilization of the machines is increased as, for example, experiments over the weekend are possible.
- The digital camera now allows an exact geometrical superimposition of sequential pictures; therefore, the error of measurement in sedimentation runs to measure complex multi-modal molecular weight distributions is reduced dramatically.
- The analysis of Schlieren lines is performed by newly developed highly sophisticated software. By this the achievable precision and resolution of Schlieren optics is increased considerably.

Physical and optical setup

The scheme of the physical and optical setup of the modified Beckman XL equipped with the newly developed Schlieren optical system is shown in cross-section in Fig. 1.

The arrangement of the optical elements is basically as in the model E Schlieren optics. The white light flash lamp illuminates the Schlieren slit. The light then passes on through the collimating lens, a glass prism and then enters the vacuum chamber through a vacuum-sealed window. After passing the sample-containing cell the light is focused by the condensing lens through the second vacuum-sealed window onto the plane of the phase plate. The newly setup optical arm contains the camera lens and the cylindrical lens. The light then falls onto the charge-coupled-device (CCD) camera, which is connected to a computer for further processing.

Modifications of the Optima XL

As already described [8] certain modifications had to be added to the Optima XL:

- An optical bench containing the light source and Schlieren slit assembly was mounted on the XL. The flash lamp can be adjusted by an *x-y-z* stage that it is put onto.
- A horizontal hole was drilled and a tube containing the collimating lens and the prism was added into the heat sink.
- The vacuum chamber was modified by drilling in the two holes that allow the mounting of vacuum-sealed windows and by adding a rack that carries the condensing lens.
- The CCD camera and the remaining optical elements were mounted on a scaffold above the vacuum chamber as indicated in Fig. 1.

A picture of the physical setup as realized in our laboratory is shown in Fig. 2.

Fig. 1 Scheme of the setup of the new XL-SO II analytical ultracentrifuge: flash lamp (*1*), Schlieren slit (*2*), collimating lens (*3*), glass prism (*4*), XL drive (*5*), heat sink (*6*), vacuum-sealed windows (*7*). After passing the sample-containing cell the light is focused by the condensing lens (*9*). Vacuum chamber (*10*), phase plate (*11*), camera lens (*12*), cylindrical lens (*13*), charge-coupled-device camera (*14*)

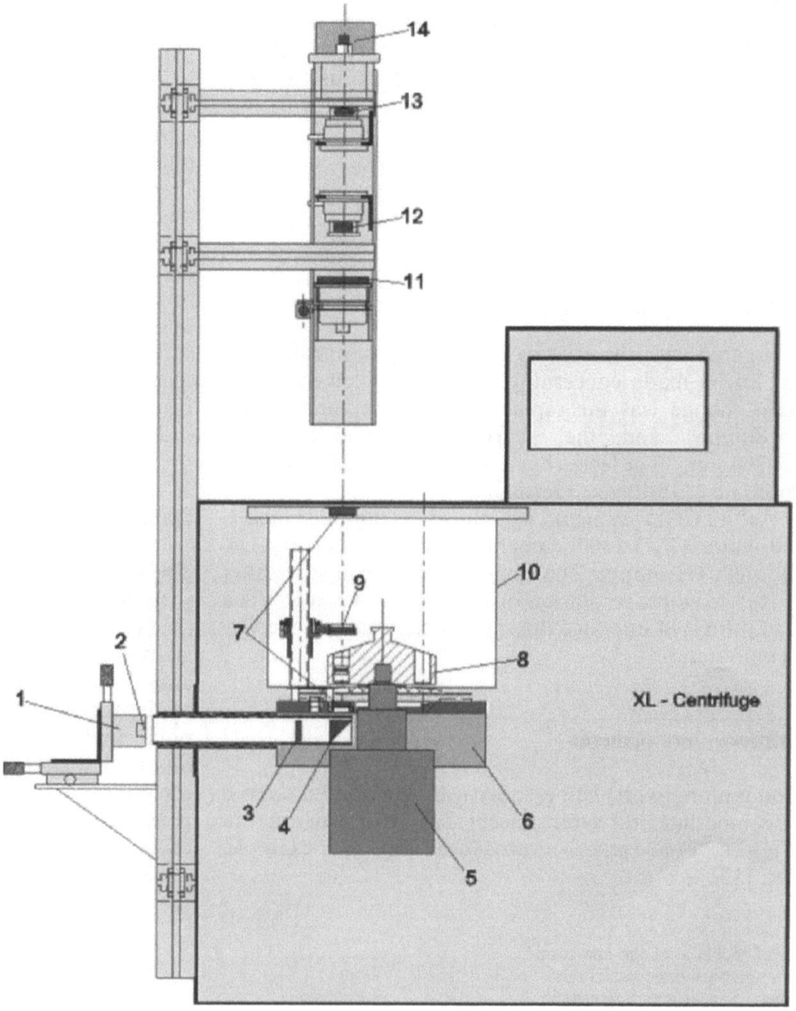

Light source

In principle, either the light source, the detector or an element in the optical light-path has to be triggered in Schlieren optical systems of AUCs [9, 10, 11]. For our optical setup we chose to trigger the light source mainly because of the high reliability of the stroboscopic light source and the possibility to control the light source by the newly developed multiplexer software. The ease to create superimposed pictures of more than one cell is another reason for triggering the light source. In each centrifugation run there is one reference cell per rotor.

◄

Fig. 2 Photographic picture of the setup of the new XL-SO II as realized in our laboratory

This cell is used to perform radial calibration during centrifugation by superimposing the Schlieren picture of the reference cell and the Schlieren picture of the cell containing the samples examined. This allows the measurement of up to seven different samples in one centrifugation run.

As a light source a Cathodeon C 82007 xenon flash lamp (LOT/oriel, Darmstadt, Germany) was chosen (flash sequence 0–100 Hz, flash energy 1 J).

Optical system and camera

Compared with the existing system only minor changes had to be made concerning the optical elements. The main change was the replacement of the camera lens ($f = 80$ mm) and the achromatic cylindrical lens ($f = 200$ mm). The lenses have been purchased from Linos Photonics, Göttingen, Germany.

As a CCD camera we chose a digital black+and-white CCD 1300 camera (VDS Vosskühler, Osnabrück, Germany). The camera allows exposure times of 10^{-5} se with a resolution of 1,280 × 1,024 pixels. Via a digital RS-644 interface the camera can be controlled by a computer.

Software developments

Going along with the new physical and optical setup a new modular software concept has been worked out (Fig. 3). For earlier approaches see, for example, Ref. [12].

Data acquisition

Data acquisition was realized via a Labview-based software tool using a National Instruments DAQ board (National Instruments, Austin, USA). The triggering signal was created by a homemade system. For that a photoelectric cell was added to the bottom of the vacuum chamber. The photoelectric cell consists of a continuously light emitting diode (LED) and a fast photodiode. The light of the LED is reflected by a mirror that is fitted on the base of the analytical rotor. By this alignment each revolution of the rotor generates one sharp electrical pulse. This signal is imported by the data acquisition software and used for the calculation of the actual rotor speed and for the calculation of the actual cell positions. From that the times the flash lamp has to be pulsed in order to illuminate the desired cells are calculated.

Data reduction

The CCD camera supplies Schlieren optics photographs with a resolution of 1,280 × 1,024 pixels as bitmap (BMP) files. These BMP files are reduced semiautomatically into X, Y files with $X = R =$ absolute rotor radius and $Y = dn/dR$. Semiautomatically means that the radius of the meniscus, the radius of the bottom and the start point and the end point of the Schlieren curve are digitized manually, whereas all of the Schlieren curves are digitized automatically. In the case of sedimentation runs (program VelScan) all Schlieren curves at different times are automatically digitized and transformed into X, Y files.

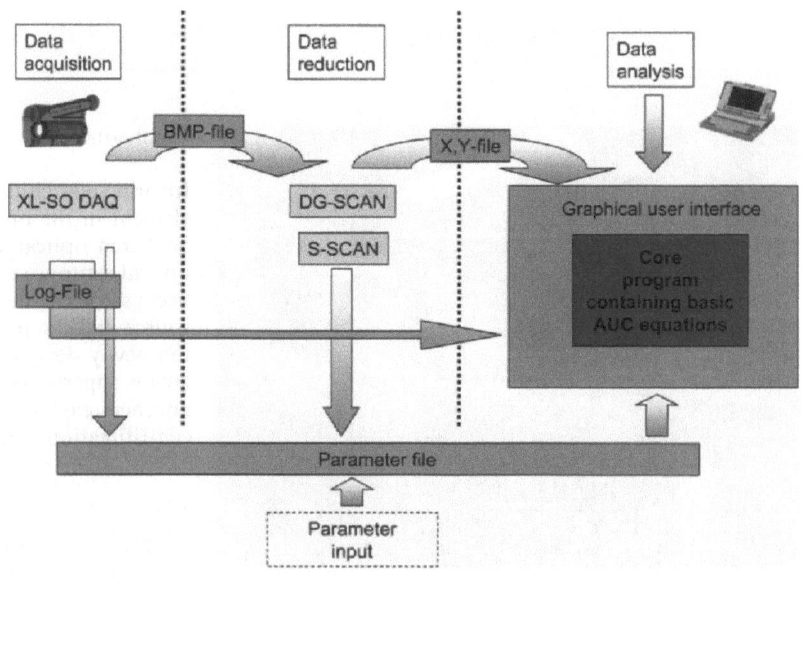

Fig. 3 Scheme of the new modular software concept. Three steps are clearly distinguished: picture recording, editing of the pictures and transforming them into analyzable data and finally analyzing the data with regard to the desired physical information

Data analysis

The data analysis was performed with the programs DGXLS and VelXLS. The program DGXLS evaluates static and dynamic density gradients by calculating and plotting the density of the dissolved or dispersed particles as a function of the rotor distance. In the case of static density gradients this job is done according to (1) the Hermans–Ende method, (2) a calibration method using particles of known density, and (3) mass conservation with respect to the index of refraction as a function of the rotor distance in the cell (refractometry) [14].

The program VelXLS evaluates sedimentation runs, flotation runs, and runs with a synthetic boundary. The following physical quantities can be determined: all kinds of sedimentation coefficients, sedimentation coefficient distributions, diffusion coefficients, particle sizes, particle size distributions, all kinds of molar masses, molar mass distributions and virial coefficients.

Example measurements

In the following some examples from industrial research and development are presented to demonstrate the quality and the versatility of the new Schlieren optical system.

Sedimentation velocity run

Two pictures from the sedimentation velocity run of a bovine serum albumin sample ($c = 4$ g/l) are compared in Fig. 4. The run was performed at 40,000 rpm with 25-mm monosector cells.

The Schlieren picture after 100 min taken with the old reflex-camera-based photographic system is shown in Fig. 4a. In contrast Fig. 4b displays the Schlieren picture recorded with the new digital camera (also after 100 min). Both pictures are superimpositions of the Schlieren pictures of the corresponding sample cells and the reference cell. Whereas in the reflex camera system approximately 300 flashes were needed to generate a picture of this quality, with the new digital camera system 5–15 flashes of the xenon flash lamp are sufficient. This, of course, allows us to follow faster processes via Schlieren optics.

From the detailed comparison of the two pictures taken with the two different optical systems it can be concluded that equivalence of the optical systems is reached.

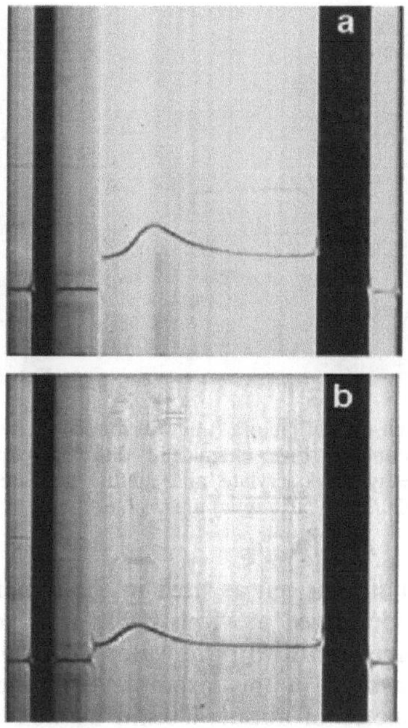

Fig. 4 Schlieren picture taken with the old reflex-camera-based photographic system (after 100 min); **b** Schlieren picture recorded with the new digital camera (after 100 min)

Dynamic density gradient

The next example makes use of the ability of Schlieren optics to resolve steep refractive indices as well as of the ability of the new digital camera system to follow fast processes. The Schlieren optical picture (after 20 min) taken from an experiment using a special synthetic boundary technique (a so-called dynamic density gradient) is shown in Fig. 5. The technique has been described elsewhere [3]. The sample measured in this example contained 11 differently composed model latex samples. The latex samples were prepared by copolymerization of ethylhexyl acrylate and methyl acrylate monomers with varying relative concentrations [poly(methyl acrylate) = 0/10/20/30/40/50/60/70/80/90/100% w/w] resulting in different densities of the latex samples [ρ(samples) = 0.980/1.000/1.021/1.043/1.066/1.089/1.114/1.167/1.196/1.225 g/cm^3]. From these 11 different samples equal portions were mixed and a dynamic density gradient experiment was performed on the resulting mixture. Within the density range covered by the dynamic density technique (1.00–1.10 g/cm^3) we expect to find five different latex samples according to the densities just given. And in fact Fig. 5 shows all of the expected Schlieren optical signals indicating the presence of chemically differing samples. The absolute density values accessible from the analysis of this picture via the procedure described earlier match the theoretical values exactly so they do not have to be repeated here.

Synthetic boundary experiment

The last example in this work demonstrates another special synthetic boundary type of experiment. In this

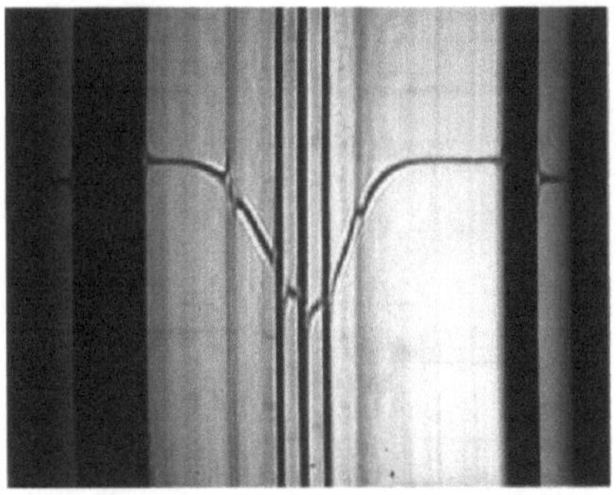

Fig. 5 Schlieren optical picture (after 20 min) of a dynamic density gradient experiment. Five chemically differing copolymer latex fractions of ethylhexyl acrylate and methyl acrylate [poly(methyl acrylate) = 10/20/30/40/50% w/w) are visible

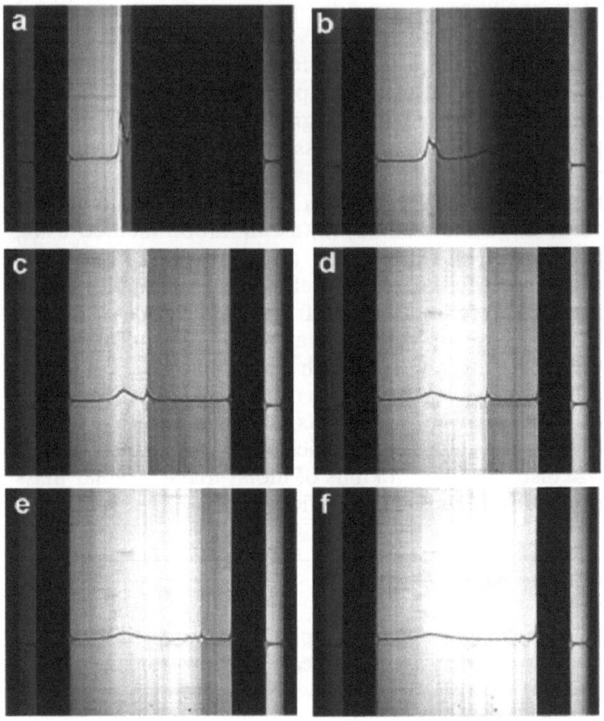

Fig. 6 Schlieren pictures of a synthetic boundary run of a synthetic polymer latex sample taken at different times after the overlaying occurred: **a** 1 min, **b** 3 min, **c** 13 min, **d** 28 min, **e** 35 min, **f** 43 min.). The experiment was performed at 40,000 rpm (Philpot angle 70°)

technique a similar cell as used in the dynamic density experiment described previously was used to overlay a polymer dispersion with pure water at 40,000 rpm. The sample measured in this example contained particles with a diameter of about 400 nm as well as particles with a diameter of about 25 nm. The Schlieren picture taken 1 min after the overlaying occurred is shown in Fig. 6a. The experiment is usually performed at high latex concentrations making the sample turbid at the beginning of the experiment. Owing to the relatively high rotor speed the fast sedimenting 400-nm latex particles contained in the sample examined are forced to the bottom of the cell. This can be seen in Fig. 6b (after 3 min). The sedimentation of the 25-nm species indicated by the shift of the turbid band towards the bottom of the cell can be followed in the Schlieren pictures taken after 13 min (Fig. 6c), 28 min (Fig. 6d), and 35 min. (Fig. 6e). The last Schlieren picture of this series (Fig. 6f, after 43 min) shows a flat Schlieren peak at the radial position where the overlaying had occurred. This peak indicates the presence of nonsedimenting low molecular weight polymers in the sample. By analyzing the peak area, the amount of the nonsedimenting portion of the sample can be quantified (in this case approximately 3% w/w).

Conclusion

In this paper a new Schlieren optical systems for AUCs was described. It is based on the Schlieren optics of the model E AUC that was expanded and further developed by introducing a new optical alignment that allows the online recording of Schlieren optical pictures using a digital camera. Going along with the development of the physical setup, a new software concept has been worked out and realized. In future developments a more automated processing of the data reduction step is desirable [13]. The outcome of this is a fast and powerful universal detector system combining high optical resolution with comfortable manageability of the detector operation. This has been demonstrated with examples from industrial research and development measurements.

Acknowledgements The authors thank W. Mächtle for his great contributions in setting up the precursor Schlieren optical system and fruitful discussions during the setup of the new one. K. Reindel is acknowledged for his serious contributions to the installation of the camera and the data acquisition procedure. Thanks go to K. Vilsmeier for discussion of the software concept and for exemplary measurements. F. Hörner is acknowledged for help with the physical setup.

References

1. Cölfen H, Völkel A Prog Colloid Polym Sci 127:
2. Mächtle W (2000) In: Meyers RA (ed) Encyclopedia of analytical Chemistry. Wiley, Chichester, pp
3. Mächtle W (1992) In: Harding SE, Rowe AJ, Horton JC (eds) Analytical ultracentrifugation in biochemistry and polymer science. Royal Society of Chemistry, Cambridge, pp 147–175
4. Scholtan W, Lange H (1972) Kolloid-Z Z Polym 250:782–796
5. Laue TM, Anderson AL, Demaine PD (1994) Prog Colloid Polym Sci 94:74–81
6. Lloyd PH (1974) Optical methods in ultracentrifugation, electrophoresis and diffusion. Clarendon Press, Oxford
7. Schmidt B, Riesner D (1992) In: Harding SE, Rowe AJ, Horton JC (eds) Analytical ultracentrifugation in biochemistry and polymer science. Royal Society of Chemistry, Cambridge, pp 176–207
8. Mächtle W (1999) Prog Colloid Polym Sci 113:1–9
9. Mächtle W, Klodwig U (1979) Makromol Chem 180:2507–2511
10. Mächtle W, Klodwig U (1976) Makromol Chem 177:1607–1612
11. Kisters D, Borchard W (1999) Prog Colloid Polym Sci 113:10–13
12. Clewlow AC, Errington N, Rowe AJ (1997) Eur Biophys J 25:311–317
13. Klodwig U, Mächtle W (1989) Colloid Polym Sci 267:1117–1126
14. Mächtle W, Lechner MD (2002) Prog Colloid Polym Sci 119:1–10

Progr Colloid Polym Sci (2004) 127: 26–30
DOI 10.1007/b94247

A. Straatmann
T. Windhues
W. Borchard

Effects of acetylation on thermodynamic properties of seaweed alginate in sodium chloride solutions

A. Straatmann · T. Windhues
W. Borchard (✉)
Fakultät für Naturwissenschaften,
Universität Duisburg–Essen,
47048 Duisburg, Germany
e-mail: borchard@uni-duisburg.de
Fax: +49-203-3792110

Abstract Alginates are polysaccharides consisting of mannuronate and guluronate units. While bacterial alginate is acetylated, seaweed alginate does not contain acetyl groups. This is one of the main differences between algal and bacterial alginates. The acetyl groups play an important role in the physical–chemical properties of the alginates. To determine the influence of acetyl groups on the properties of alginates in solutions, seaweed alginates were acetylated according to a procedure described by Skjåk-Bræk et al. Effects of the acetylation on the size, the interactions and the shape of the macromolecules were investigated amongst other methods with analytical ultracentrifugation. Though the molar mass of the alginate hardly increases after acetylation, a size-increasing effect on the alginate spheres is observed. In very dilute solutions positive second virial coefficients are observed for acetylated and nonacetylated alginate. Acetylation increases the second virial coefficients of alginate solutions, indicating better interaction of acetylated alginate with the water molecules than nonacetylated alginates.

Keywords Alginate · Acetylation · Aggregation · Virial coefficients

Introduction

Alginates are used in a wide range of industrial applications as a gelling and thickening agent [1]. For medical purposes they have been applied as immobilization material for cells [2]. They can be isolated from different sources. Commercial alginates are extracted from brown seaweeds, which can be harvested throughout the year [1]. Alginates are also produced by some bacteria, like *Pseudomonas aeruginosa* and *Azotobacter vinelandii.* Alginate from the bacterium *P. aeruginosa* is known to cause lung infections in patients with cystic fibrosis [3].

Alginates are copolymers of α-L-guluronate (G) and its epimer β-D-mannuronate (M). The monomers are distributed as homogeneous blocks (MM or GG) and in alternating sequences (GM) as sketched in Fig. 1. The distribution of the different units depends on the source of the alginate [4].

Alginates from bacteria vary from seaweed alginate in different ways. Bacterial alginates have much higher molecular masses than commercial alginates [5]. In contrast to seaweed alginates bacterial alginates contain acetyl groups. Some of the mannuronate residues are acetylated in the O-2 or the O-3 position, as shown in Fig. 2 [6].

The acetyl group protects the mannuronate residues from epimerisation to guluronate and therefore controls the monomer sequence of the polymer [6]. The lack of G blocks in the alginate from *P. aeruginosa* might be affected by the acetyl groups [7]. Though the function of the acetyl groups concerning the biological synthesis of alginate has been widely investigated over the last few years, its influence on the physicochemical properties of alginate solutions has hardly been discussed.

An increased chain extension of alginates after acetylation in sodium chloride solutions ($c = 0.1$ mol/l) was

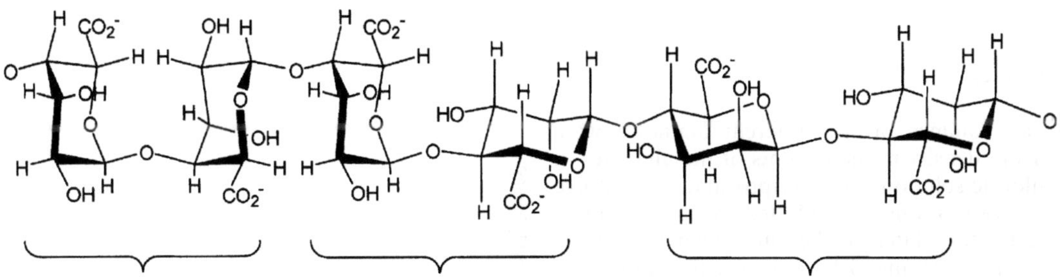

Fig. 1 Structure of alginate sequences: mannuronate unit (*M*), guluronate unit (*G*)

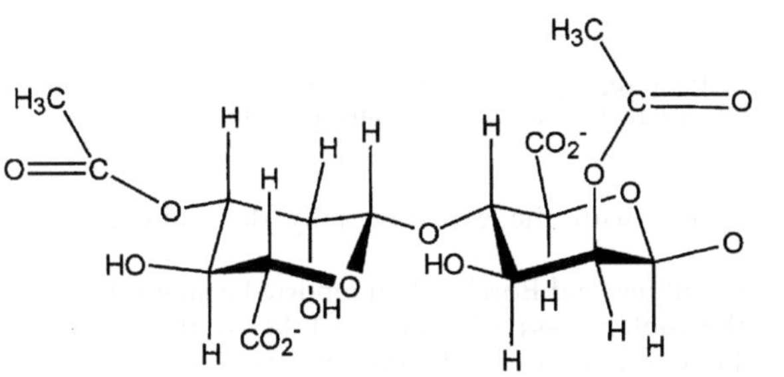

Fig. 2 Structure of acetylated M units

found by Skjåk-Bræk et al. [8] using viscosity measurements. Recently the effects of acetylation on thermodynamic properties of alginate in physiological sodium chloride solution was investigated using light scattering techniques. For the alginate Manucol DM an increase of the radius of gyration after acetylation was determined, while at the same time a slight decrease of the degree of polymerisation was detected [9]. Slightly negative virial coefficients were reported for both the acetylated and the nonacetylated seaweed alginate. The second virial coefficient became less negative after acetylation, indicating better interaction of the acetylated alginate with the solvent than nonacetylated alginates.

In this contribution the thermodynamic properties of a seaweed alginate and the same seaweed alginate after chemical acetylation are determined by means of analytical ultracentrifugation. The differences in the properties of these alginates are discussed and the results compared with the results from static light scattering measurements.

Experimental

Acetylation

The acetylation of Manucol DM (Monsanto) was performed according to a slightly modified procedure described by Skjåk-Bræk et al. [10]. To a calcium chloride solution (0.1 mol/l) 500 ml alginate solution (2 g/l) was added dropwise. The alginate beads were left in the solution overnight, the supernatant fluid was then decanted and 500 ml pyridine was added to the precipitate before storage for 1 day at room temperature. After decantation of the pyridine, the alginate beads were covered with a mixture of pyridine and acetic anhydride (1:1). The reaction mixture was stored at room temperature for 12 h and the excess pyridine/acetic anhydride mixture was then decanted. The remaining precipitate was washed with 250 ml acetone and 250 ml deionised water before dissolving it in 250 ml sodium ethylenediaminetetraacetate (0.05 mol/l). The solution was dialysed for 24 h against 5 l sodium chloride solution (0.1 mol/l) and deionised water and the alginate was freeze-dried afterwards. The acetyl content was determined by a photometric assay according to the method of Hestrin [11].

The degree of acetylation was 0.48 acetyl groups per uronate unit.

Ultracentrifuge measurements

For the ultracentrifuge runs the alginate was dialysed against water and freeze-dried. The sodium alginate was dissolved in 0.14 mol/l sodium chloride solution and stirred overnight to prepare a stock solution (0.25 wt%). The stock solution was diluted to prepare concentrations between 0.05 and 0.25 wt%.

The measurements were performed with an analytical ultracentrifuge equipped with Schlieren optics (Beckman). A six-hole An-G rotor with sector-formed cells (sector angle 4°) was used. The velocity runs were performed at 30,000 rpm and the sedimentation equilibrium runs were performed with a velocity of 18,000 rpm at 20 °C.

The sedimentation coefficients were determined from the movement of the hypothetical sharp boundary according to the method of Goldberg [12]. The diffusion coefficients were derived from the height–area ratio of the moving boundary.

The apparent molar mass was calculated from the sedimentation equilibrium by using the following equation [6]:

$$\int \rho_2(1 - \bar{V}_2^0\rho)r\omega^2 dr = \mathrm{RT}\left(\frac{1}{M_{\mathrm{w,app}}}\rho_2 + \mathrm{A}_2{}^*\rho_2^2\right),$$

where ρ_2 is the partial density or mass concentration, $\bar{V}_2^0$ is the specific volume of the pure polymer, ρ is the density of the solution, r is the distance from the axis of rotation, ω is the angular velocity, R is the ideal gas constant, T is the thermodynamic temperature, M_w is the weight-average molar mass and A^*_2 is a second virial coefficient.

The reciprocal apparent molar mass was then plotted versus the partial density to obtain the weight-average molar mass from the intercept and the second virial coefficient of the osmotic pressure from the slope.

The densities of the alginate solutions were measured with a DMA 60 density meter from Paar (Graz, Austria) with two comparative measuring cells.

Results

Hydrodynamics

The specific volumes of the different alginates were obtained from density measurements of the alginates in sodium chloride solutions. The reciprocal density at five different concentrations was plotted versus the mass fraction and extrapolated to the pure polymer. For the seaweed alginate Manucol DM a specific volume of $\bar{V}_2^0 = 0.48$ l/g was obtained, which is in the range which is known for alginates [13]. After acetylation the partial specific volume increases to $\bar{V}_2^0 = 0.70$ g/l This can be explained by sterical effects of the acetyl groups. By replacing the small hydroxyl group with the bigger acetyl group the polymer chain is expanded [8]. This leads to a decrease in density and therefore an increase in the specific volume.

The specific volume of the acetylated alginate is in the same range as the specific volume of a bacterial alginate from *P. aeruginosa*, which has a value of $\bar{V}_2^0 = 0.76$ l/g [5]. This indicates further that the acetylation causes a widening of the coils.

This widening effect obtained from the density measurements is in agreement with the larger radii of gyration determined with light scattering techniques for the acetylated alginate [9]. Theoretical considerations of Stokke and Smidsrod [14] confirmed this size increase. They reported that the conformational freedom of the polymer chain is decreased probably owing to a hindered rotation of the polymer resulting in an increase in the coil size.

Sedimentation equilibrium

For alginate solutions a limiting concentration was reported above which aggregates are formed. This concentration was determined by a step which occurred in the Schlieren curve at the limiting concentration [5]. Such behaviour was not found for the acetylated alginate, though a slight decrease in the slope of the Schlieren curve could be detected near the bottom of the cell. This indicates that no microgels are formed and the aggregation tendency of the alginate is decreased after acetylation.

For Manucol DM a molar mass of 130,000 g/mol was determined [5]. The molar mass of the acetylated alginate is with 140,000 ± 20,000 g/mol the same if the margin of error is considered. Since owing to the acetylation, hydroxyl groups are exchanged for acetyl groups a slight increase of the molar mass is expected. Considering the degree of acetylation of 0.48 acetyl groups per uronate unit the molar mass should increase by 12 wt%. This is equal to a molar mass of 145,000 g/mol, which is consistent with the experimentally determined molar mass.

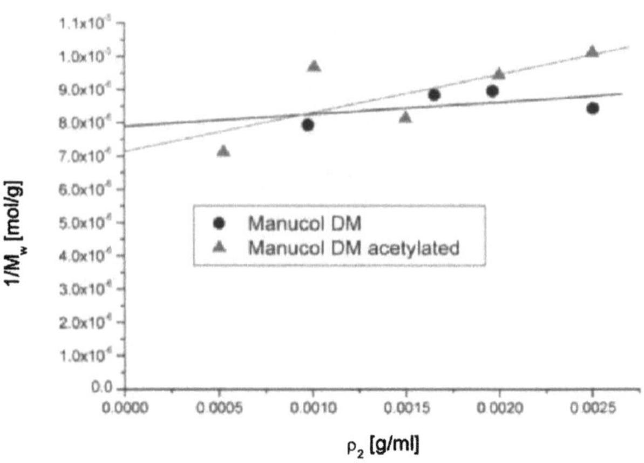

Fig. 3 Reciprocal apparent molar mass of acetylated and nonacetylated alginate versus the mass concentration of the alginates

Windhues and Borchard [9] reported a slight decrease of the molar mass of Manucol DM from 128,000 to 97,000 g/mol after acetylation, and the degree of acetylation was determined to be 0.37. This decrease cannot be verified here and might be due to different conditions during acetylation.

The reciprocal molar mass of the acetylated and nonacetylated alginate is plotted against the mass concentration of the polymer in Fig. 3. It can be seen that both plots have a positive slope, but the slope of the acetylated alginate is greater. From these slopes the second osmotic virial coefficients A_2 were calculated, leading to values of $A_2 = (2 \pm 2)\times10^{-4}$ ml mol/g^2 for the nonacetyled alginate and $(6 \pm 3)\times10^{-4}$ ml mol/g^2 for the acetylated alginate. This increase further indicates that the aggregation tendency of the alginate is decreased by acetylation. Though the acetyl group is hydrophobic in contrast to the hydroxyl group, aqueous solution is a better solvent for the acetylated alginate. The acetyl group sterically hinders the aggregation of the molecules. Furthermore, the hydroxyl group can bind together via hydrogen bonds, which is impossible for the acetyl groups.

The increase of the virial coefficient after acetylation is in agreement with the results of Windhues and Borchard [9]. Though they found negative virial coefficients in both cases, the virial coefficient of the acetylated alginate was less negative. This indicates better interaction of the acetylated alginate with the solvent. The light scattering measurements were performed in the concentration range between 1 and 5 g/l, whereas the ultracentrifugal measurements were performed in the concentration range from 0.5 to 2.5 g/l. Former results from sedimentation equilibrium for different algal alginate solutions

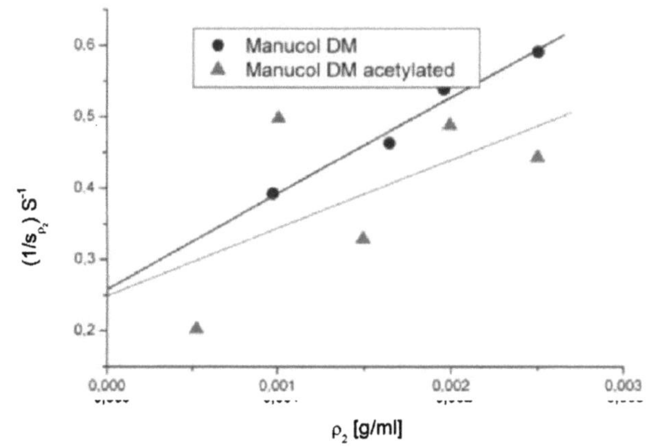

Fig. 4 Reciprocal apparent sedimentation coefficients of acetylated and nonacetylated alginate versus the mass concentration of the alginates

showed that the virial coefficient became negative above 2.5 g/l [5]. Therefore, the differences in the results obtained by these methods are probably due to the different concentration range.

Sedimentation velocity

The Schlieren peaks of both the nonacetylated and the acetylated alginate are widened during sedimentation. The sedimentation coefficients were therefore calculated from the hypothetic boundary according to the method of Goldberg [12]. The limiting sedimentation coefficients were $s_2=3.9\pm0.4$ S before acetylation and $s_2=4.0\pm2.0$ S after acetylation. The difference is within the error range of the results and the sedimentation coefficient is therefore unchanged after acetylation. A plot of the apparent sedimentation coefficients versus the mass concentration of the polymer is illustrated in Fig. 4.

The diffusion coefficients were obtained using the height–area ratio. For the nonacetylated alginate a diffusion coefficient of $(1.0\pm0.4)\times10^{-7}$ cm^2/s was calculated, which is slightly higher than the diffusion coefficient reported for other alginates, which ranged from 0.3×10^{-7} to 0.9×10^{-7} cm^2/s [13]. The higher diffusion coefficient of Manucol DM is expected because of the method. Peak-broadening is partly caused by polydispersity. This effect cannot be separated when calculating the diffusion coefficient. The limiting diffusion coefficient increased to $(4.0\pm0.9)\times10^{-7}$ cm^2/s after acetylation. Considering that the radius of gyration increased after acetylation, a decrease of the diffusion coefficient is expected. Probably the acetylation leads to a higher polydispersity of the alginate, which leads to a larger overestimation of the diffusion coefficient than for macromolecules with lower polydispersity. The acetylation might not be homogeneous, leading to different molar masses of the individual chains.

Conclusion

The larger radii of gyration and the higher specific volume of the acetylated alginate in comparison to the nonacetylated alginate indicate that the acetyl groups have a widening effect on the alginate macromolecules. The acetylation decreases the tendency for aggregation in dilute solutions and aqueous sodium chloride solution is a better solvent for the acetylated alginate than for the nonacetylated alginate, which can be concluded from the increased second virial coefficients of the osmotic pressure.

Using sedimentation equilibrium the molar mass could be detected from measurements in the dilute regions and positive virial coefficients were obtained. Owing to the higher concentrations compared to samples used for sedimentation equilibrium negative virial coefficients were found with light scattering techniques.

The increased diffusion coefficient of the acetylated alginate is probably caused by higher polydispersity after acetylation. The polydispersity leads to increased peak-broadening and an apparent diffusion coefficient, which is higher than the true value.

Acknowledgements. The authors are participants of the research group Physical Chemistry of Biofilms and thank the Deutsche Forschungsgemeinschaft for financial support.

References

1. Rehm BHA, Valla S (1997) Appl Microbiol Biotechnol 48:281
2. Gutowska A, Jeong B, Jasionowski M (2001) Anatom Rec 263:342–349
3. Hanlon GW, Denyer SP, Oliff J, Ibrahim L (2001) Appl Environ Microbiol 67:2746
4. Rees DA, Welsh EJ (1977) Angew Chem 89:228
5. Straatmann A, Borchard W (2002) Prog Colloid Polym Sci 119:64
6. Christensen BE (1999) In: Wingender J, Neu TR, Flemming HC (eds) Microbial extracellular polymeric substances. Springer, Berlin Heidelberg New York, p 143
7. Schürks N, Wingender J, Flemming HC, Mayer C (2002) Int J Biol Macromol 30:105
8. Skjåk-Bræk G, Zanetti P, Paoletti S (1989) Carbohydr Res 185:131

9. Windhüs T, Borchard W (2003) Carbohydr Polym 52:47
10. Skjåk-Bræk G, Paoletti S, Gianferrara T (1988) Carbohydr Res 185:1
11. Hestrin SJ (1949) J Biol Chem 180:249
12. Goldberg RJ (1953) J Phys Chem 57:194–202
13. Wedlock DJ, Fasihuddin BA, Phillips GO (1986) Int J Biol Macromol 8:57
14. Stokke B, Smidsrod O (1993) Carbohydr Polym 22:57

Progr Colloid Polym Sci (2004) 127: 31–47
DOI 10.1007/b98011

Helmut Cölfen
Antje Völkel

Analytical ultracentrifugation in colloid chemistry

Abstract Analytical ultracentrifugation (AUC) has proven to be a valuable tool for the study of colloids since its invention by Svedberg, who initially used it for the particle size characterization of gold colloids. The character of colloids raises analytical questions, which are often different from those which are commonly addressed for biopolymers; therefore, AUC analysis of colloidal systems differs from that of well-defined proteins. Colloids are often extremely polydisperse owing to an inherent tendency to aggregate and thus have to be treated as interacting polydisperse systems. Furthermore, colloids are often hybrid systems (e.g. inorganic–organic) and have a particle size and density distribution. A further complicating feature is that many colloids, especially inorganic species, are extremely small, so diffusion broadening of boundaries can make AUC analysis of such systems difficult. This overview paper highlights which answers AUC can give for colloidal systems despite the previously mentioned obstacles, which restrict the use of many common AUC evaluation techniques for proteins. Four principal examples are presented and discussed: (1) Pt and ZrO_2 colloids as examples for the determination of particle size distributions of extremely small colloids in the nanometre range; (2) an interacting binary mixture of reactive latices with a protein, which leads to aggregation as an example for the quantitative assessment of colloid aggregation phenomena; (3) iron oxides in a carrageenan microgel, coated metal particles and a growing inorganic mineral inside an organic aggregate as organic–inorganic hybrid particles, as well as ferritin, an organic–inorganic hybrid colloid which has a particle size and density distribution as an example for the combination of AUC with other size-dependent but density-independent methods like flow-field-flow fractionation to yield further insight into the nature of the different oligomers; and (4) synthetic boundary crystallization ultracentrifugation as an example of how fast processes like crystal growth can be transferred to a spatial resolution of species which can be analysed by the much slower AUC technique. These examples illustrate how common problems in colloid chemistry can be solved by AUC and which pitfalls and future perspectives exist.

Keywords Colloids · Particle size distribution · Nanoparticles · Aggregation · Analytical ultracentrifugation

H. Cölfen (✉) · A. Völkel
Department of Colloid Chemistry,
Max Planck Institute of Colloids and Interfaces, MPI Research Campus Golm,
14424 Potsdam, Germany
e-mail: coelfen@mpikg-golm.mpg.de
Tel.: +49-331-5679513
Fax: +49-331-5679502

Introduction

Colloids are a class of particles and molecules which gained increasing importance throughout the last decade with the advent of nanochemistry/physics. Although important basic work on these systems was done at the beginning of the last century, only the development of improved analytical methods allowed a more precise understanding and development of colloidal systems. Colloids are commonly defined as molecules/particles with a size in the range 1–1,000 nm [1], corresponding to molar masses in the range 10^4–10^{12} g/mol. Thus, they represent an intermediate state of matter between atoms and the macroscopic solid state, which is called the mesoscopic regime. This includes polymers and biopolymers at the lower size end, DNA, viruses and micelles/vesicles as well as the important class of latices, microgels and inorganic colloids spanning the whole colloidal range. A main feature of colloids is their large surface-to-volume ratio, which can be exploited for applications like catalysis, adsorption and release. Furthermore, colloids show size-dependent electronic and optical properties, especially if they are very small. Examples for this are the different colours of gold colloids and the size-dependent colours and electronic properties of semiconductors (quantum size effect). Thus, it becomes clear that the particle size distribution (PSD) is a fundamental quantity in colloid chemistry. Analytical ultracentrifugation (AUC) has been recognized to be a powerful tool for the determination of PSDs since the days of its invention by Svedberg. In fact, the ultracentrifuge was first used for the determination of PSDs before its virtue for the analysis of polymeric systems was discovered. This is nicely illustrated by the early work of Svedberg from 1923 to 1925 [2, 3, 4, 5] before he performed the first ultracentrifuge investigations on proteins [6]. After the tremendous success of the ultracentrifuge for the analysis of polymeric systems, including the proof that polymers exist [6], its application for the analysis of colloidal systems became more or less forgotten for decades. Only in industrial laboratories this technique was still applied with great success for the determination of PSDs, mainly of latices [7, 8, 9, 10, 11, 12, 13]. However, the technique is suitable to address many more colloidal systems and their associated problems. This overview paper is based on our own work and is intended to give a brief exemplary survey of the common problems one encounters in the ultracentrifugal analysis of colloidal systems which are often much different from those of relevance in the analysis of proteins or other biopolymers and intends to point out ways towards their solution.

Common problems encountered during the AUC of colloids

AUC of colloids differs in many respects from that of polymeric systems, so special emphasis is laid upon the special problems associated with these systems and their solutions:

- Colloids may exhibit extremely broad particle size and associated sedimentation coefficient distributions. There is a danger that big aggregates or small impurities are not detected.
- Colloids can aggregate or grow during centrifugation (concentration-dependent aggregation).
- The density of hybrid colloids is unknown to access particle size.
- Electrostatic stabilization complicates analysis owing to charge contributions.
- Particle polydispersity in size, shape, density and hydration.
- High particle density often makes density gradients or density variation methods impossible.
- Colloids are often multicomponent mixtures.

Some of these problems can be nicely addressed by AUC, whereas others are more difficult or even impossible to solve by AUC alone. For example, one typical property of many colloidal systems (especially those of inorganic nature) is their very small particle size (smaller than 10 nm). Whereas light scattering commonly fails owing to the too low particle size and the associated low scattering intensity combined with the often disturbing light absorbance of these particles and electron microscopy lacks statistical significance owing to the limited number of particles detected [14], AUC can be favourably used as it detects all particles even down to the smallest sizes. An example can be given for the Au_{55}-cluster, which was reported to be a defined and monodisperse species and is of special interest owing to its size at the transition between a metal and a molecule, which should lead to special physical properties [15]. However, an AUC investigation of this cluster reveals an at least bimodal PSD and Gauss fitting to the experimental sedimentation coefficient distribution (*s* distribution) ot the correlated PSD indicated a third species (Fig. 1a) [16].

Although the AUC distribution is not diffusion-corrected, meaning that the distribution is broader than the real one owing to significant boundary-broadening by diffusion of such small particles [17], the resolution of the PSD is in the angstrom range. If this information is compared with that in Fig. 1a, which shows the corresponding transmission electron microscopy (TEM) results, the higher accuracy of the PSD from AUC becomes obvious although both distributions

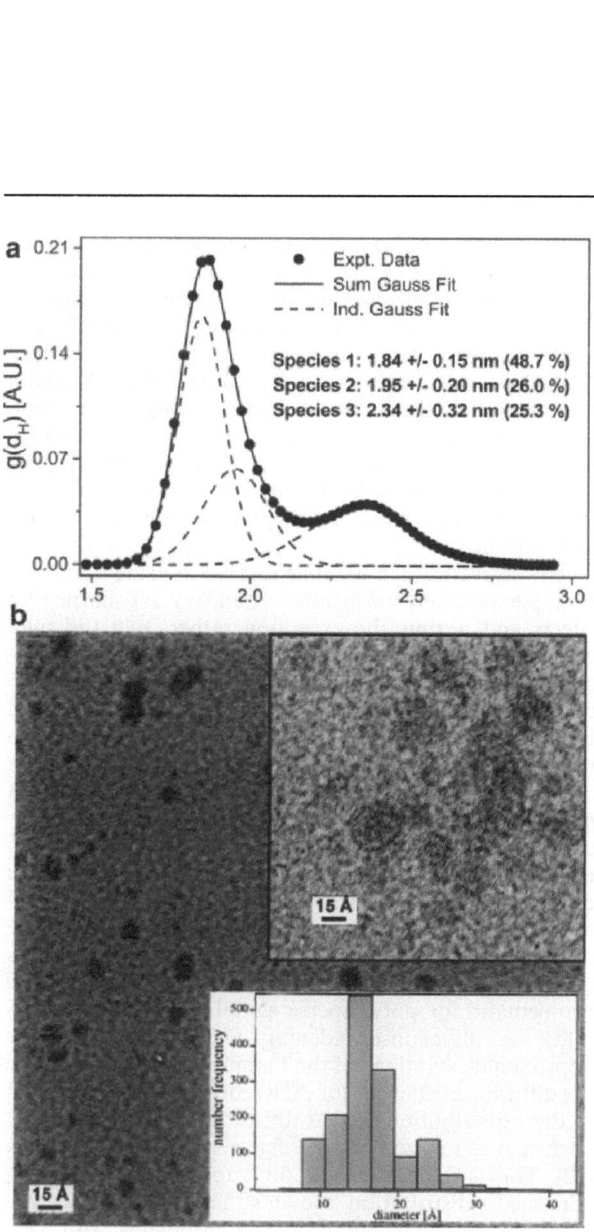

Fig. 1 a Particle size distribution of $Au_{55}[P(Phe_3)]_{12}Cl_6$ clusters as investigated by analytical ultracentrifugation (*AUC*) (25 °C, 60,000 rpm, $\lambda = 500$ nm). Re-evaluated data from Ref. [16]. **b** Transmission electron microscopy (*TEM*) investigation taken from Ref. [16]. Reproduced with permission of the American Chemical Society

qualitatively agree. However, TEM provides important information about the particle shape, so the hydrodynamic equivalent diameters in Fig. 1a are the real diameters.

Another common feature of colloids is their polydispersity owing to their inherent tendency to aggregate or caused by the nucleation process. This has the consequence that the *s* distribution can be extremely broad, spanning decades of sedimentation coefficients. An example is given in Fig. 2 for aggregated $BaSO_4$. This has the consequence that a single speed is not sufficient

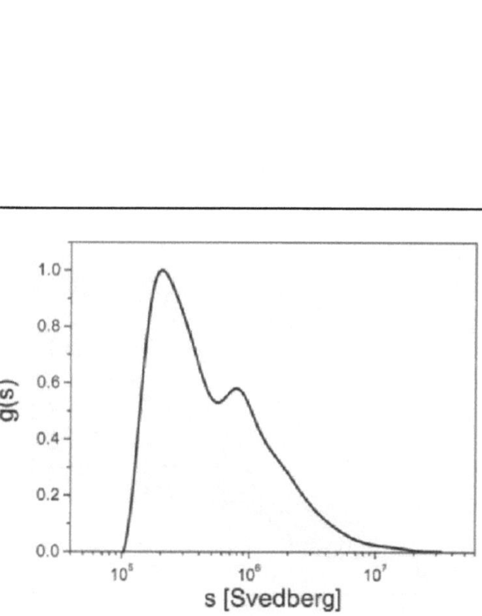

Fig. 2 Sedimentation coefficient distribution of aggregated $BaSO_4$ colloids

for the characterization of such colloids. Instead, speed profiles have to be applied which proved to be successful for the characterization of latices with broad PSDs. With such speed profiles, the entire colloidal range can be accessed in a single experiment or if increased accuracy is needed in two experiments. This was shown for model latex mixtures [10, 12].

Another problem is the light scattered by the colloids if absorption or turbidity detection are used. Here, a Mie correction has to be applied, which has the potential problem that this correction can amplify uncertainties or artefacts in the small particle size range. Thus, it would be much more advantageous to apply refractive index detection, but this is not yet possible with the necessary speed profiles. An elegant workaround was very recently developed by Müller [18], who overlaid parts of the *s* distribution obtained at several speeds to a master curve, thus allowing refractive index detection to be applied even for very broad distributions.

Angstrom-resolved PSDs

AUC is especially useful for the investigation of very small colloids with sizes smaller than 10 nm. In that case, a conventional XL-A or XL-I analytical ultracentrifuge can be applied using a single speed, which is usually 60,000 rpm. The problem of diffusion boundary broadening is circumvented in rare cases by self-sharpening effects of the different particle boundaries in the mixture. A good example for this effect are AUC measurements on Pt colloids which were quenched during growth by addition of excess stabilizer [19] (Fig. 3).

The raw data can be easily converted to an *s* distribution, which can then be converted to the PSD using

$$d_H = \sqrt{\frac{18\eta s}{\rho_P - \rho_S}},$$

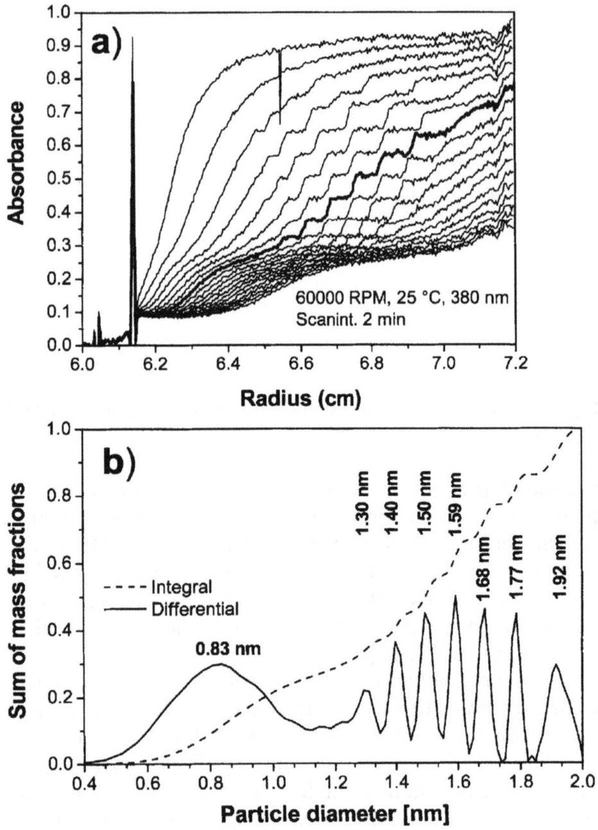

Fig. 3a, b AUC on Pt colloids in a methanol–acetic acid mixture quenched during growth by excess stabilizer. **a** Experimental raw data (25 °C, 60,000 rpm, scan interval 2 min). **b** Integral and differential particle size distribution for the *bold* scan in **a**. Taken from Ref. [19] with permission of Springer-Verlag

where η is the solvent viscosity, ρ_P is the particle density and ρ_S is the solvent density. This modified Svedberg equation applies a hard-sphere approach and is only valid for solid particles.

If one looks at the raw data in Fig. 3a, one can see that even after a very short time, a step profile with sharp steps develops, which indicates multiple monodisperse species. If one looks at the PSD (Fig. 3b), one observes two main features:

- The detected particles are very small (down to 8.3 Å).
- The baseline resolution between the different species is extremely high (1 Å).

If one knows the crystal structure of the colloids from wide-angle X-ray scattering and the spherical particle shape from TEM/high-resolution TEM (HRTEM), one can convert the maxima in the PSD to atom numbers, which are as low as 21 Pt atoms for the 0.83-nm species [19]. It is most remarkable that the difference between the larger species also corresponds to about 21 Pt atoms, so the PSD in Fig. 3b allows conclusions about the particle growth mechanism to be drawn. The smallest stable species appears to be the spherical 21-atom cluster, which coalesces with other particles to build the larger defined particles. This is supported by HRTEM observations, which show multiple lattice fringes in a single particle hinting at a coalescence process [20]. Furthermore, the PSD from AUC is in qualitative agreement with that from TEM, the latter showing a much lower statistical accuracy displaying only a bimodal distribution.

However, the high baseline resolution achieved in this example was likely the result of boundary self-sharpening effects and is thus the exception rather than the rule. Normally, one would expect significant diffusion broadening for such small particles. There are several ways to correct for this. One classical way is the so-called van Holde–Weischet analysis, which extrapolates the sedimentation profiles to infinite time to yield a diffusion-corrected *s* distribution [21]. However, this extrapolation procedure does not usually exactly yield the diffusion-corrected *s* distributions and is more useful as a diagnostic tool. Another possibility is to calculate the diffusion broadening from the known particle size and to subtract this effect from the *s* distribution to yield the diffusion-corrected *s* distribution [17]. Whereas this procedure is useful for monodisperse samples, it becomes problematic for polydisperse samples. The third possibility for diffusion-broadening correction is to apply approximate solutions of the Lamm equation to estimate the diffusion coefficient for each sedimentation coefficient in the *s* distribution so that diffusion broadening can be corrected as realized in the software SEDFIT by Schuck [22]. This is the best of the tools to obtain a diffusion-corrected *s* distribution known so far, and it works well in most cases but can also have problems with very polydisperse samples, resulting in artificial peaks. However, to exclude such artificial peaks in the *s* distribution, it is good advice to overlay the diffusion-corrected and noncorrected *s* distributions to check for equal maxima, similar tailing, etc.

A practical example: ZrO_2 catalyst film deposition

ZrO_2 has important applications as a strongly acidic catalyst; therefore, it is desirable to design coatings with a high surface area, which means homogeneous coatings of very small nanoparticles. One possibility to achieve this is the application of self-assembled monolayers, which on the one hand homogeneously assemble on a substrate and on the other hand provide sticking functional groups for ZrO_2. When ZrO_2 nanoparticles were prepared via a hydrolysis reaction of $Zr(SO_4)_2$ at 70 °C, the resulting film showed cracks

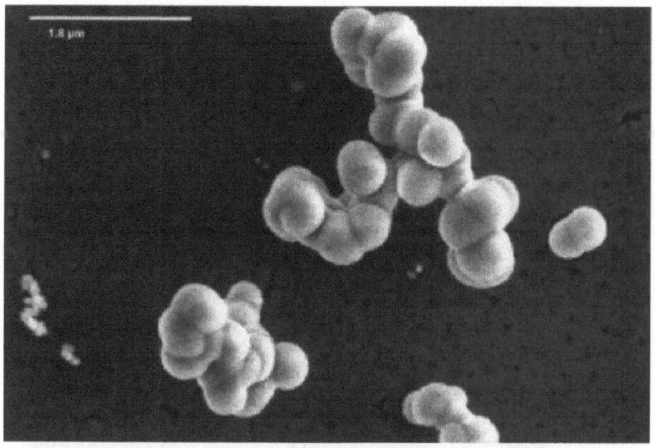

Fig. 4 ZrO_2 film deposited on a self-assembled monolayer via $Zr(SO_4)_2$ hydrolysis at 70 °C. The cracks and large particles are clearly visible. Taken from Ref. [23] with permission of the American Chemical Society

and, furthermore, the solution particle growth was not controlled, so particles in the micron range formed and led to an inhomogeneous film as shown in Fig. 4 [23].

AUC which was performed on samples where the reaction was quenched by cooling the reaction mixture with liquid nitrogen and subsequent rethawing revealed fast nanoparticle growth even in the early growth stages (Fig. 5).

It is evident that four different early growth stages can be observed by multiple Gauss fits to the sedimentation coefficient distributions:

- Species 1: $r_H = 0.30$ (0.43) nm.
- Species 2: $r_H = 0.55$ (0.80)–0.60 (0.87) nm.
- Species 3: $r_H = 1.04$ (1.50)–1.10 (1.61) nm.
- Species 4: $r_H = 1.58$ (2.29) nm.

However, their exact particle size cannot be determined according to Eq. (1) as the particle density is unknown, ranging between 3.22 g/ml for $Zr(SO_4)_2$ (particle sizes in parentheses) and 5.66 g/ml for ZrO_2. As numerous complexes can be formed in the hydrolysis reaction before the first crystalline particle is formed, only a density range can be used as a limit for the particle size calculations. This is a limitation which often occurs in colloid analysis with the AUC, especially for hybrid colloids and very small particles, as for the very small colloids, the bulk density cannot be used anymore owing to the dominance of the particle surface with its solvation effects or a different surface structure from that of the bulk. On the other hand, hybrid colloids consist of several components, so they can even exhibit a density distribution (see the Organic–inorganic hybrid colloids with size and density polydispersity section). Nevertheless, the particle size

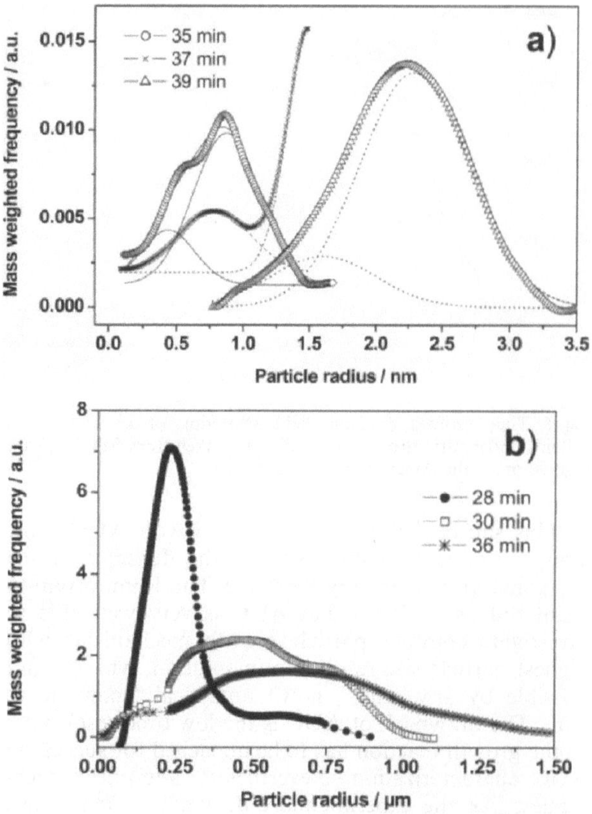

Fig. 5a, b Particle size distributions of ZrO_2 colloids at 70 °C at different reaction times. Particle size calculated with $\rho = 3.22$ g/cm^3. **a** After quenching in liquid nitrogen and re-thawing and **b** evolution of particle size distribution with time, representative of the transition from a clear to a cloudy medium. The distributions were not corrected for diffusion broadening. Note that the different reaction times in **a** and **b** are due to the different heating rates in these experiments. Taken from Ref. [23] with permission of the American Chemical Society

range is often rather small like in the previous example, so comparisons with sizes of precursor species reported in the literature can be made, giving an insight into the reaction mechanism. In the previous example, it can be seen that the resolution of AUC is again very high and in the angstrom range. The smallest resolved species has a radius of just 0.3–0.4 nm, which corresponds to the $[Zr_4(OH)_8(H_2O)_{16}]^{8+}$ complex described in the literature with a radius of gyration found by small-angle X-ray scattering of between 3.8 and 5 Å [24, 25, 26]. This tetrameric complex further oligomerises to the octameric species $[Zr_8(OH)_{20}(H_2O)_{24}Cl_{12}]$ with a reported radius of gyration of 6 Å [24], which is in very good agreement with species 2 in Fig. 5. The octamer then oligomerises to higher oligomers (species 3), which will precipitate (species 4).

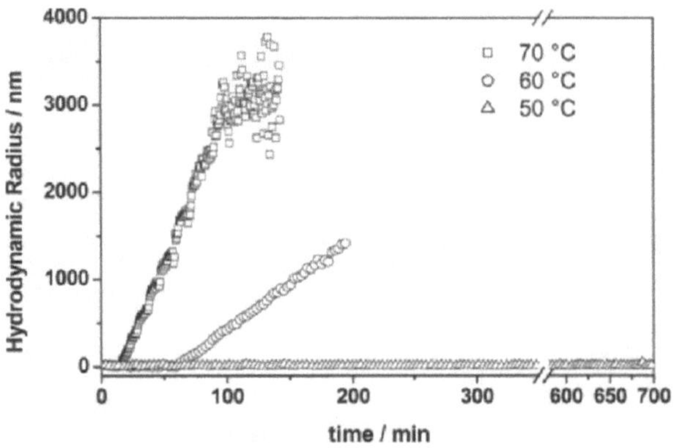

Fig. 6 Time-resolved dynamic light scattering of growing ZrO_2 colloids under various reaction conditions. Taken from Ref. [23] with permission of the American Chemical Society

This shows that PSDs determined with AUC show almost atomic resolution even if the densities of the inorganic species are only moderate. This is important to point out as it shows that AUC is very well able to investigate complex particle growth mechanisms with highest particle size resolution in solution, which is not possible by any other known analytical technique to date. The drawback of AUC is the low time resolution, so the growth reaction has to be quenched for successful AUC characterization. Nevertheless, alternative techniques, like the experimentally demanding small-angle X-ray scattering using synchroton radiation, can on the one hand give a very high time resolution in the observation of the particle growth processes well in the millisecond range [27], but lack the fractionation capability, which can be problematic as in the previous case of multiple growth species only average values can be determined.

After the growth process on the nanometre scale had been elucidated, the further growth was followed up to the micron scale (Fig. 5b). Here, it could be shown that the particles grow very fast to the micron range within a few minutes and the PSD gets extremely broad.

However, AUC alone cannot solve the problem of the elucidation of an experimental procedure for the controlled particle growth owing to the lack in time resolution, so dynamic light scattering (DLS) was applied, which shows a time resolution in the second range but could not detect the very small nanoparticles. However, the particle growth could be nicely observed under various experimental conditions as shown in Fig. 6.

From the DLS curves, it became obvious that an "inhibition" period was observed where no particle growth could be detected as the particles were too small for DLS detection as well as a temperature-dependent

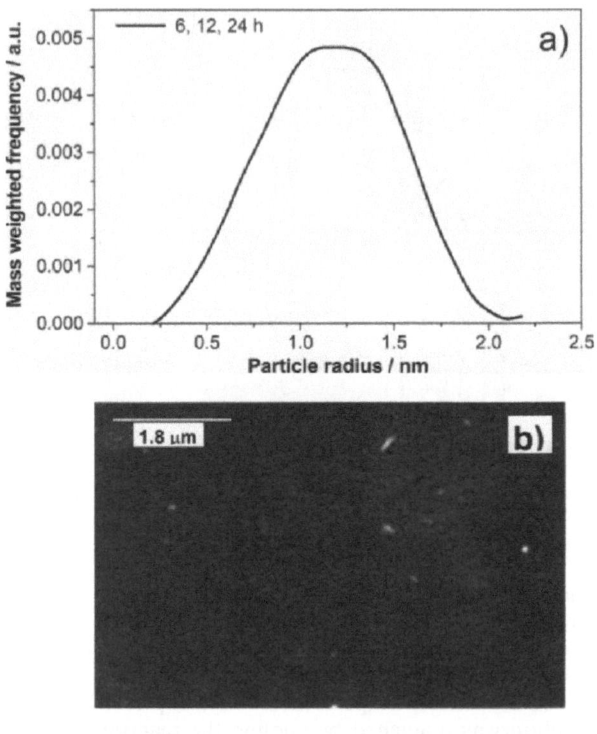

Fig. 7 a AUC-derived ZrO_2 particle size distribution obtained for the reaction temperature of 50 °C and **b** ZrO_2 thin film deposited on a self-assembled monolayer at 50 °C, 24-h deposition time. Taken from Ref. [23] with permission of the American Chemical Society

linear growth rate. This experiment revealed fast growth rates of 34 nm/s at 80 °C reaction temperature, 11 nm/min at 70 °C, 0.7 nm/s at 60 °C and no detectable growth at 50 °C; thus, 50 °C is either suitable to produce very small nanoparticles, which do not grow but which are also not detectable by DLS, or no particles are produced at all. To address this question, AUC was performed on the reaction mixture with 50 °C reaction temperature. The result clearly shows that around $r_H = 1{,}16$ nm big particles are grown (Fig. 7a).

This species is remarkably similar to species 3 in Fig. 5. In addition, it was checked if growth occurred with prolonged reaction times of 6, 12 and 24 h by AUC and the resulting PSDs were identical. This showed that the undesired nanoparticle growth could be inhibited right after the particle nucleation by applying a reaction temperature of 50 °C. Thin films deposited on a self-assembled monolayer clearly showed the absence of larger particles and cracks like in Fig. 4 (Fig. 7b), proving that the control of the solution-based particle nucleation can lead to the desired homogeneous thin catalyst film with enhanced reactivity [23]. This experimental outcome underlines the importance of highly resolved PSDs from AUC.

Turbidimetric immunoassays as a model example for a polydisperse interacting system

A common feature of colloidal systems is their inherent tendency to aggregate unless sufficient particle stabilization is provided. To demonstrate that AUC is very well suited for the investigation of such systems in comparison to other commonly applied techniques for particle size measurement like TEM or light scattering, we investigated a model system for a polydisperse aggregating system. Here a mixture of two monodisperse latices (127 and 221 nm) was set up with 92 wt% of the small latices [28]. The latices were coated with antibodies for C-reactive protein (CRP), where the smaller particles were coated with an antibody of low reactivity and the bigger ones with a higher reactive antibody. Such coated latex mixtures are applied as turbidimetric assays with an enhanced dynamic range for the detection of antigen concentrations via simple turbidity measurements [29] (Fig. 8).

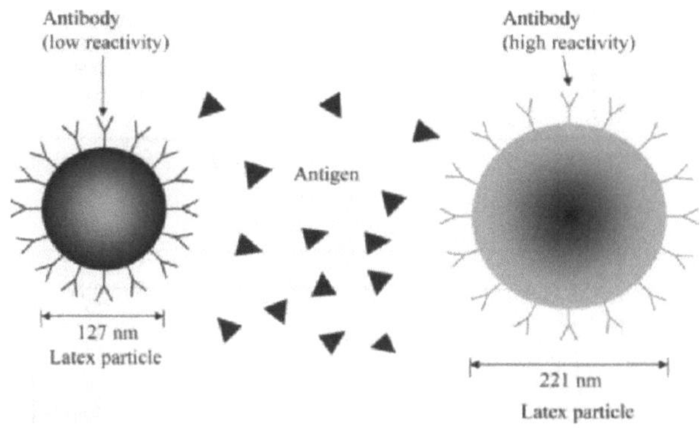

Fig. 8 Schematic representation of a turbidimetric assay based on two latices, which are coated with antibodies of different reactivity (see text)

To achieve a linear turbidity response with increasing antigen concentrations, the large latices with the highly reactive antibody have to aggregate first at low antigen concentrations, yielding enhanced sensitivity in the low antigen concentration range owing to the higher turbidity of the larger particles. At increasing antigen concentrations, the smaller particles then have to aggregate and, in addition, have to detect high antigen concentrations. The measurement problem was to detect in a quantitative manner if the previously described concept works, so an aggregating polydisperse system had to be characterized with respect to its PSD. TEM investigations indicated

Fig. 9 Representative TEM pictures of 7 coated with antibodies for C-reactive protein (*CRP*) latex particles after agglutination with **a** 0, **b** 4, **c** 25 and **d** 156 mg/l CRP. *Scale bar* 500 nm. Taken from Ref. [28] with permission of the American Chemical Society

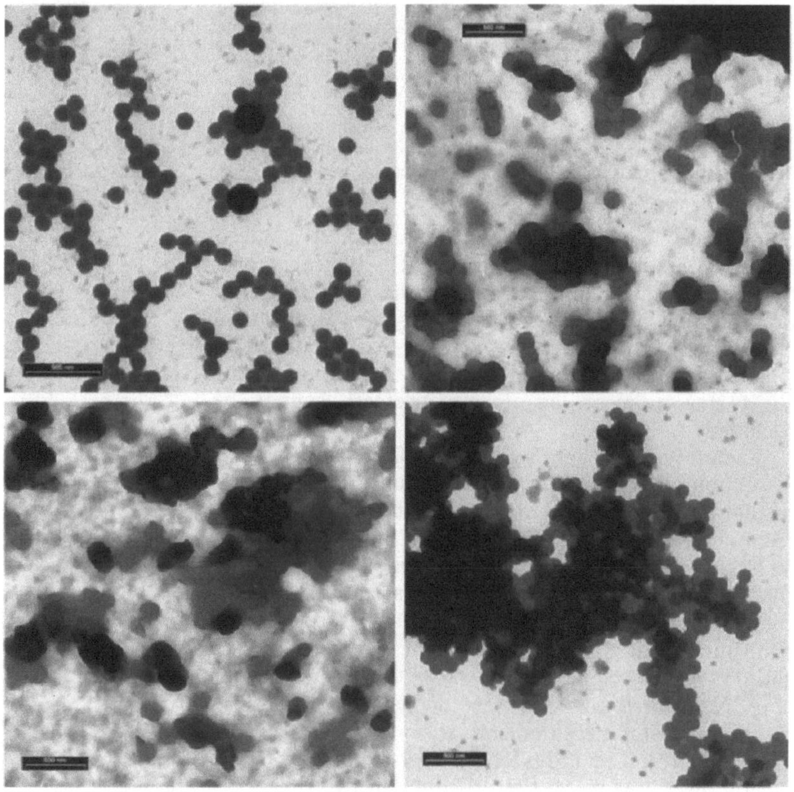

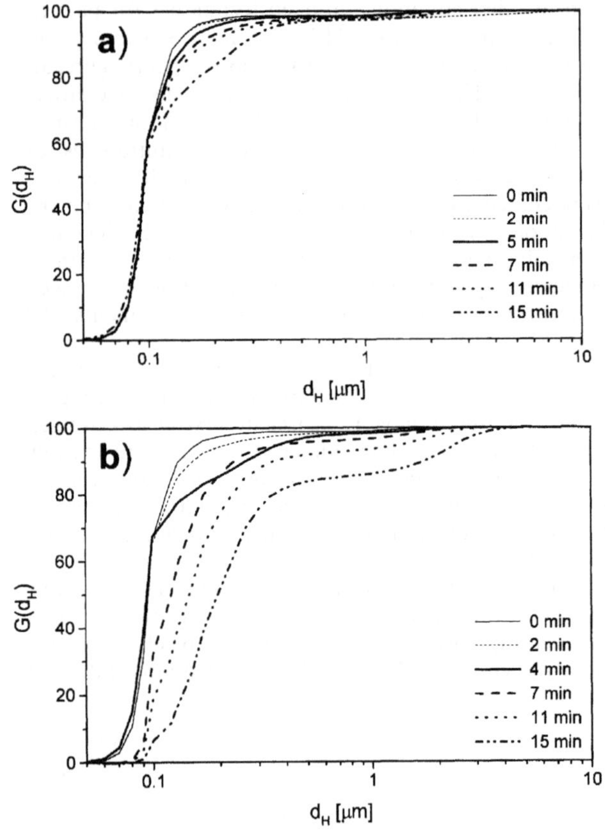

Fig. 10 Integral particle size distribution, $G(d_H)$, from a static light scattering kinetic measurement with **a** 50 and **b** 171 mg/l. Taken from Ref. [28] with permission of the American Chemical Society

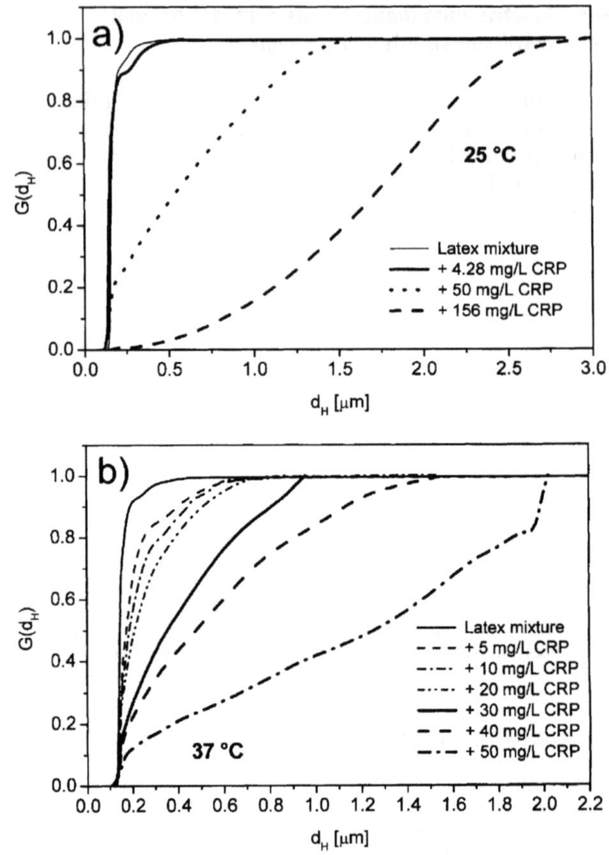

Fig. 11 Mass-weighted integral AUC particle size distributions of an agglutinated latex mixture at different CRP concentrations: **a** after 10 min at 25 °C and **b** incubated for 50 min at 37 °C. Taken from Ref. [28] with permission of the American Chemical Society

that at low CRP concentrations, the larger latices indeed aggregate first and that at higher CRP concentrations, the smaller ones follow, but besides the visual information on the aggregate morphologies, no quantitative information could be derived from this investigation (Fig. 9). In addition, the samples had to be dried for the TEM measurement, so drying artefacts cannot be excluded.

Therefore, static light scattering (SLS) was applied as a solution technique with a rather good time resolution so that the aggregation kinetics could be monitored. The results are shown in Fig. 10. It can be seen that the aggregation kinetics can be nicely followed, indicating that the larger particles indeed aggregate first. However, the resolution of the PSD is low as can be seen for the distributions at 0 min, where 92 wt% 127-nm latices should be detected next to 8 wt% 221-nm latices (Fig. 10). This is a consequence of the lack of fractionation in the SLS measurement.

Highly resolved PSDs could be obtained by AUC where the 92 wt% of the 127-nm latices was nicely

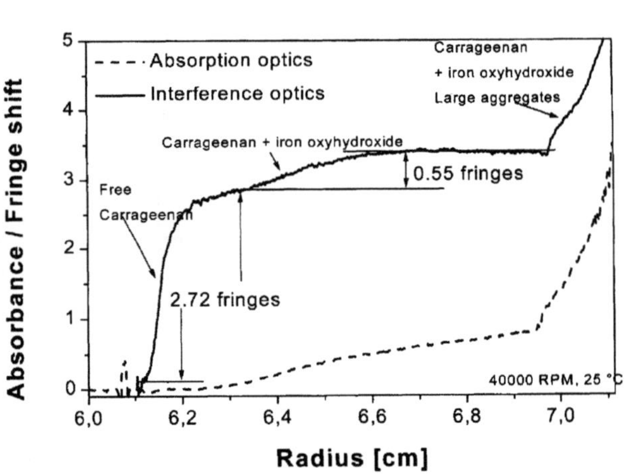

Fig. 12 Iron oxyhydroxide in κ-carrageenan microgels. Detection of the different species by a combination of UV–vis and refractive index detection. Reproduced from Ref. [30] with kind permission of Springer-Verlag

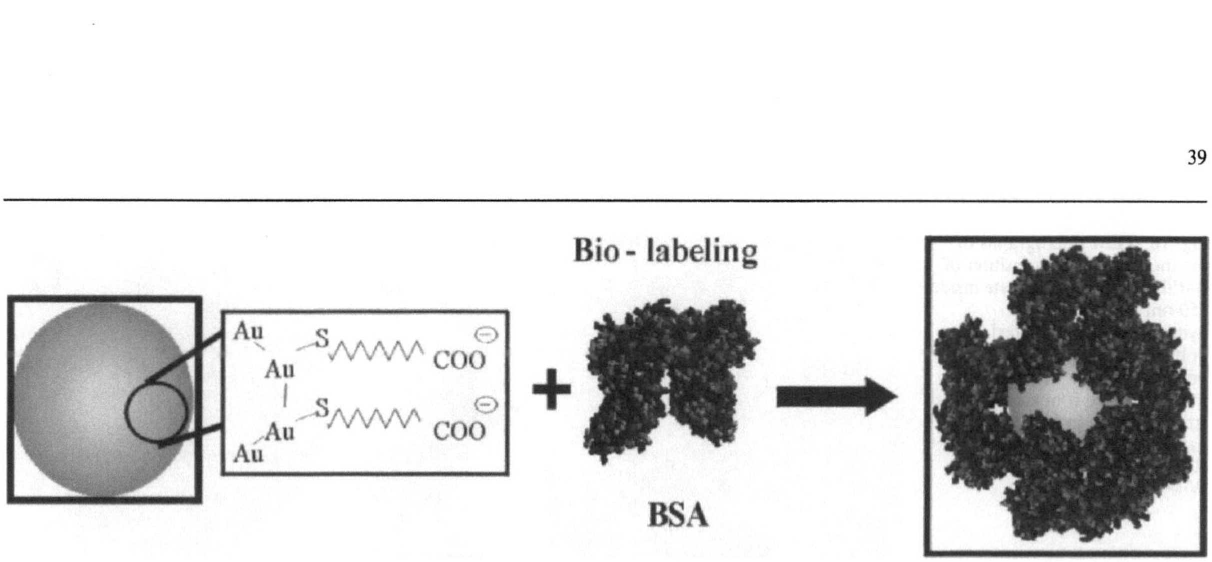

Fig. 13 Schematic presentation of the bovine serum albumin (*BSA*) coating of gold nanoparticles using 11-mercaptoundecanoic acid as a phase compatilizer

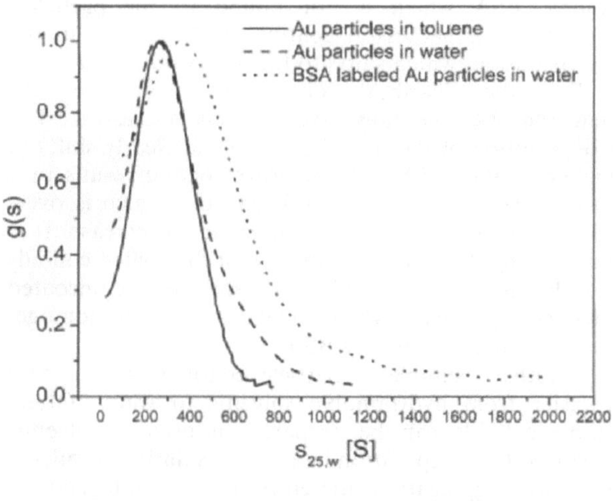

Fig. 14 Sedimentation coefficient distributions (not diffusion corrected) of gold particles with different coatings corrected to water at 25 °C [32]. Hydrophobic gold in toluene (*solid line*), hydrophilic mercaptoundecanoic acid capped gold particles in water (*dashed line*), BSA-labelled gold particles in water (*dotted line*). The peak width extending the 0-S limit is a result of diffusion broadening

Fig. 15 Photograph of five diluted nanoparticle samples before and after phase transfer into water from their reaction mixtures (toluene) using mercaptoundecanoic acid. Silver (*A*, *B*), gold, prepared according to Ref. [33] (*C*, *D*), platinum (*E*, *F*), gold, prepared according to Ref. [34] (*G*, *H*), palladium (*I*, *J*). Reproduced from Ref. [32] with permission of Wiley-VCH

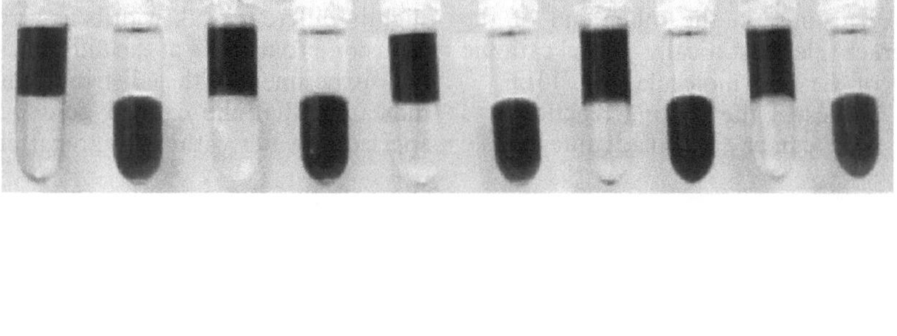

resolved after the PSDs were corrected for Mie scattering. Also, it could be shown that the larger particles aggregate first (Fig. 11). Overall, these results were in good agreement with the model of a diffusion-controlled aggregation process [28].

This demonstrates that AUC is well able to study aggregating systems with a high particle size resolution, but with the drawback of low kinetic resolution as the temperature equilibration time for an AUC experiment is more than 10 min, even with a pretempered rotor. SLS (or alternatively DLS) on the other hand can yield good kinetic resolution but lacks the capability of high particle size resolution. Nevertheless, both techniques compliment each other in a favourable way (see also the section A practical example: ZrO_2 catalyst film deposition). However, if very broad PSDs with particle sizes spanning the whole colloidal range from 1 nm to a few microns result from aggregation processes, scattering techniques are likely to fail and a fractionating technique like AUC is the only realistic option to obtain quantitative PSDs.

Organic–inorganic hybrid colloids with size and density polydispersity

Sample homogeneity/number of components

In many fields of polymer or colloid chemistry, complicated complexes or hybrid colloids are synthesized. It is often desirable to have a quick and convenient check for the efficiency of the reaction as well as a check of sample homogeneity. Sedimentation velocity experiments can very advantageously be used here owing to the fractionation without any stationary phase as is demonstrated

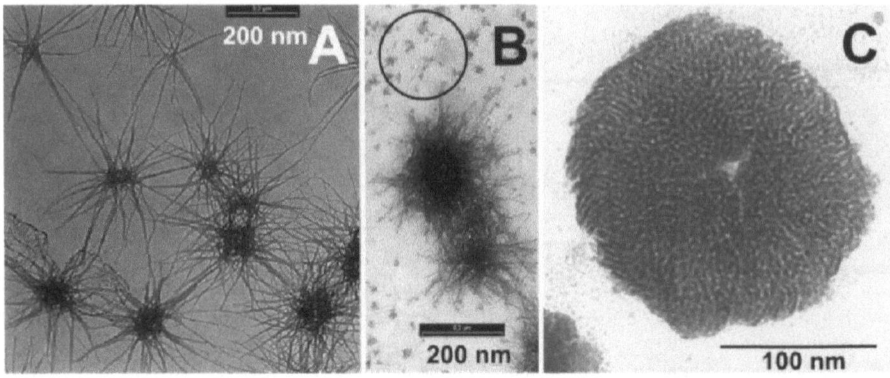

Fig. 16 TEM micrographs of the morphological transition of starlike calcium phosphate inside 150-nm aggregates of a hydrophobically modified poly(ethylene oxide)-*block*-poly(methacrylic acid) copolymer to a more compact spherical structure. The polymer aggregates are not visible. Taken from Ref. [35] with permission of Wiley-VCH

for the case of iron oxyhydroxide particles partly stabilized in a self-assembling microgel of κ-carrageenan (Fig. 12) [30]. If one component absorbs light (FeOOH) and the second does not (κ-carrageenan), one can selectively detect the absorption of the iron oxide in the polymeric superstructures with absorption optics, whereas all components together are detected by the simultaneously used Rayleigh interference optics. Here it must be stated that most coloured inorganic compounds have a high extinction coefficient and are thus almost exclusively detected with the UV–vis absorption optics, whereas the simultaneously applied interference optics almost exclusively represents the local polymer concentrations which can be converted to the real concentrations via the refractive index increment and the known refractive index of the solvent.

From Fig. 12, three species are clearly quantitatively detected: Free unbound κ-carrageenan, κ-carrageenan plus iron oxyhydroxide microgel and larger cross-linked aggregates. The efficiency control of chemical reactions was also reported for complicated complexes between oppositely charged polyelectrolytes [31]. In addition, the species detected can be further characterized by their sedimentation coefficient distribution.

Particle coating and liquid phase transfer

An s distribution can yield insight into the success of chemical reactions forming hybrid colloids. An example is 11-mercaptoundecanoic acid capped further gold colloids, which were coated with bovine serum albumin (BSA) [32]. The thiol is used as a phase compatibilizer as the thiol groups bind chemically to the gold surface, whereas the carboxy groups on the other end of the molecule can interact electrostatically with cationic amino acid residues of the BSA molecule (Fig. 13).

It is of primary interest if the coating reaction was successful. For this, AUC can be applied advantageously and the raw experimental data can reveal the amount of soluble BSA which is not bound to the particles. However, the successful coating is also visible in the s distribution, which shifts to higher s values (Fig. 14).

From the s distributions in Fig. 14, it can be deduced that the BSA coating reaction was successful. The s distribution of the coated gold colloid clearly shifts to higher s values although the BSA coating results in a density decrease of the overall particle, which is over-compensated by the particle molar mass increase. It is noteworthy that the s distribution of the coated colloids is only slightly broadened with respect to the uncoated precursor particles, which indicates a rather homogeneous particle coating by BSA.

Another example is particle coatings for a phase transfer from hydrophobic solvents to water. Often, stable colloids can be prepared in organic solvents, whereas their preparation in aqueous solutions results in undesired aggregation; therefore, it is advantageous to synthesize the colloids in organic solvents equipping them with the appropriate stabilizer for a subsequent phase transfer to aqueous solutions. The gold colloids can serve as an example for this. They were prepared in toluene, then capped with mercaptoundecanoic acid, which enabled the coated particles to dissolve in water [32]. The same is possible with other metal nanoparticles prepared according to the methods in Refs. [33, 34] (Fig. 15).

For the proof of successful phase transfer, it is of crucial importance to show that the particles do not aggregate upon redispersion in another phase. This can again easily be monitored by the investigation of the colloid s distributions in both aqueous and organic solvents (Fig. 14, solid and dashed lines). Despite the slightly different peak widths as a result of the different diffusion broadening at the different speeds applied in the two experiments, the distributions are very similar, indicating that the capped gold particles can be redispersed in water without aggregation.

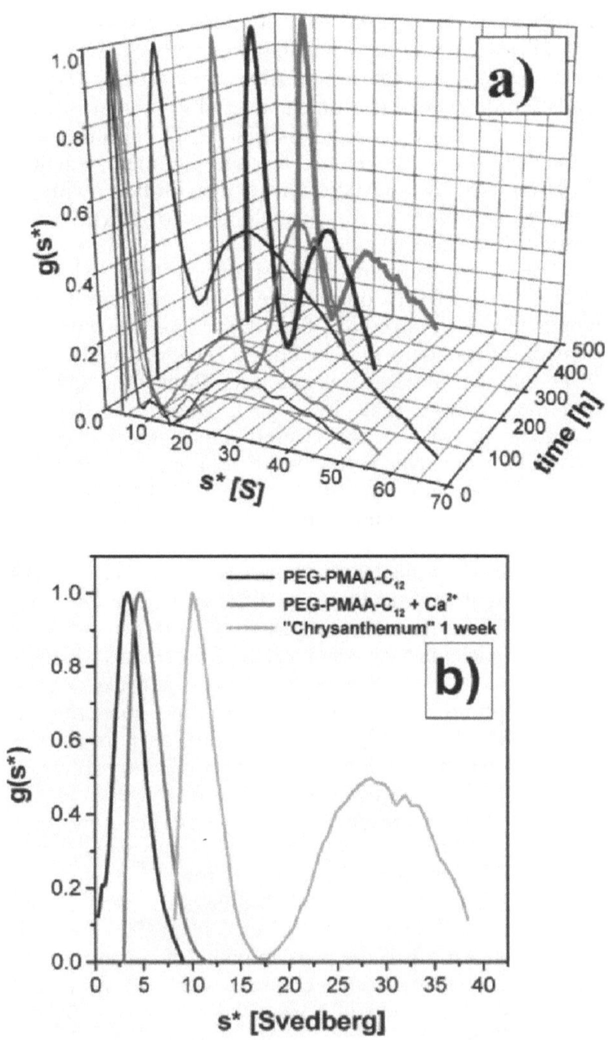

Fig. 17 a) Time-dependent sedimentation coefficient distributions (normalized) of poly(ethylene oxide)-*block*-poly(methacrylic acid)-C_{12}/calcium phosphate aggregates at pH 4.0, with a polymer-to-Ca ratio of 2:1 [35]. Taken from Ref. [35] with permission of Wiley-VCH. b) Sedimentation coefficient distributions of the block copolymer, the polymer plus Ca^{2+} and the final hybrid colloid

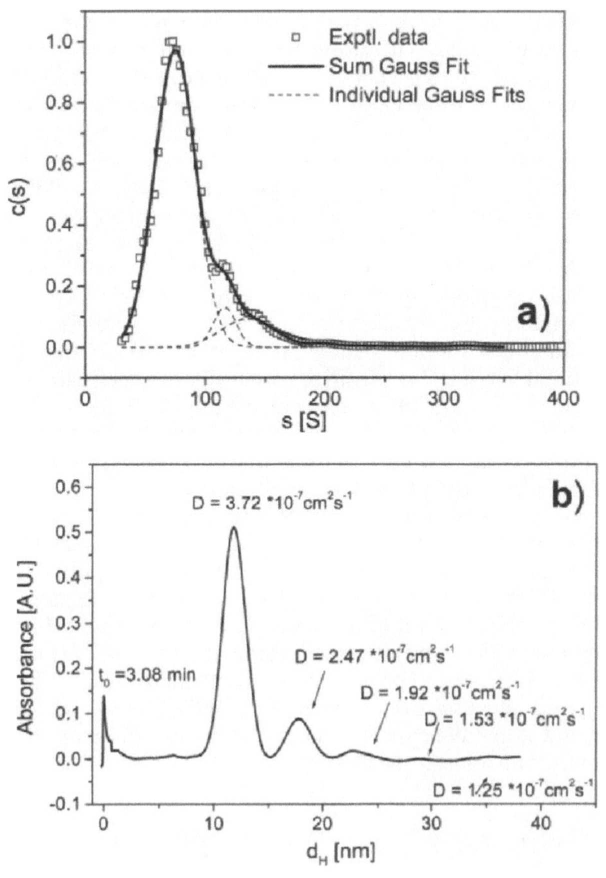

Fig. 18 a Diffusion corrected *s* distribution from AUC. Run parameters 10,000 rpm, 25 °C, λ=410 nm (99 absorption data sets evaluated). **b** flow-field flow fractionation elugram after conversion to the particle size distribution. λ=255 nm, cross-flow 1.97 ml/min, elution flow 0.53 ml/min, Elution liquid 0.3 g/l NaCl+0.08 g/l Tween 20 [39]

Particle growth/transformation

Sometimes, colloids – especially organic–inorganic hybrid colloids or micelles/vesicles, etc. – exhibit morphological transitions, which can sometimes be triggered by external stimuli. An intriguing example is the formation of starlike calcium phosphate inside 150-nm aggregates of a hydrophobically modified poly(ethylene oxide)-*block*-poly(methacrylic acid) copolymer [35]. The long whiskers can be partly as thin as 2–3 nm and up to 1,000-nm long. This creates a very high surface area and it is not astonishing that this situation is not thermodynamically stable but subject to a slow morphological transition to a more compact particle (Fig. 16). However, it was not clear if these complex structures were present in solution or were just drying artefacts, so time-dependent sedimentation velocity experiments were performed (Fig. 17).

It can be clearly seen that after the start of the hybrid colloid formation, only a monomodal *s* distribution is detected, whereas after a few hours, a second species can be seen at higher *s* values. The low-*s* species corresponds to the starlike structures with low density and high friction in Fig. 17a, whereas the higher-*s* species is the compact, denser structure in Fig. 17c. It is remarkable that the time evolution of the relative amount of the high-*s* species has a maximum after which macroscopic precipitation is observed in the system, so the amount of the compact species is decreased again. Therefore, the structural evolution of the hybrid colloids could be characterized as starlike structures, transforming to the more compact spherical structure, which finally aggregates and precipitates. Although the particle size could not be accessed as the particle density is unknown, the *s* distributions allow the characterization of the morphological transition. In fact, DLS measurements revealed that the average hybrid particle size remained that of 150 nm of the initial polymer aggregate. This means that the observed changes in the *s* distribution in Fig. 14 mainly reflect a density distribution besides changes in the frictional ratio. That AUC is sensitive to density changes can be seen in Fig. 17, where the *s* distribution of the initial polymer aggregates shifts to higher *s* values upon Ca^{2+} addition with an unchanged shape of the distribution. Addition of the phosphate counterion yields the hybrid colloid, which after time shows a bimodal *s* distribution, although DLS shows a constant particle size of 150 nm. For such a system, a global analysis approach would be of benefit, and this has already been realized on the SEDFIT program platform by combining DLS, sedimentation velocity and sedimentation equilibrium [36]. Such an approach could potentially yield the hybrid particle size, density and molar mass, which could allow conclusions about the composition of the hybrid colloids to be drawn.

Analysis approaches for organic–inorganic hybrid systems

The need for global analysis approaches is evident for almost all organic–inorganic hybrid systems as they usually exhibit a PSD superimposed with a density distribution and their densities often exceed 2 g/ml, so they cannot be accessed by means of the traditional density gradient techniques, which showed great success for the analysis of complex latex systems [12]. Also, the density variation method [37, 38] cannot be applied here, as the common solvent densities are too low to exhibit a large enough difference for the high-density colloids. An example where a combined analysis approach can be successfully applied is ferritin, which was investigated in terms of the *s* distribution by AUC and the diffusion coefficient distribution by flow-field flow fractionation (Fl-FFF) [39]. Ferritin is an iron storage protein and is a hollow sphere (outer diameter 12.5 nm, inner diameter 8 nm) consisting of 24 protein subunits of 18,000–24,000 g/mol [40] with a molar mass of the iron-deficient apoferritin of 450,000 g/mol as determined by sedimentation equilibrium [41]. Ferritin is also an oligomerizing system which can incorporate varying iron levels up to 4,500 iron atoms per capsule [42, 43] depending on the iron level in the blood where iron is continuously incorporated or released from the ferritin capsules and, in addition, it was reported that Fe^{3+} which is not yet incorporated into mature ferrihydrite particles can be transferred between ferritin capsules [43]. In view of this, one has to expect a paucidisperse PSD reflecting the concentrations of the defined ferritin oligomers, which is superimposed by a particle density distribution owing to possible variations in the iron loading or ferrihydrite/protein hydration.

Considering the *s* and *D* distributions from AUC and Fl-FFF, respectively, one can see differences at first sight which reflect the previous facts (Fig. 18). The particle-density-independent *D* distribution (proportional to the PSD via the Stokes–Einstein law) shows the defined oligomer species (Fig. 18, right), where the ferritin monomer size of 11.9 nm is found in nice agreement with that of 12.5 nm given in the literature [44]. In contrast, the density-dependent *s* distribution (Fig. 18, left) just indicates different oligomers and only peak deconvolution can lead to a picture resolving the different oligomers.

This clearly indicates that both distributions in Fig. 18 are not identical and the only difference between AUC and Fl-FFF for the same sample is that AUC also depends on the particle density whereas Fl-FFF does not. Hence, the results in Fig. 18 clearly reflect a particle density distribution and PSD. In view of this, conventional sedimentation analysis is likely to fail, as it is usually based upon constant density or partial specific volume, $\bar{v}$. Here, a global analysis has clear advantages and the relatively cheap computer power available nowadays facilitates this kind of analysis as already partly realized in the SEDFIT package of Schuck [36]. In the following, we present a rather crude global analysis approach by combining the distribution results from Fl-FFF and AUC with respect to the peak maxima of the individual species detected by both techniques although it is clear that global evaluation of both distributions could yield further insight. The results are presented in Table 1.

From the *s* and *D* values given in Table 1, the particle-density-independent buoyant molar mass, M_b, can be calculated and it can be seen that it is relatively constant between 500 and 580 kDa for a monomer–trimer. Deviations arise for the higher oligomers but as their amounts are very small (about 2 wt% as detected by Fl-FFF, which shows more confidence owing to the better resolved distribution), they have to be treated with care in the further analysis, at least with respect to the sedimentation analysis, which is based on a fitting procedure of the experimental sedimentation profiles [22]. In view of this, it is indicated that the ferritin oligomers have a constant $\bar{v}$ indicating a constant average iron loading per capsule of the different oligomers. Nevertheless, the smeared *s* distribution in Fig. 18a, compared with the *D* distribution (Fig. 18b) indicates the $\bar{v}$ distribution although it cannot be too big, as the different oligomers can still be distinguished despite the smearing.

Table 1 Sedimentation coefficients, *s*, diffusion coefficients, *D*, hydrodynamic diameters, d_H, and buoyant molar mass, M_b, of each of the ferritin oligomers from analytical ultracentrifugation (*AUC*) and flow-field flow fractionation (*Fl-FFF*) measurements

	Monomer	Dimer	Trimer	Tetramer	Pentamer
s (S) at 25 °C (AUC)	74.6	116.3	135.7	195.3	317.0
$D\times10^7$ (cm^2/s) at 25 °C (Fl-FFF)	3.72	2.47	1.92	1.53	1.25
d_H (nm) (Fl-FFF)	11.9	17.7	22.7	28.9	35.1
M_b per monomer (g/mol)	496,900	583,300	583,700	790,700	1,256,700
Oligomer amount (wt%) from AUC[a]	82.50	6.17	10.46	0.64	0.23
Oligomer amount (wt%) from Fl-FFF[a]	80.70	14.63	2.59	0.47	1.61

[a] The relative oligomer amounts were calculated on the basis of a proportionality between the absorption signal and the ferritin concentration implying a constant ferrihydrite content for every capsule [39]

In principle, the molar mass of each oligomer could now be calculated from the buoyant molar mass with the measured $\bar{v} = 0.598$ ml/g. If this is done starting with the monomer, a molar mass of 1,242,000 g/mol is derived, which is much higher than that of apoferritin $\bar{v} = 0.74699$ ml/g 450,000 g/mol as determined by sedimentation equilibrium in Ref. [45]) plus the maximum possible iron loading of 4,500 atoms [42, 43], resulting in a molar mass increase of 432,200 g/mol calculated via the molar mass of ferrihydrite (5 $Fe_2O_3{\cdot}9H_2O$, M_r= 960.46 g/mol). This results in a maximum total ferritin molar mass of about 880,000 g/mol, a value much lower than that found on basis of the measured $\bar{v}$, indicating that the measured $\bar{v}$ is significantly wrong. This is most likely a result of the fact that it could not be measured in dialysis equilibrium owing to the danger of ferrihydrite dissolution. Therefore, absolute molar masses of the oligomers cannot be calculated with the present data sets. In such cases, the buoyant molar masses could be determined in solvents of different density by sedimentation velocity in combination with Fl-FFF to yield $\bar{v}$ of the different oligomers via the zero buoyant molar mass, or possibly even the $\bar{v}$ distribution, which would be important information for ferritin. This approach would be similar to the sedimentation equilibrium approaches for the $\bar{v}$ determination of supramolecular systems via the zero buoyant molar mass, where the $\bar{v}$ can hardly be measured via conventional density measurements [46]. However, sedimentation equilibrium cannot be applied here, as $\bar{v}$ of each oligomer is desired. Another potential approach for $\bar{v}$ determination could be density gradients, but here problems can be expected owing to the very low ferritin $\bar{v}$ and in the assignment of $\bar{v}$ to the different oligomers in a $\bar{v}$ distribution. Nevertheless, even if $\bar{v}$ remains unknown, the present data sets allow further conclusions, which are summarized in Table 2.

Assuming a constant $\bar{v}$ and frictional ratio, which neglects the small contributions of asymmetry to the frictional coefficient, the s_n/s_1 or the D_1/D_n ratio of a hard-sphere aggregate with oligomerization degree *n* should show an $n^{2/3}$ dependence reflecting the $M^{2/3}$ dependence. This is indeed found to be in good agreement for *s* and *D* with the exception of the pentamer *s* value, which is determined unreliably (Table 1). Another conclusion which can be drawn from the good agreement with the simple aggregation model is that the ferritin oligomers adopt a globular solution structure and can be hydrodynamically treated as a sphere.

Table 2 Ferritin data from AUC and Fl-FFF in comparison to a hard-sphere aggregation model. Index *n* indicates the degree of oligomerization [39]

	Monomer	Dimer	Trimer	Tetramer	Pentamer
Hard-sphere aggregation	1.00	1.59	2.08	2.52	2.92
s_n/s_1	1.00	1.56	1.82	2.62	4.25
D_1/D_n	1.00	1.51	1.94	2.43	2.98

In summary, the combined Fl-FFF and AUC results could give much more insight into the nature of this complex organic–inorganic hybrid material than each of the techniques would allow on its own. Not only the amount of oligomer (best characterized by Fl-FFF, as AUC oligomer results rely on peak deconvolution of a continuous distribution) but also the *s* and *D* values and the buoyant molar mass of each oligomer could be evaluated. As the analysis was only based on the detected oligomer peak maxima, the benefit of a global analysis integrating the whole AUC and Fl-FFF distributions together with molar mass information from sedimentation equilibrium becomes obvious. The need of a global analysis for colloidal systems is further emphasized by the fact that many colloids are stabilized by organic additives, so they have to be treated as organic–inorganic hybrid colloids once their size is very small, so surface properties become important. This has been found for colloids with sizes smaller than 10 nm, but depending on the system, much larger colloids also have to be considered in this respect. Nevertheless, for very small colloids, the problem remains that DLS often cannot be used as a density-independent method for the determination of the particle size, so Fl-FFF may become increasingly important for the global analysis of such systems.

Synthetic boundary crystallization ultracentrifugation

In colloid chemistry, particle nucleation and growth processes are of particular interest as they determine the particle size and its distribution and with it the colloid properties. Numerous methods are available to study nucleation and growth processes but the particle sizes of such small particles are very difficult to determine in solution. Besides AUC, small-angle X-ray scattering can be applied and this method has a high time resolution if synchrotron radiation is applied to study growth kinetics [27]. The disadvantages of this method are the lack of fractionation, so only average particle sizes can be determined and the need for synchrotron radiation owing to the usually low particle concentrations in solution. Therefore, AUC is a potentially powerful choice for the investigation of nucleation and growth processes and it was already shown (see sections Angstrom-resolved PSDs and A practical example: ZrO_2 catalyst film deposition) that AUC can yield PSDs with almost atomic resolution. However, most of the nucleation and growth processes are fast, so they cannot be investigated by AUC if they cannot be quenched successfully. Therefore, synthetic boundary crystallization ultracentrifugation was developed [47, 48]. The basis for this synthetic boundary method is the so-called active enzyme centrifugation, where an enzyme is layered upon a layer of substrate using a Vinograd-type synthetic boundary cell and the reaction products are monitored. In synthetic boundary crystallization ultracentrifugation, a small amount (usually 15 μl) of a reactant solution is layered onto about 350 μl of a second reactant solution so that a crystal is rapidly formed at the sharp interface between the two solutions (Fig. 19).

The particles formed then have the possibility to sediment/diffuse away from the reaction boundary, where their further growth is quenched owing to the lack of the second reactant, or they can diffuse back into the reaction boundary, where they can grow further to a second growth generation. These particles again have the option to quench their growth or grow further and so forth until the minor reactant is used up. This is the case after a few seconds, which is important as this prevents extensive reactant interdiffusion and a resulting broadening of the reaction boundary. This also implies that only fast nucleation and growth processes can be investigated by this AUC method, which usually is the case for inorganic crystals if sufficient supersaturation can be provided. After the complete consumption of the first reactant, the particles sediment according to their size, density and shape and as the sedimentation pathway is much longer than the initial thickness of the reaction boundary, the small smearing effect by possible different particle generation spots inside the thin reaction zone is negligible; therefore, a PSD can be determined via the *s* distribution. By using synthetic boundary crystallization ultracentrifugation, a time distribution of different particle growth species is converted to a radial distribution, so the low time resolution of AUC experiments does not matter anymore. Furthermore, the automatic quenching enables us to detect even the earliest growth species, including subcritical complexes.

The first application of synthetic boundary crystallization was reported for the test of different stabilizers on the growth of CdS particles [48]. The results in Fig. 20 indicate differences in the apparent PSDs for the CdS particles, which were stabilized with chemically very similar thiol stabilizers. Again, a resolution in the angstrom range could be achieved.

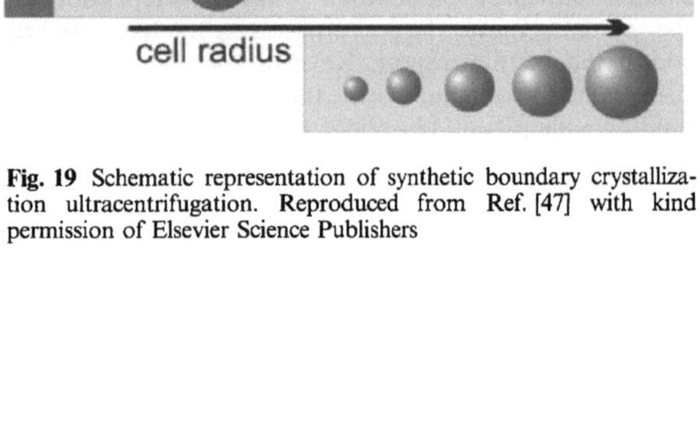

Fig. 19 Schematic representation of synthetic boundary crystallization ultracentrifugation. Reproduced from Ref. [47] with kind permission of Elsevier Science Publishers

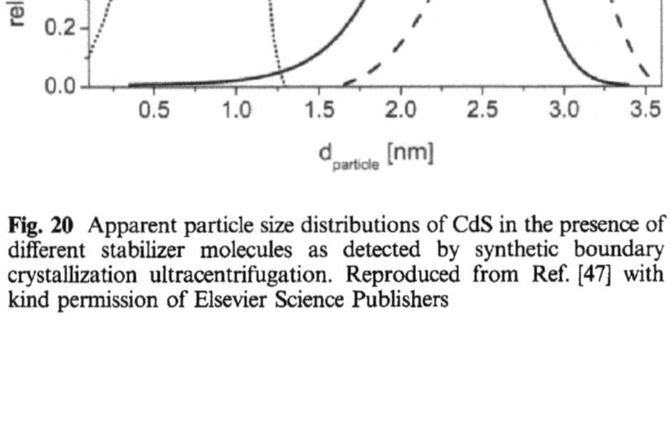

Fig. 20 Apparent particle size distributions of CdS in the presence of different stabilizer molecules as detected by synthetic boundary crystallization ultracentrifugation. Reproduced from Ref. [47] with kind permission of Elsevier Science Publishers

However, it has to be noted that the PSDs in Fig. 20 only have apparent character as they were calculated with the density of bulk CdS of 4.82 g/ml as a result of the unknown density of the small CdS colloids. It is well known that the density of very small colloids can differ from that of the bulk materials owing to the increased specific surface area where solvation effects or structural effects become predominant over the bulk crystal properties. This is reflected in the fact that the density of CdS stabilized with thioglycerine was determined to be 3.2 g/ml via a combined AUC, DLS and UV–vis study [49]. In addition to the particle density, the stabilizer shell thickness could be calculated to be 0.2 nm by applying a core–shell model [49] and was in good agreement with literature data [50], showing the high resolution of the AUC experiments. This again reflects the need for global analysis to gain more information on colloidal systems. Nevertheless, although the PSDs in Fig. 17 have only apparent character, the influence of the stabilizer becomes obvious. The particle density and/or the particle size are significantly influenced by the stabilizer, with thiogylcerine showing the biggest effect. These studies indicated the potential benefit of synthetic boundary crystallization ultracentrifugation for the study of nucleation and growth processes, but difficulties still remain. Up to now, only light-absorbing colloids could be investigated by this approach as the refractive index detection via Rayleigh interference optics so far yields uninterpretable results [49]. This is a result of the complex sum refractive index changes upon crystallization in a synthetic boundary cell, which do not only depend on the refractive index change of the crystal forming ions upon crystal formation, but furthermore on the rapidly diffusing counterions as well. As a result, the detected fringe patterns are sigmoidal curves, which are very similar to the conventional sedimentation velocity boundaries. Even their time development throughout the experiment is similar. Therefore, the experimental technique needs to be improved to correct for this. One possibility could be the use of a similar amount of counterion solution, which is overlaid in the reference sector so that the resulting interference traces would be similar to that in differential sedimentation experiments permitting the observation of the crystallizing species. If interference optics could be used as a detection system in synthetic boundary crystallization ultracentrifugation, the scope of the crystal systems for investigation by this technique could be tremendously increased, including all systems of scientific and industrial relevance.

Conclusion

The examples here show that AUC is a versatile and powerful technique for the analysis of colloids, which was already realized by Svedberg, the pioneer of AUC. Nevertheless, although the pioneering colloid work of Svedberg was published 80 years ago, the range of possible methods for colloid analysis is by far not yet fully explored (see also Future perspectives). The beauty of AUC for colloid analysis lies in the possibility to investigate colloids with sizes over the entire colloidal range, even in one experiment, when applying speed profiles, the fractionation of complex mixtures with s values ranging over decades, the high statistical accuracy as every particle is detected and the almost atomic resolution for the smallest colloids so that crystal nuclei and even subcritical clusters can be resolved. New methods like synthetic boundary crystallization centrifugation have their special virtues. For example, this method especially allows the observation of crystal growth processes by transformation of a time profile into a spatial profile. Nevertheless, colloids can have various unfavourable properties, which makes them problematic to investigate. Examples are coupled polydispersity in particle size, density, shape and/or charge or simply aggregation. Especially the charge should not be neglected as many colloids are stabilized by charge, but the sedimentation of charged colloidal systems has so far only rarely been considered theoretically with respect to the quantitative charge determination [51]. This implies that the analysis of such complex colloidal systems outlined here cannot be performed with AUC alone, although the s distribution often yields a lot of information. Thus, global analysis combining AUC with the information from other techniques, like DLS or Fl-FFF, which yield the diffusion coefficient or the diffusion coefficient distribution, is highly promising for the analysis of complex colloidal systems. The virtue of an AUC/DLS global analysis for the determination of size and shape distributions of macromolecules has already been demonstrated [36] and, hopefully, in the future many other techniques can be included, like Fl-FFF, ζ-potential measurements, axial ratio distributions from electron microscopy or atomic force microscopy, so that whole distributions can be obtained for quantities like size, shape, density and charge.

Future perspectives

More than 10 years ago, Mächtle [52] wrote a visionary paper about the future requirements for modern analytical ultracentrifuges. It is amazing that only a few parts of the requirements outlined have been realized so far in the commercial XL-I, although their potential benefit is obvious. Other requirements are still only realized in a few laboratories (multiple detection optics, automatic online data analysis, application of speed profiles) and especially these requirements are important if not essential for the analysis of complex colloids. It is important to obtain as much analytical information as possible simultaneously, so multiple detection systems have to be applied (Schlieren/interference, UV–Vis, SLS, turbidity and others) in order to unravel the complex folded distributions of size, shape, density and charge. Even existing optical systems, like UV–vis optics, should be improved as they can give further information. For example, fast fibre-based UV–vis optics with a charge-coupled-device array spectrophotometer, which we are currently designing (P. Maciejewska, L. Börger, H. Cölfen, W. Mächtle unpublished results), can yield a whole spectrum consisting of 800 wavelengths in a few milliseconds. This means that such a detector is not only a fast new potential turbidity detector with varying sensitivity owing to the wavelength dependence of light scattering, but could also be used to determine the wavelength dependence of turbidity, which can yield the particle size. As the same spectrum could yield the particle size via the wavelength dependence of turbidity and the sedimentation coefficient from the time when the spectrum was taken, the application of a hard-sphere model could yield the particle density via a modified Svedberg equation. This is just an example of what can be expected from future efforts in detector development.

As colloids are often very polydisperse, the application of speed profiles is essential. If the detectors can be designed in a way that the data acquisition rate is in the millisecond-to-second range, which is possible with modern techniques, the speed profiles can be designed to be very fast to suppress diffusion broadening of the sedimenting boundaries. So a future sedimentation velocity experiment could be designed in a way that a fast acceleration profile from 0 to 60,000 rpm is used and the experiment is finished after 15 min.

The increasing amount of information from multiple AUC detection systems as well as the useful combination with other analytical techniques, to increase the information content about complex systems, make global analysis essential. Owing to increasingly complex colloidal systems and the increasing importance of hybrid systems (organic–organic, inorganic–inorganic, organic–inorganic) or functional colloids with programmed self-assembly properties which are potential candidates for size, shape, density and charge polydispersity, global analysis will be the most promising option to characterize such systems in terms of whole distributions of these quantities, which are of great importance.

Despite better detection systems and analysis approaches, the future could show the establishment of new ultracentrifuge methods. The versatility of AUC is amazing and it is still possible after 80 years of application of this technique to set up new methods like, for example, a synthetic boundary technique for the kinetic study of membrane growth processes [53]. Further actual examples include pH gradients [54], which can potentially be used for particle charge determination or observation of the smallest crystallizing species. Therefore, the number of physicochemical quantities accessible by AUC will further increase in the future. This may help to lead to a broader application of AUC than realized at the moment, with methods focussing on biopolymers and could extend it to a general method also for the industrially and scientifically important fields of synthetic polymers, latices, pigments, functional colloids, nucleation and crystal growth studies and so on.

Acknowledgements We have to thank numerous people for their contributions to this work. M. Antonietti for the fine working environment and useful discussions. B. Zilske, G. Lucas L. Börger, T. Pauck and K. Schilling for their skillful AUC measurements. H. Schnablegger is acknowledged for the DLS studies on ZrO_2, H. Wachernig for the SLS on antibody-coated latices. We also thank C. Göltner for the TEM micrographs of the antibody-coated latices. G.A. Braun, F. Caruso, S. Eda, A. Fischer, D.I. Gittins, J. Kaufmann, U. Kobold and D. Rapoport are acknowledged for the supply of interesting colloidal samples. D.I. Gittins is also thanked for making graphical material available to the authors. Finally, we thank the Max Planck Society for the financial support of this work.

References

1. Everett DH (1992) Grundzüge der Kolloidwissenschaft. Steinkopff, Darmstadt
2. Svedberg T, Nichols JB (1923) J Am Chem Soc 45:2910
3. Svedberg T, Rinde H (1923) J Am Chem Soc 45:943
4. Svedberg T, Rinde H (1924) J Amer Chem Soc 46:2677
5. Svedberg T (1925) Kolloid Z Zsigmondy Festschr Erg-Bd Zu 36:53
6. (a) Svedberg T, Fahraeus R (1926) J Am Chem Soc 48:430; (b) Svedberg T, Nichols JB (1926) J Am Chem Soc 48:3081
7. Scholtan W, Lange H (1972) Kolloid Z Z Polym 250:782
8. Mächtle W (1984) Makromol Chem 185:1025
9. Mächtle W (1988) Angew Makromol Chem 162:35
10. Müller HG (1989) Colloid Polym Sci 267:1113
11. Mächtle W (1992) Makromol Chem Makromol Symp 61:131
12. Mächtle W (1992) In: Harding SE, Rowe AJ, Horton JC (eds) Analytical ultracentrifugation in biochemistry and polymer science. The Royal Society of Chemistry, Cambridge, p 147
13. Müller HG, Herrmann F (1995) Prog Colloid Polym Sci 99:114
14. Lange H (1995) Part Part Syst Charact 12:148
15. Schmid G, Pfeil R, Boese R, Bandermann F, Meyer S, Calis GHM, van der Velden JWA (1981) Chem Ber 114:3634
16. Rapoport DH, Vogel W, Cölfen H, Schlögl R (1997) J Phys Chem B 101:4175
17. Lechner MD, Mächtle W (1999) Prog Colloid Polym Sci 113 37
18. Müller HG Prog Colloid Polym Sci 127: 9–13

19. Cölfen H, Pauck H (1997) Colloid Polym Sci 275:175
20. Braun GA (1998) PhD thesis. RWTH Aachen
21. van Holde KE, Weischet WO (1978) Biopolymers 17:1387
22. Schuck P (2000) Biophys J 78:1606
23. Cölfen H, Schnablegger H, Fischer A, Jentoft FC, Weinberg G, Schlögl R (2002) Langmuir, 18:3500
24. Singhal A, Toth LM, Lin JS, Affholter K (1996) J Am Chem Soc 118:11529
25. Singhal A, Toth LM, Beaucage G, Lin JS, Peterson J (1997) Colloid Interface Sci 194:470
26. Toth LM, Lin, JS, Felker LK (1991) J Phys Chem 95:3106
27. Bolze J, Peng B, Dingenouts N, Panine N, Narayanan T, Ballauff M (2002) Langmuir 18:8364
28. Cölfen H, Völkel A, Eda S, Kobold U, Kaufmann J, Puhlmann ACD, Wachernig H (2002) Langmuir 18:7623
29. Eda S, Kaufmann J, Roos W, Pohl S (1998) J Clin Lab Anal 12:137
30. Jones F, Cölfen H, Antonietti M (2000) Colloid Polym Sci 278:491
31. Karibyants N, Dautzenberg H, Cölfen H (1997) Macromolecules 30:7803
32. Gittins DI, Caruso F (2002) Chem Phys Chem 3:110
33. Brust M, Bethell D, Schiffrin DJ, Kiely CJ (1995) Adv Mater 7:795
34. Wilcoxon JP (1992) US Patent 5,147,841
35. Antonietti M, Breulmann M, Göltner C, Cölfen H, Wong KKW, Walsh D, Mann S (1998) Chem Eur J 4:2493
36. Schuck P (2002) Size and shape distributions of macromolecules in solution by global analysis of sedimentation and dynamic light scattering. 8–11 June 2002. Lecture on advances in AUC & Hydro 2002. Autrans, France
37. Mächtle W (1984) Makromol Chem 185:1025
38. Müller HG, Herrmann F (1995) Prog Colloid Polym Sci 99:114
39. Cölfen H, Völkel A (2003) Eur Biophys J (2003) 32:432
40. Harrison PM, Artymiuk PJ, Ford GC, Lawson DM, Smith JMA, Treffry A, White JL (1989) In: Mann S, Webb J, Williams RJP (eds) Biomineralisation; chemical and biochemical perspectives. VCH, Weinheim, p 257
41. Pauck T, Cölfen H (1998) Anal Chem 70:3886
42. Proulxcurry PM, Chasteen ND (1995) Coord Chem Rev 144:347
43. Bauminger ER, Harrison PM, Hechel D, Nowik I, Treffry A (1991) Proc R Soc Lond Ser B 244:211
44. Mann S (1986) Chem Unserer Zeit 20:69
45. Tziatzios C, Durchschlag H, Sell B, van den Broek JA Mächtle W, Haase W, Lehn JM, Weidl CH, Eschbaumer C, Schubert D, Schubert US (1999) Prog Colloid Polym Sci 113:114
46. Börger L, Cölfen H, Antonietti M (2000) Colloids Surf A 163:29
47. Börger L, Cölfen H Progr (1999) Colloid Polym Sci 113:23
48. Börger L (2000) PhD thesis. University of Potsdam
49. Weller H, Schmidt HM, Koch U, Fojtik A, Baral S, Henglein A, Kunath W, Weiss K, Diemann E (1986) Chem Phys Lett 124:557
50. Stigter D (1980) J Phys Chem 84:2758
51. Mächtle W (1991) Prog Colloid Polym Sci 86:111
52. Wandrey C, Bartkowiak A (2001) Colloids Surf A 180:141
53. Lucas G, Börger L, Cölfen H (2002) Prog Colloid Polym Sci 119:11

Progr Colloid Polym Sci (2004) 127: 48–53
DOI 10.1007/b98012

C. Tziatzios
D. Schubert
P. Schuck
C. R. D. Lancaster
R. B. Gennis
B. Barquera

The state of association of the Na^+-translocating reduced nicotinamide adenine dinucleotide:quinone oxidoreductase in detergent solution – an ultracentrifugation study

C. Tziatzios (✉) · D. Schubert
Institut für Biophysik, J.W. Goethe-Universität, Haus 74, Theodor-Stern-Kai 7, 60590 Frankfurt am Main, Germany
e-mail: tziatzio@biophysik.uni-frankfurt.de
Tel.: +49-69-63015837
Fax: +49-69-63015838

P. Schuck
Division of Bioengineering and Physical Sciences, ORS, National Institutes of Health, Bethesda, MD 20892–5766, USA

C.R.D. Lancaster
Max-Planck-Institut für Biophysik, Marie-Curie-Strasse 15, 60439 Frankfurt am Main, Germany

R.B. Gennis · B. Barquera
Department of Biochemistry, University of Illinois at Urbana-Champaign, 600 South Mathews Street, Urbana, IL 61801, USA

Abstract Na^+:reduced nicotinamide adenine dinucleotide:quinone reductase (Na^+-NQR) is a redox-driven sodium pump found in some bacterial respiratory chains. The oligomeric state of Na^+-NQR from *Vibrio cholerae* was studied by sedimentation velocity and sedimentation equilibrium experiments in the analytical ultracentrifuge. Sedimentation velocity analysis of the purified enzyme in solutions of the nonionic detergent *n*-dodecyl-β-D-maltoside (Dm) revealed the presence of a nearly homogeneous protein population. From its sedimentation and diffusion coefficient, and considering reasonable amounts of Dm bound by the enzyme, it is shown that the component corresponds to monomeric Na^+-NQR. This result is corroborated by sedimentation equilibrium experiments, performed under conditions of density matching for the bound detergent. No influence of NaCl on the sedimentation behaviour of Na^+-NQR was detected. The amount of the protein-bound detergent was found to be about 0.57 g Dm per gram of protein.

Keywords Membrane protein · State of association · Sedimentation velocity analysis · Sedimentation equilibrium analysis

Introduction

The respiratory chains of some bacteria contain a unique redox-driven sodium pump, Na^+:reduced nicotinamide adenine dinucleotide (NADH):quinone reductase (Na^+-NQR) [1, 2, 3]. Most of these bacteria are marine bacteria; some are pathogenic [4, 5, 6, 7]. Na^+-NQR constitutes the gateway for electrons into the respiratory chain, accepting reducing equivalents from NADH and donating them to the quinone pool [1]. The energy from the redox reaction is used to pump sodium from the interior to the exterior of the cell, thus establishing a sodium gradient across the membrane, which is used for metabolic work [8, 9]. In the bacteria, the enzyme represents an essential part of the sodium cycle and thus plays an important role in their physiology [10, 11].

Na^+-NQR from three *Vibrio* species has been isolated and partly characterized. In all, the enzyme is made up of six different subunits, A–F, one of them (F) containing binding sites for NADH, flavin adenine dinucleotide, and a 2Fe–2S center. Subunits B and C each contain a covalently bound flavin mononucleotide. The enzyme also contains riboflavin at an unknown location [6, 7, 12, 13]. However, information concerning the secondary, tertiary and quaternary structure of the enzyme is still lacking. This even holds for a relatively gross property of the protein complex, its state of association.

Several respiratory complexes exist as dimers, for example, the bc_1 complex [14] and fumarate reductase [15]. In the case of the bc_1 complex, the dimer is mechanistically important [14]. Determining whether Na^+-NQR is a monomer or dimer may thus help to understand the mechanism by which the enzyme operates. The present paper aims at determining the quaternary structure of Na^+-NQR, using the enzyme from *Vibrio cholerae*, a protein of molar mass 216,000 g/mol (as calculated from its gene sequence [7]). The methods used are sedimentation velocity and sedimentation equilibrium analysis of the purified active enzyme in solutions of the mild, nonionic detergent *n*-dodecyl-β-D-maltoside (Dm). It is generally accepted that the complex size thus determined will correspond to the size of the complex in the intact membrane [16, 17].

Materials and methods

Materials

The recombinant, histidine-tagged Na^+-NQR complex was prepared as previously reported, the last purification step being affinity chromatography using a Ni-nitrilotriacetic acid resin [7]. Afterwards, the enzyme was dialysed against 50 mM sodium phosphate (pH 8), 100 mM NaCl, 5% (w/v) glycerol, 0.05% (w/v) Dm, and 10 mM ethylenediaminetetraacetate (EDTA), frozen in liquid nitrogen and kept at –80 °C until needed. For most of the analytical ultracentrifugation experiments it was transferred, by tenfold dilution with appropriate buffer, into 50 mM sodium phosphate (pH 8), 100 mM NaCl, 7 mM EDTA, 0.05% (w/v) Dm, and 3.3% (w/v) glycerol. For those sedimentation equilibrium experiments performed under conditions of density-matching for the detergent, the sample was brought, by dilution and dialysis, to 50 mM sodium phosphate (pH 8), 100 mM NaCl, 1 mM EDTA, 0.05% (w/v) Dm, 95% D_2O/5% H_2O, and 29% (w/v) sucrose [14]. The influence of the NaCl concentration on the sedimentation behaviour of Na^+-NQR was investigated by sedimentation velocity experiments in buffers containing 50 mM potassium phosphate (pH 8), 1 mM EDTA, 0.05% (w/v) Dm, and 0 or 300 mM NaCl. The samples were transferred into these buffers by dialysis for more than 20 h.

Dm was purchased from Anatrace, Ohio, USA. All other chemicals were from Sigma or Calbiochem and were of analytical grade (if available).

Analytical ultracentrifugation

Experiments were performed using a Beckman Optima XL-A ultracentrifuge and an An-50Ti rotor at a temperature of 4 °C. Absorbance versus radius data, $A(r,t)$ or $A(r)$, were measured at 460 nm. Sedimentation velocity runs used double-sector charcoal-filled Epon centrepieces of 12-mm optical pathlength. The sample and reference volumes were 300 and 320 μl, respectively. The rotor speed was 35,000 rpm. The $A(r,t)$ data were collected at time intervals of 4.5 min, with a radial step size of 0.001 cm and three averages per scan. The initial protein concentration was 5.6 g/l. The evaluation used the program Sedfit applying direct boundary modelling with discrete Lamm equation solutions for ideal and nonideal sedimentation [18, 19] or the van Holde–Weischet analysis [20, 21, 22]. For the sedimentation equilibrium experiments, six-channel charcoal-filled Epon centrepieces ($d = 12$ mm) were used with sample and reference volumes of 130 and 135 μl, respectively. The initial protein concentration was 1 g/l. The rotor speed was 13,000 or 17,000 rpm. Owing to the high viscosity of the solutions, equilibrium was reached only after approximately 90 h. (To test the stability of the purified Na^+-NQR during equilibrium sedimentation, the enzyme was incubated in buffer containing 30% sucrose at 4 °C for 1 week. After this time, no significant loss of activity was observed). The $A(r)$ data were collected with a radial step size of 0.001 cm and 30 averages per scan. The evaluations used the program Discreeq [23].

The partial specific volume, $\bar{v}$, of Na^+-NQR in aqueous buffers was calculated according to Durchschlag [24] as 0.739 ml/g. Under density-matching conditions it was corrected (1) for the influence of the sucrose and for H–D exchange to $\bar{v} = 0.779$ ml/g according to the procedure in Ref. [14], and (2) for the contribution of protein-bound phospholipid [17, 25]. The latter correction was based on a phospholipid average molar mass of 800 g/mol and $\bar{v}$ of (0.970 ± 0.020) ml/g [26]. The $\bar{v}$-value for Dm under those conditions is 0.83 ml/g [14]; in sucrose-free buffer it is 0.814 ml/g [27, 28]. The densities of the buffers were measured with a Paar DMA 02 digital density meter. Buffer viscosities were calculated using the software Sednterp [29].

Other methods

Analytical gel filtration

A Superose 6 column (2.4 ml) was used in a SMART purification system (Pharmacia) with a two-wavelength absorbance monitor. The running conditions were as reported in Ref. [7]. The following commercially available proteins were run as standards: bovine serum albumin (67 kDa), aldolase (158 kDa), and ferritin (440 kDa), using the same conditions as for Na^+-NQR.

Protein concentration

The protein content of the sample was determined photometrically, based on the conversion $A^{1cm}_{g/l}(280\ nm) = 1.16$, as calculated from the amino acid composition of the protein [7] according to the method in Ref. [29]. The conversion factor does not consider the protein-bound cofactors and may thus slightly overestimate protein concentration.

Phospholipid content

Protein-bound phospholipid in the samples was determined by the Bartlett assay [30], after extensive dialysis against 0.2% $C_{12}E_9$ in water.

Results

Gel filtration experiments

As a first approach to understand the aggregation state of Na^+-NQR, we performed analytical gel filtration on the purified enzyme. More than 90% of the absorbance (at 280 nm) eluted in a narrow, symmetrical peak. The remainder was found in a small shoulder at the trailing edge of the main peak (5%) and at the void volume (less than 3%). The apparent molar mass (including bound detergent) of the material in the main peak, which apparently consisted of homogeneous Na^+-NQR, was calculated from the elution volume to be about 354 kDa (data not shown). It is thus reasonable to suggest that Na^+-NQR is a monomer; however, owing to the known problems of determining molar masses of intrinsic membrane proteins by gel filtration [31], more conclusive data have to be added to confirm this conclusion.

State of association of Na^+-NQR from sedimentation velocity analysis

Experimental sedimentation velocity data on Na^+-NQR, together with curves fitted to them by application of the sedimentation coefficient distribution method [$c(s)$ method] [32], are shown in Fig. 1a. The continuous distribution model results in a fit of very good quality (root-mean-square, rms, error 0.0082 optical density units). Even visual inspection of the experimental sedimentation profiles reveals the presence of sample heterogeneity, since two distinct boundaries can be identified (Fig. 1a). The calculated $c(s)$ distribution (Fig. 1b) is relatively simple: it shows a major, relatively sharp peak at 6.1 S and another sharp but small peak at 3.5 S. The ratio of the areas under the two peaks is approximately 20. Since the relative area under a peak corresponds to the relative loading concentration of the respective species, it follows that the major peak accounts for approximately 95% of the total absorbance at 460 nm. Considering that a small amount of aggregated material would have been removed from the solution already during acceleration of the rotor, the result is fully consistent with the gel filtration data.

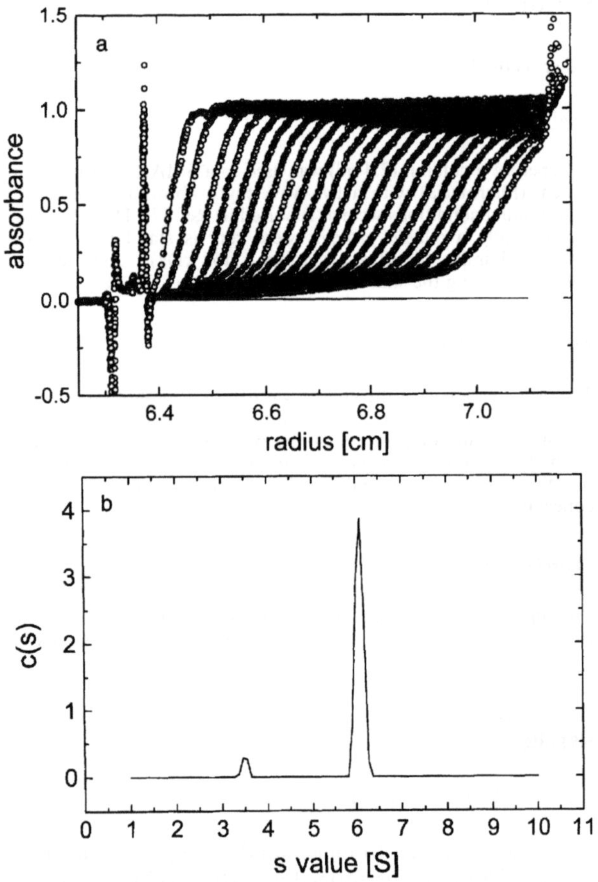

Fig. 1a, b A sedimentation velocity experiment with Na^+:reduced nicotinamide adenine dinucleotide:quinone reductase (Na^+-NQR). For clarity only every second data set is shown. **a** Measured sedimentation velocity profiles $A(r,t)$ versus radius r (*open circles*) and best-fit distribution from the Lamm equation model for continuous distributions (*solid lines*). **b** Best-fit sedimentation coefficient distribution $c(s)$

The sedimentation velocity data $A(r,t)$ (Fig. 1a) were analysed in more detail using the method of van Holde and Weischet. It has been shown that this method provides a useful tool for detecting concentration-dependent sedimentation. At a resolution of 25 divisions per boundary (leading to a rms deviation of 0.0173 optical density units) the apparent sedimentation coefficient for each boundary fraction was calculated and plotted against $t^{-1/2}$ (Fig. 2a). s values were determined by extrapolation to infinite time (Fig. 2a), defining the integral sedimentation coefficient distribution, $G(s)$ (Fig. 2b). As can be deduced from the analysed boundary fractions and in agreement with the results from the continuous distribution analysis, the s value of the main component is 6.1 S; the fractions corresponding to lower protein concentration confirm the presence of a slower component (Fig. 2a). The data are also indicative of a small concentration dependency of the s value of the main component, since the point of intersection of the extrapolated lines in Fig. 2a is not at infinite time but is shifted to the right. This is supported by the shape of the $G(s)$ distribution (Fig. 2b): s values corresponding to the upper boundary fractions, i.e. to higher sample concentrations, are lower than those corresponding to lower boundary fractions, resulting in a slightly negative slope of the $G(s)$ profile [20, 21].

Since the results described suggest that at least the main peak represents a homogeneous component, we performed Lamm equation modelling using two discrete components, with starting s values as found by the continuous model [32]. The fit was of similar quality as that obtained with the $c(s)$ method (rms error 0.0083 optical density units); the s values found were 6.07 and 3.58 S, with $D = 1.76 \times 10^{-7}$ cm^2/s for the 6 S-peak. The limits of error for the latter value are, however, relatively large: it could be varied by approximately ±10% without significantly increasing the sum of the squared residuals of the fit. The buoyant molar mass,

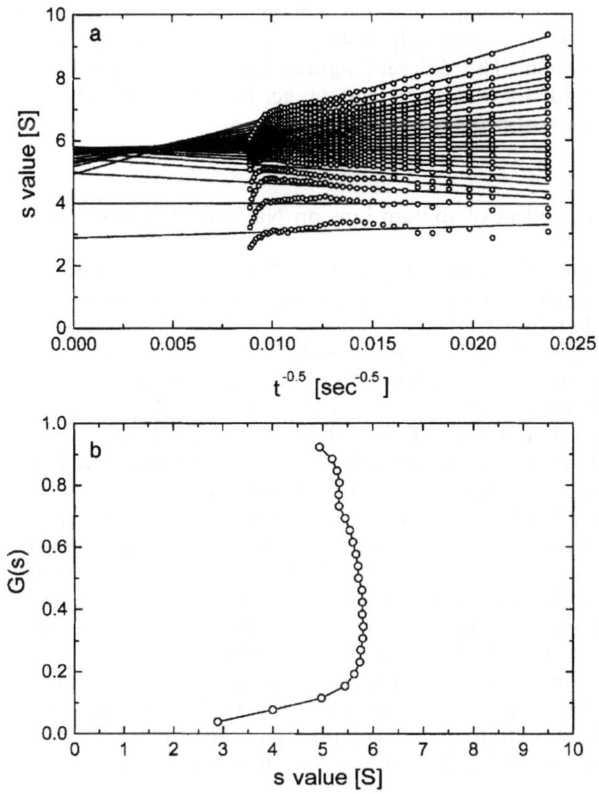

Fig. 2a, b van Holde–Weischet analysis of the sedimentation velocity data shown in Fig. 1a. The number of boundary fractions was 25; the *s* values shown are those for fractions 1–24. **a** Time dependency of the apparent sedimentation coefficients of the boundary fractions (*open circles*) and linear extrapolation to infinite time (*solid lines*). **b** Corresponding sedimentation coefficient distribution *G*(*s*)

$M_{\text{eff,c}} = M_c(1 - \bar{v}_c\rho_0)$, for the main peak, as calculated from its *s* and *D* via the Svedberg equation, is $(78{,}000 \pm 8{,}000)$ g/mol.

In order to account for possible hydrodynamic nonideality, as indicated by the van Holde–Weischet analysis, an alternative model with two discrete components was used, which includes hydrodynamic interactions, leading to a concentration-dependent sedimentation coefficient of the major component [33]. This led to a better quality of fit (rms error 0.00657 optical density units) and a best-fit value of $D_0 = 2.17 \times 10^{-7}$ cm^2/s for the major fraction with $s_0 = 6.35$ S and $k_s = 0.0128$ l/g. Models including terms for thermodynamic nonideality did not improve the fit. The $M_{\text{eff,c}}$ value calculated from s_0 and D_0 at infinite dilution was lower than that from the ideal sedimentation model, with a best-fit value of 66,500 g/mol. However, $M_{\text{eff,c}}$ values up to 71,100 g/mol could be obtained by increasing k_s up to 0.0075 l/g (which is consistent with the range of values empirically observed for globular proteins [33, 34]), with an only slight decrease in the quality of the fit. The *s* values are more consistent among the different models, with our estimate for the most likely value of s_0 being (6.22 ± 0.12) S.

In order to identify the oligomeric state of the major component, we compared the measured $M_{\text{eff,c}}$ values with theoretical ones calculated for (1) a Na^+-NQR monomer with 0.5/0.6...1.0 g of Dm bound per gram of protein, and (2) a Na^+-NQR dimer without bound detergent, using the $\bar{v}$-values given earlier, $M_1 = 216$ kDa and $\rho_0 = 1.020$ g/ml. Calculation 1 with (0.68 ± 0.20) g detergent per gram of protein led to an $M_{\text{eff,c}}$ value in the range of that from *s* and *D*. On the other hand, calculation 2, which completely neglects the contribution of bound detergent to $M_{\text{eff,c}}$ and thus will underestimate it, led to $M_{\text{eff,c}} = 106{,}600$ g/mol, i.e. much higher than the figure calculated from *s* and *D* from either model, with or without hydrodynamic nonideality. Thus, the 6 S-peak clearly has to be assigned to monomeric Na^+-NQR.

With respect to the relative amount of detergent bound to the Na^+-NQR monomer, the figure of 0.68 g Dm per gram of protein only holds in the case of ideal sedimentation. A value of 0.36 g/g results if the calculation is based on $M_{\text{eff,c}} = 66{,}500$ g/mol, as obtained when considering nonideal sedimentation.

State of association of Na^+-NQR from equilibrium sedimentation

To confirm the results, sedimentation equilibrium experiments with Na^+-NQR were performed under conditions of density-matching for the detergent [25, 35], using the procedure described in Ref. [36]. The *A*(*r*) distributions could be fitted with high precision under the assumption that the protein is homogeneous (Fig. 3). (Apparently, at the rotor speeds used in the experiments the minor component, see earlier, did not sediment significantly and thus could be accounted for by an increased baseline position). The best-fit figure for the effective molar mass of the protein, $M_{\text{eff,p}}$, from a one-component fit was well defined at $(13{,}400 \pm 500)$ g/mol. This figure, however, includes the contributions of the protein-bound phospholipid. We found (with three different samples) that the samples contain, per mole of protein, 18 mol of phosphorus, only 4 mol of which can be accounted for by the phosphorus content of the enzyme-bound cofactors [13]. The remaining 14 mol of phosphorus, when considered as representing phospholipid, will contribute (-1900 ± 400) g/mol to $M_{\text{eff,c}}$. With the corresponding correction and $\bar{v}_p = 0.779$ ml/g, $\rho_0 = 1.2055$ g/ml, we obtain $M_p = (251{,}000 \pm 15{,}000)$ g/mol, which is higher than the molar mass of monomeric Na^+-NQR calculated from the gene sequence by 16%. Without the correction, the M_p value obtained would be 220,000 g/mol and thus virtually identical to the calculated

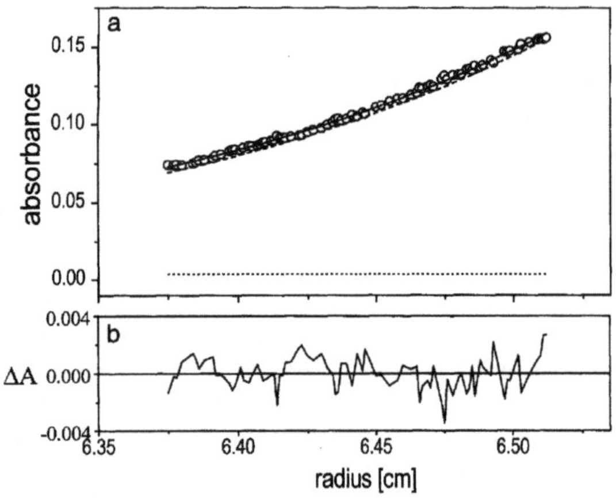

Fig. 3a,b Sedimentation equilibrium analysis of Na^+-NQR. **a** Experimental $A(r)$ data (*open circles*); curve fitted to them assuming the presence of only one component (*solid line*); calculated contribution to $A(r)$ of the monomeric Na^+-NQR (*dashed line*) and of the baseline (*dotted line*). **b** Local differences between fitted and experimental data

figure for the monomer (considering H–D exchange). Interpreting the M_p value of 251,000 g/mol as a weight-average molar mass of a monomer/dimer mixture would indicate a dimer content of the sample of approximately 14% (w/w). There is, however, no corresponding dimer peak in the $c(s)$ profiles described earlier (Fig. 1b). This may indicate that the correction applied for protein-bound phosphorus is too large, possibly owing to trapped phosphate in the sample (which originally contained 50 mM phosphate). In any case, the result supports the conclusion from sedimentation velocity analysis that the observed homogeneous protein represents monomeric Na^+-NQR.

To obtain more information on the bound detergent, sedimentation equilibrium experiments with Na^+-NQR were also performed in H_2O-based buffer ($\rho_0 = 1.014$ g/ml), without density-matching for the bound detergent. Again, the experimental sedimentation equilibrium distributions could be fitted with high precision assuming homogeneity of the sample. The best one-component fit to the data yielded a figure for $M_{\mathrm{eff,c}}$ of (71,100 ± 3,000) g/mol (data not shown). The combination of this effective molar mass with that determined under conditions of density-matching for the bound Dm, neglecting bound phospholipid, yielded a molar mass for the bound detergent of $M_d = (97{,}000 \pm 18{,}000)$ g/mol, which corresponds to (0.45 ± 0.08) g Dm per gram of protein and is thus lower than the value derived from the sedimentation velocity experiments. Of course, the state of association of the protein particle can also be determined from the $M_{\mathrm{eff,c}}$ value obtained without density-matching, in the same way as done from the sedimentation velocity data. In the present case it turns out that $M_{\mathrm{eff,c}}$ is consistent with a protein monomer/detergent complex only, again with approximately 0.45 g Dm bound per gram of protein. The different figures for the latter parameter yield, for the mean and range, (0.57 ± 0.12) g Dm per gram of protein.

The effect of sodium ions on Na^+-NQR sedimentation

Since Na^+-NQR represents a Na^+ pump, its conformation may depend on whether or not sodium is bound. We therefore studied the effect of NaCl on the oligomeric state of Na^+-NQR by sedimentation velocity experiments in samples containing potassium phosphate buffers with 0 or 300 mM NaCl (see Materials and methods). The experimental data were analysed by using Lamm equation modelling for two noninteracting discrete species. The s value of the main component was 6.84 S or 6.35 S in 0 and 300 mM NaCl, respectively. When corrected for standard conditions [34], the results were $s_{20,w} = 11.47$ and 11.39 S, respectively. Converting, in the same way, the sedimentation coefficient of 6.07 S measured in sodium phosphate buffer containing 100 mM NaCl leads to $s_{20,w} = 11.65$ S. Obviously, the sedimentation coefficients observed in the range 0–300 mM NaCl are very close to each other, indicating that sodium ions have no significant influence on the shape or the state of association of the protein.

Discussion

In the present paper, we have unequivocally shown that Na^+-NQR from *V. cholerae*, isolated and studied in solutions of the nonionic detergent Dm, is a homogeneous population of the monomeric "minimal complex" of six subunits. This conclusion is based both on sedimentation velocity and sedimentation equilibrium experiments in the analytical ultracentrifuge; it is further supported by data from gel filtration. Since, in general, the state of association of intrinsic membrane proteins in solution of appropriate mild, nonionic detergents is thought to correspond to that in the intact membrane [16, 17], our result also reports the enzyme's state of association in the *V. cholerae* membrane. Thus, the association behaviour of the enzyme is simple and does not allow the subtle subunit–subunit interactions known from some dimeric components of the respiratory chain (e.g. the cytochrome bc_1 complex [14]).

With respect to the sedimentation equilibrium analysis performed, we have shown recently that the classical technique of density-matching [17, 25, 27, 28, 35] can be successfully applied also with Dm, despite the fact that (1) the density of the detergent is rather high and its matching requires high additive concentrations [14], and

(2) owing to the similarity between $1/\bar{v}_p$ and the matching solvent density, any error in $\bar{v}_p$ will be strongly magnified with respect to the error in M. However, since the high additive concentration will lead to changes in $\bar{v}_p$, it is mandatory that appropriate corrections for this effect are applied. For the experimental conditions used in the present report, these corrections were determined in a previous study [14]. In addition, it is obvious from our data that, owing to the large difference between phospholipid and Dm density, knowledge of the amount of protein-bound phospholipid is a prerequisite for accurate determinations of molar protein mass.

The sedimentation velocity data, collected without using density-matching, have turned out to be at least as useful in determining the state of association of Na^+-NQR as the results from equilibrium sedimentation. Sedimentation velocity experiments, combined with Lamm equation fitting, thus represent a valuable tool also when studying membrane proteins in detergent solution, as also shown by Musatov et al. [22] using different software [20, 21]. Our results indicate that moderate hydrodynamic nonideality can be taken into account. As compared with equilibrium sedimentation, sedimentation velocity analysis has the obvious advantage of requiring less time for completing a run and by this also being suitable for relatively unstable proteins. In addition, it will give valuable information on the purity of the protein sample. The much smaller contribution of protein-bound phospholipid to the molar complex mass, as compared with conditions of Dm density-matching, represents another advantage. On the other hand, sedimentation velocity analysis will not yield absolute values for the particle mass if the amount of protein-bound detergent is not known. This was uncritical in the present study (which even allowed quantifying detergent binding) but will be distinctly disadvantageous in others. We have tried to overcome this disadvantage by performing the sedimentation velocity experiments under conditions of density-matching for the detergent. Our hitherto existing experience is, however, disappointing, at least when using sucrose as a cosolute acting as a "densifier". It remains to be seen whether the up-to-now unidentified problems can be solved.

Acknowledgements We thank A. Roth for technical assistance with the gel filtration experiments and J.A. van den Broek for help with the phosphorus determinations.

References

1. Unemoto T, Hayashi M (1993) J Bioenerg Biomembr 25:385
2. Tokuda H (1989) J Bioenerg Biomembr 21:693
3. Dimroth P (1991) BioEssays 13:463
4. Kogure K (1998) Curr Opin Biotechnol 9:278
5. Hayashi M, Nakayama Y, Unemoto T (1996) FEBS Lett 381:174
6. Zhou W, Bertsova YV, Feng B, Tsatsos P, Verkhovskaya ML, Gennis RB, Bogachev AV, Barquera B (1999) Biochemistry 38:16246
7. Barquera B, Hellwig P, Zhou W, Morgan JE, Häse CC, Gosink K, Nilges M, Bruesehoff PJ, Roth A, Lancaster CRD, Gennis RB (2002) Biochemistry 41:3781
8. Dibrov PA, Kostyrko VA, Lazarova RL, Skulachev VP, Smirnova IA (1986) Biochim Biophys Acta 850:449
9. Dibrov PA, Lazarova RL, Skulachev VP, Verkhovskaya ML (1986) Biochim Biophys Acta 850:458
10. Häse CC, Barquera B (2001) Biochim Biophys Acta 1505:169
11. Häse CC, Fedorova ND, Gaperin MY, Dibrov PA (2001) Microbiol Mol Biol Rev 65:353
12. Hayashi M, Nakayama Y, Unemoto T (2001) Biochim Biophys Acta 1505:37
13. Barquera B, Zhou W, Morgan JE, Gennis RB (2002) Proc Natl Acad Sci USA 99:10322
14. Mayer G, Anderka O, Ludwig B, Schubert D (2002) Prog Colloid Polym Sci 119:77
15. Lancaster CRD, Kröger A, Auer M, Michel H (1999) Nature 402:377
16. Helenius A, Simons K (1975) Biochim Biophys Acta 415:29
17. Tanford C, Reynolds JA (1976) Biochim Biophys Acta 457:133
18. Schuck P, Perugini MA, Gonzales NR, Howlett GJ, Schubert D (2002) Biophys J 82:1096
19. Solovyova A, Schuck P, Costenaro L, Ebel C (2001) Biophys J 81:1868
20. Demeler B, Saber H, Hansen JC (1997) Biophys J 72:397
21. Demeler B, Saber H (1998) Biophys J 74:444
22. Musatov A, Lopez JO, Demeler B, Osborne JP, Gennis RB, Robinson NC (1999) FEBS Lett 457:153
23. Schuck P (1994) Prog Colloid Polym Sci 94:1
24. Durchschlag H (1986) In: Hinz H-J (ed) Thermodynamic data for biochemistry and biotechnology. Springer, Berlin Heidelberg New York, pp 45–128
25. Reynolds JA, Tanford C (1976) Proc Natl Acad Sci USA 73:4467
26. Steele JCH, Tanford C, Reynolds JA (1978) Methods Enzymol 48 part F:11
27. Schubert D, Tziatzios C, van den Broek JA, Schuck P, Germeroth L, Michel H (1994) Prog Colloid Polym Sci 94:14
28. Tziatzios C, Schuck P, Schubert D, Tsiotis G (1994) Z Naturforsch C Biosci 49:220
29. Laue TM, Shah BD, Ridgeway TM, Pelletier SL (1992) In: Harding SE, Rowe AJ, Horton JC (eds) Analytical ultracentrifugation in biochemistry and polymer science. Royal Society of Chemistry, Cambridge, pp 90–125
30. Bartlett GR (1959) J Biol Chem 234:466
31. Schaegger H (1994) In: Schaegger H, von Jagow G (eds) A practical guide to membrane protein purification. Academic, New York, pp 23–57
32. Schuck P (2000) Biophys J 78:1606
33. Rowe AJ (1992) In: Harding SE, Rowe AJ, Horton JC (eds) Analytical ultracentrifugation in biochemistry and polymer science. Royal Society of Chemistry, Cambridge, pp 394–406
34. Cantor CR, Schimmel PR (1980) In: Biophysical chemistry, part II. Freeman, New York, pp 591–641
35. Schubert D, Schuck P (1991) Prog Colloid Polym Sci 86:12
36. Mayer G, Ludwig B, Müller H-W, van den Broek JA, Friesen RHE, Schubert D (1999) Prog Colloid Polym Sci 113:176

Progr Colloid Polym Sci (2004) 127: 54–60
DOI 10.1007/b98013

C. Tziatzios
A.A. Precup
B.G.G. Lohmeijer
L. Börger
U.S. Schubert
D. Schubert

Dimerization of monofunctionalized poly(ethylene oxide) via metal–ligand interactions and hydrogen bonds

C. Tziatzios · D. Schubert
Institut für Biophysik, Johann Wolfgang Goethe-Universität, Frankfurt am Main, Germany

A.A. Precup · B.G.G. Lohmeijer
U.S. Schubert (✉)
Laboratory of Macromolecular Chemistry and Nanoscience, Eindhoven University of Technology and Dutch Polymer Institute, Eindhoven, The Netherlands
e-mail: u.s.schubert@tue.nl
Tel.: +31-40-2474083
Fax: +31-40-2451036

L. Börger
Polymerforschung, BASF AG, Ludwigshafen, Germany

Abstract Supramolecular chemistry utilizes noncovalent bonds for assembling large ordered structures. Among the interactions used, metal–ligand interactions and hydrogen bonds play important roles. We have studied the degree of association attainable with these two interactions by quantifying dimer formation between suitable bond-forming blocks which were covalently linked to a poly(ethylene oxide) chain (or a dye-labelled derivative). Dimer formation was monitored by sedimentation equilibrium analysis in the analytical ultracentrifuge. The partial specific volume, $\bar{v}$, of the compounds was also determined by equilibrium sedimentation, using the "buoyant density method" and the tetrahydrofuran/propylene carbonate solvent pair. Up to complete dimerization of the molecules was observed when complex formation was based on Fe(II)–terpyridine interactions. In contrast, under our experimental conditions, quadruple hydrogen-bonded arrays led to only 20–50% dimerization, with surprisingly low yields in apolar solvents and significant yields even in solvents also capable of participating in hydrogen-bond formation. With the compound studied, dimer formation was found to be largely irreversible.

Keywords Supramolecular chemistry · Metal–ligand interactions · Hydrogen bonding · Sedimentation equilibrium analysis · Partial specific volume

Introduction

The essence of supramolecular chemistry is to assemble, via noncovalent bonds, large and in many cases highly regular structures from tailor-made, relatively small units [1]. The incorporation of such units into macromolecules is currently of major interest for the engineering of tailor-made macromolecules which, owing to the reversibility of the bonds involved, may act as "smart" or switchable materials: Under suitable conditions, the polymer molecules may associate, via their bond-forming units, in two or three dimensions, while changing external parameters, such as temperature, may lead to complete cancellation of these associations and thus have a major impact on the material properties [2]. Since, in principle, nearly unlimited combinations of different small "functional" units and macromolecules are possible, realization of the assembly principle described offers promising chances in materials research.

Among the noncovalent interactions applicable for the construction of supramolecular architectures, metal–ligand interactions [3] and hydrogen bonding [4] play a prominent role (both of them often being assisted by van der Waals interactions). In the present work, we have compared the relative binding strength of (1) a metal–ligand interaction based on Fe(II) ions and terpyridine ligands [3] and (2) a hydrogen-bonding array based on ureido-pyrimidinone units [4]. The bond-forming groups were linked to one of the chain ends of poly(ethylene oxide) (PEO) with a degree of polymerization of about 70, and their association into dimers (plus eventual higher oligomers) was quantified by means of sedimentation equilibrium analysis in the analytical ultracentrifuge. All molecules studied were detected by absorption optics; in the case of the quadruple hydrogen-bonded PEO this was made possible by coupling a dye label to the other chain end.

Experimental

Syntheses

Compound 1 (PEO_{70}-[Fe]-PEO_{70})

For details concerning synthesis and characterization of the Fe(II) complex and the free ligand, α-methoxy-ω-(4′-oxoterpyridinyl)-PEO, see Ref. [5]. The structures are shown in Fig. 1. Selected analytical data: Free ligand: UV/vis (H_2O) λ(ε) 278 (13,200), 234 (17,000); ^{1}H NMR (500 MHz, $CDCl_3$, δ) 8.68 (dt, 2H, H6;H6″), 8.61 (dt, 2H, H3;H3″), 8.04 (s, 2H, H3′;H5′), 7.85 (td, 2H, H4; H4″), 7.34 (ddd, 2H, H5;H5″), 4.40 (t, 2H, tpyOCH_2), 3.93 (t, 2H, tpyOCH$_2$CH_2), 3.83–3.45 (m, PEO backbone), 3.38 (s, 3H, OCH_3). Fe(II) complex: UV/vis (H_2O) λ(ε) 556 (8,400), shoulder at 506 (5,700), 361 (5,300), 315 (31,000), 271 (39,200), 245 (39,200); ^{1}H NMR (400 MHz, CD_3CN, δ) 8.57, 8.51, 7.93, 7.20, 7.12, 4.80, 4.14, 3.83–3.39, 3.34; IR (attenuated total reflection, cm^{-1}) 2,883, 2,714, 1,616, 1,563, 1,551, 1,467, 1,455, 1,425, 1,414, 1,360, 1,342, 1,280, 1,241, 1,147, 1,099, 1,060, 961, 947, 840, 793, 758, 731, 699, 665.

Compound 2 (hydrogen-bond forming)

The synthesis and structure of compound 2 are shown in Fig. 2.

Preparation of α-dye-ω-boc-PEO

The dye, 4-(5-dimethylaminonaphtalene-1-sulphonyl)phenylisothiocyanate (40.5 mg, 0.105 mmol), and of α-amino-ω-boc-PEO (300 mg, 0.088 mmol) were dissolved in CH_2Cl_2 and stirred at room temperature for 24 h. The solvent was evaporated in vacuo and the crude product was purified by size-exclusion chromatography. Yield: 286 mg, 90%.

^{1}H NMR (400 MHz, $CDCl_3$, δ) 1.45 (s, 9H, $H_{t\text{-}but}$), 2.88 (s, 6H, N(CH_3)$_2$), 3.65 (m, 390H, H_{poly}), 5.10 (s_{broad}, 1H, H_{NH}), 6.94 (s_{broad}, 1H, H_{NH}), 7.04 (m, 3H, H_{arom}), 7.19 (m, 3H, H_{arom}), 7.44–7.61 (m, 2H, H_{arom}), 8.18–8.53 (m, 2H, H_{arom}).

Preparation of α-dye-ω-amino-PEO

α-Dye-ω-boc-PEO (190 mg, 0.052 mmol) was dissolved in 3.0 ml CH_2Cl_2. Trifluoracetic acid (1.0 ml) was added and the reaction mixture was stirred for 2 h at room temperature. A 2.0-ml aliquot of a 4% NaOH aqueous solution was added and the mixture was extracted four times with 5 ml CH_2Cl_2 each time. The organic fractions were combined and dried over Na_2SO_4. The solvent was removed in vacuo and the crude product was purified by size-exclusion chromatography. Yield: 164 mg, 90%.

Fig. 1 Scheme for the preparation of a poly(ethylene oxide) dimer held together by a *bis*-terpyridine/iron(II) complex

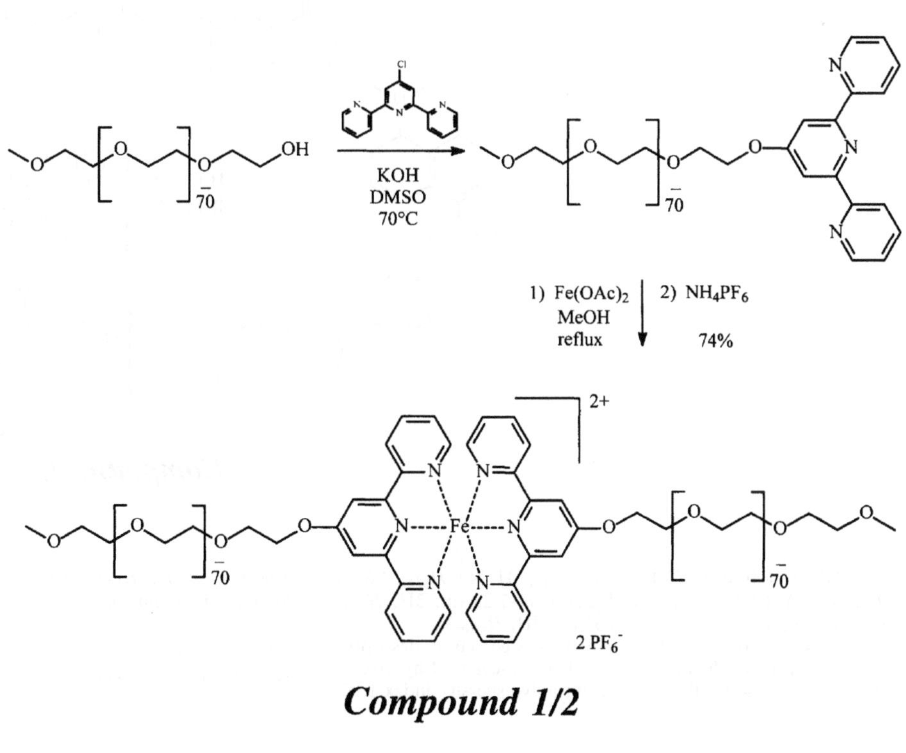

Fig. 2 Scheme for the preparation of dye-labelled poly(ethylene oxide) bearing a self-complementary ureido-pyrimidinone unit which is capable of quadruple hydrogen bonding

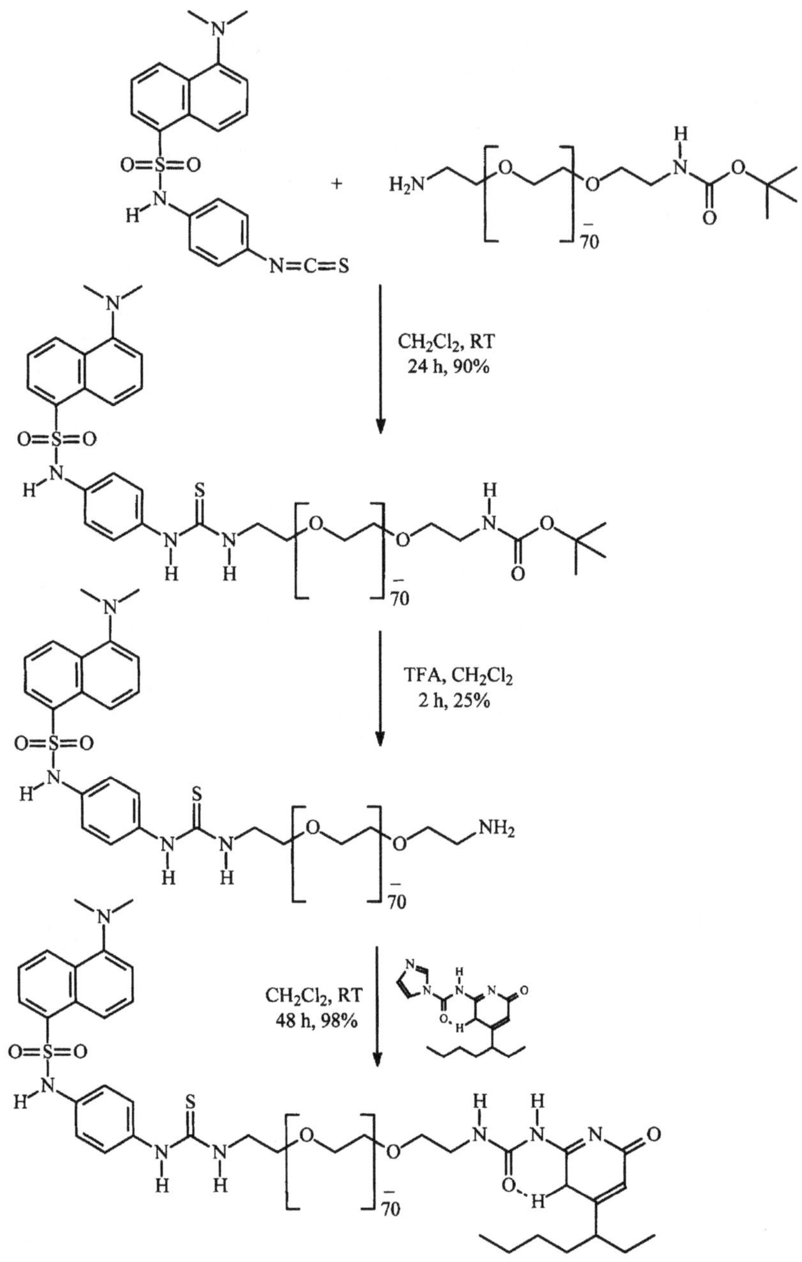

^{1}H NMR (400 MHz, $CDCl_3$, δ) 2.91 [s, 6H, $N(CH_3)_2$], 3.65 (m, 270H, H_{poly}), 6.93 (s_{broad}, 1H, H_{NH}), 6.99–7.23 (m, 3H, H_{arom}), 7.47–7.62 (m, 4H, H_{arom}), 8.22–8.54 (m, 3H, H_{arom}).

Mass spectrometry (MS) (matrix-assisted laser desorption/ionization time of flight, MALDI-TOF, α-cyano-4-hydroxycinnamic acid): $M_n = 3{,}730$; $M_w = 3{,}826$; polydispersity index 1.03.

Preparation of α-dye-PEO terminated by a ureido-pyrimidone unit (compound 2)

α-Dye-ω-amino-PEO (41 mg, 0.017 mmol) and carbodiimidazole activated 6-(2-ethylpentyl)-isocytosine (190 mg, 0.033 mmol) [4, 6,

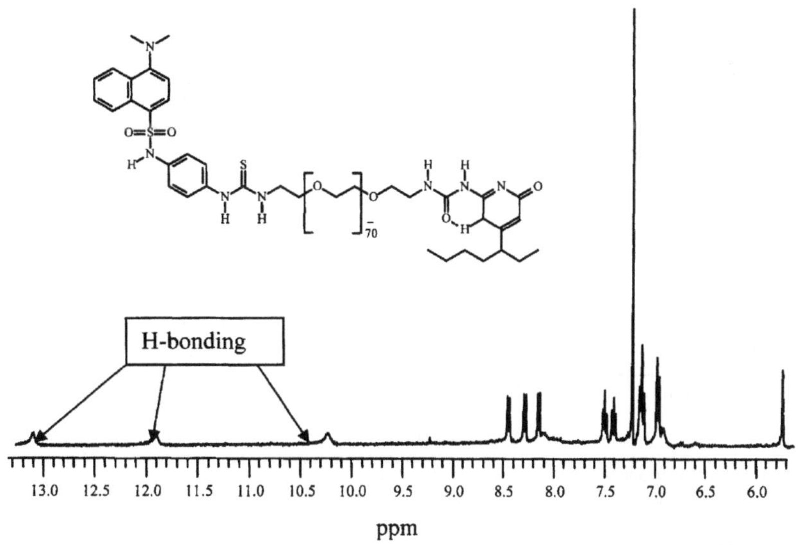

Fig. 3 Signals from the hydrogen-bonded protons and the aromatic region of the ^{1}H NMR spectra (400 MHz, $CDCl_3$) of compound 2

7] were dissolved in CH_2Cl_2 and stirred for 48 h at room temperature. The solvent was removed and the remaining crude product was purified by size-exclusion chromatography. Yield: 63 mg, 98%.

^{1}H NMR (400 MHz, $CDCl_3$, δ) 0.90–1.73 (m, 21H, H_{aliph}), 2.88 [s, 6H, N(CH_3)$_2$], 3.70 (m, 390H, H_{poly}), 5.80 (s, 1H, H_{oleph}), 7.04 (m, 3H, H_{arom}), 7.19 (m, 3H, H_{arom}), 7.451–7.61 (m, 2H, H_{arom}), 8.20–8.52 (m, 2H, H_{arom}), 10.29 (s_{broad}, 1H, H_{NH}), 11.98 (s_{broad}, 1H, H_{NH}), 13.20 (s_{broad}, 1H, H_{NH}) (see Fig. 3).

MS (MALDI-TOF, α-matrix): $M_n = 3{,}930$; $M_w = 4{,}028$.

Gel permeation chromatography (GPC) ($CHCl_3$, polystyrene standard): $M_n = 5{,}127$; $M_w = 5{,}746$.

Chemicals

4-(5-Dimethylaminonaphtalene-1-sulphonyl)phenylisothiocyanate was purchased from Molecular Probes, α-amino-ω-boc-PEO and α-hydroxy-ω-methoxy-PEO were from Shearwater, and PEO standards were from Polymer Laboratories. 1,1-Carbonyldiimidazole activated 2-amino-4-hydroxy-6-ethylpentylpyrimidine was obtained from SyMOChem. The solvents were bought from Biosolve; dimethyl sulphoxide was dried over BaO. For preparative size-exclusion chromatography, Bio-Rad SX-1 beads swollen in CH_2Cl_2 were used. All other chemicals were purchased from Sigma/Aldrich, Fluka or Merck and were used without further purification.

Analytical ultracentrifugation

Sedimentation equilibrium experiments were performed using a Beckman Optima XL-A ultracentrifuge, an An-50Ti rotor, titanium double-sector centrepieces of pathlength 12 mm (BASF), and polyethylene gaskets. The rotor speed was 35,000–40,000 rpm and the rotor temperature was 20 °C. The sample volume was 200 μl. The absorbance versus radius profiles, $A(r)$, were recorded at 315 or 555 nm (compound 1) or at 330 nm (compound 2), and those of the uncomplexed precursor of compound 1 at 280 nm. The evaluations were performed as described earlier [8], using the computer program DISCREEQ by Schuck [9, 10]. Determination of the partial specific volume, $\bar{v}$, of the compounds was achieved by application of the "buoyant density method" [8, 11, 12] and the tetrahydrofuran (THF)/propylene carbonate solvent pair [8].

Other techniques

^{1}H NMR and ^{13}C NMR spectra were recorded, at 25 °C, using a Varian Inova spectrometer at frequencies of 500 and 125 MHz, respectively, and using a Varian Mercury spectrometer at frequencies of 400 and 100 MHz. Chemical shifts are given in parts per million downfield from TMS. UV/vis spectra were recorded using a PerkinElmer Lambda 45P spectrophotometer. MALDI mass spectra were obtained using dithranol as the matrix and NaI with a PerSeptive Biosystems Voyager DE PRO spectrometer using a layer-by-layer spotting technique. Size-exclusion chromatography was performed on a 30-cm Waters Styragel HT4 column, at a flow rate of 0.5 ml/min adjusted by a Waters 1515 pump, with chloroform as the eluent and with 4% (v/v) triethylamine and 2% (v/v) 2-propanol as additives to reduce column interactions. The eluate was monitored with a refractive index detector (Waters 1414). PEO standards were used for calibration.

Results

Compound 1

The sedimentation equilibrium profiles, $A(r)$, for both the PEO–terpyridine unit and the corresponding metal–ligand complex, in the two solvents and in a mixture of them, could be fitted by single exponentials, suggesting homogeneity of the particles. Plots of the effective molar particle mass, $M_{eff} = M(1 - \bar{v}\rho_0)$, as a function of solvent density, ρ_0, yielded straight lines, suggesting independence of both M and $\bar{v}$ of the solvent [8, 11] (Fig. 4). However, for unknown reasons samples of the Fe(II) complex in 30% THF, corresponding in density to the apparently missing data point in the figure, were

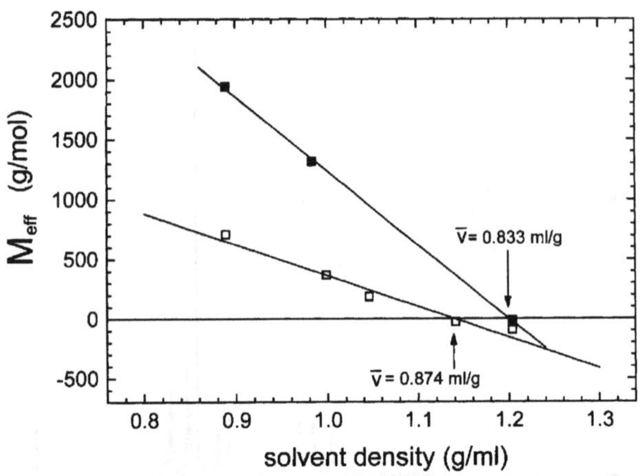

Fig. 4 Effective molar mass, M_{eff}, of the uncomplexed poly(ethylene oxide)–terpyridine unit (*open squares*) and of compound 1 (*closed squares*) as a function of solvent density. Solvent system: tetrahydrofuran (*THF*)/propylene carbonate, each solution containing 10 mM tetrabutylammonium hexafluorophosphate. Loading concentrations of the uncomplexed unit and of compound 1: approximately 0.10 and 0.26 g/l, respectively

heavily aggregated. From the zero crossover of the lines, $\bar{v}$ is obtained as (0.874 ± 0.004) ml/g for the free ligand and (0.833 ± 0.004) ml/g for the complex. Applying these figures, the M_{eff} values in THF yield, for the molar particle mass, $M = (3{,}190 \pm 150)$ g/mol for the free ligand[1] and $M = (7{,}180 \pm 300)$ g/mol for the complex. Virtually identical results were obtained from the slope of the straight lines (which equals $-M\bar{v}$ [8, 12]).

Within the limits of error of the results, these M values are identical to those calculated from the chemical structure of the compounds, 3,250 and 7,110 g/mol, respectively. (These numbers agree with experimental ones based on the NMR, MALDI-TOF-MS and GPC data obtained for the uncomplexed PEO–terpyridine; with the complex the figure includes two PF_6 counterions which cosediment with it). Thus, complex formation by the terpyridine–Fe(II)–terpyridine unit seems to be virtually complete under the experimental conditions of Fig. 4. This finding is corroborated and extended by fits to $A(r)$ using a monomer/dimer/tetramer model of self-association: more than 96% of the absorbance sediments with the molar mass of the dimer, the remainder with that of the tetramer; according to the best fit, the absorbance corresponding to the monomer is zero and may reach at most 2%. However, any nonzero figure determined in this way for the monomer content would strongly underestimate its true size since, according to Ref. [13], the molar extinction coefficient of the terpyridine–Fe(II) complex amounts to only approximately 8% of that of the terpyridine–Fe(II)–terpyridine unit. Thus, even a contribution of 1% of the terpyridine–Fe(II) complex to the total absorbance at 555 nm would correspond to approximately 11% of the Fe(II)-containing particles. Whereas the data of Fig. 4 are compatible with complete dimerization of the compound, the formation of monomers could be seen in more dilute samples, either from fits to $A(r)$ which treated the sample as a mixture of monomers and dimers, or by simply observing the strong decrease with time in sample absorbance at 555 or 315 nm (data not shown). However, according to both methods, monomer accumulation did not occur when the samples contained 1% (v/v) 2-mercaptoethanol to prevent oxidation. We suggest the following interpretation:

1. The dimeric compound 1 is not stable but is actually part of a monomer/dimer association equilibrium which is, however, strongly shifted towards the larger oligomer in the concentration range studied by us.
2. At lower compound concentration the average lifetime of the monomer is sufficiently long to allow oxidation of Fe(II) in the terpyridine–Fe unit if no precautions are taken.
3. By oxidation, the Fe-containing monomer will be removed from the association equilibrium; its concentration will increase with time (this conclusion is based on the assumption that, similar to the case of the Ru-containing analogues [14], Fe(III) does not support formation of a terpyridine–Fe–terpyridine complex).

Compound 2

The procedure used with compound 1, sedimentation equilibrium experiments in pure solvents and appropriate solvent mixtures in order to determine both $\bar{v}$ and M, was also applied to compound 2 and its precursor not capable of hydrogen bonding (Fig. 2). The measurements of $A(r)$ utilized the absorbance of the dye label covalently attached to the PEO chain. The results obtained with the precursor, in THF and mixtures of this solvent with propylene carbonate, are shown in Fig. 5. They yield $\bar{v} = 0.884$ ml/g (in agreement with results on a similar compound obtained by both the same method and digital densimetry [8]) and $M = 4{,}035$ g/mol. The latter figure exceeds the value calculated from the chemical structure by 6%, which suggests the presence of a small amount of aggregated material (most probably dimers).

[1] Care had to be taken to avoid dimerization of the compound by traces of divalent metal ions in the added salt, as repeatedly observed by us. The presence of these ions is, however, already indicated by the appearance of a peak at 320 nm in the absorbance spectra

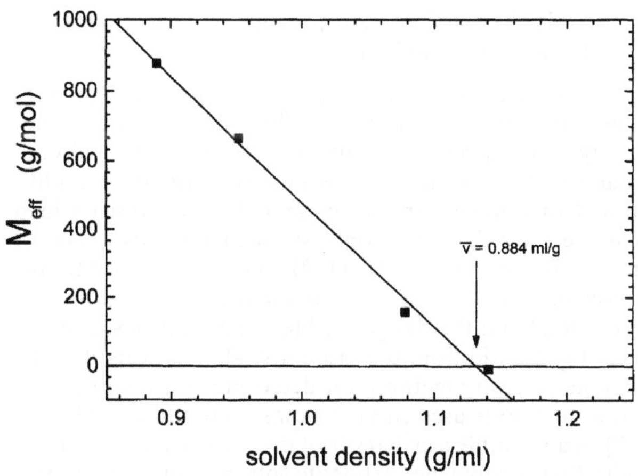

Fig. 5 M_{eff} versus solvent density, in THF/propylene carbonate mixtures, for the precursor of compound 2. Compound concentration: approximately 0.21 g/l

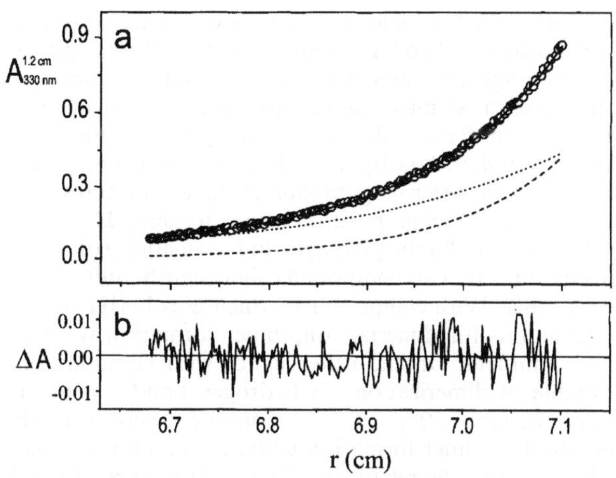

Fig. 6 **a** Sedimentation equilibrium analysis of compound 2: experimental absorbance values $A(r)$ at 330 nm (*circles*), curve fitted to them assuming the presence of monomers and dimers of the compound (*solid line*), and calculated contributions to $A(r)$ of the monomer (*dotted line*) and the dimer (*dashed line*). **b** Residuals of the fit. Initial compound concentration: approximately 0.32 g/l. Solvent: THF plus 10 mM tetrabutylammonium hexafluorophosphate. Rotor speed: 40,000 rpm

Table 1 Sedimentation equilibrium analysis of compound 2 based on a monomer/dimer model of self-association: percentage of the compound present as dimer (w/w) in different solvents. Compound concentration: approximately 0.21 g/l (except for H_2O, where data for three different samples with concentrations between 0.12 and 0.36 g/l are summarized). In addition, the $\bar{v}$ value of the compound in the respective solvent is given

Solvent	$\bar{v}$ (ml/g)	Percentage dimer
Tetrahydrofuran plus 10 mM tetrabutylammonium hexafluorophosphate	0.884	45 ± 5
$CHCl_3$ plus 10 mM NH_4PF_6	0.839	46 ± 8
H_2O plus 20 mM NH_4PF_6	0.833	20–40

In the studies on the functionalized molecule, compound 2, which were performed in THF, $CHCl_3$, and H_2O (each containing 10 or 20 mM NH_4PF_6 or the corresponding tetrabutyl salt, depending on the polarity of the solvent), the dependency of $\bar{v}$ on the solvent had to be taken into account [8, 11, 12]. With THF, the $\bar{v}$ value determined for the precursor was used (see earlier). With the other two solvents, the analyses were based on the $\bar{v}$ values determined in a previous study [8]; see Table 1.

In all sedimentation equilibrium experiments, fitting the data required the use of terms for both monomers and dimers, with the calculated relative amount of the two oligomers depending on the polarity of the solvent. A typical fit is shown in Fig. 6, and a compilation of the results in Table 1. It is obvious from the table that, under the conditions used, (1) the relative amount of dimer does not exceed approximately 50%, even in the nonpolar solvents, and (2) it unexpectedly reaches up to 40% in H_2O, where competition by the solvent molecules is thought to suppress solute–solute hydrogen bonds. (We have confirmed by analogous measurements on the precursor, which yielded dimer contents of 10% and thus only modestly larger contents than in THF, that the latter result can only to a small extent be due to nonspecific hydrophobic association or stacking via the dye component). Also unexpectedly, a decrease in solute concentration from 0.42 to 0.14 g/l did not significantly influence the relative amounts of monomer and dimer found, either in THF or in H_2O. Thus, the compound molecules do not seem to be in an association equilibrium but instead are somehow "frozen" in a certain state.

Discussion

The present paper has two objectives:

1. To compare the relative strength of two widely applied types of interactions between supramolecular building blocks when the blocks are linked to macromolecules: metal–ligand interactions and hydrogen bonding.
2. To demonstrate the potential of sedimentation equilibrium analysis in the analytical ultracentrifuge for detecting and quantifying the formation of macromolecular complexes assembled via the supramolecular building blocks mentioned, i.e. held together by noncovalent bonds. Monofunctionalized PEO derivatives capable of dimer formation were used as model compounds.

With respect to objective 1), both compounds studied apparently met two basic requirements: sufficient purity and homogeneity, and the absence of side reactions or additional reactions leading to significant unspecific aggregation. This is demonstrated by the feasibility of fitting the $A(r)$ data by a single exponential or by two exponentials describing monomer/dimer mixtures, together with baseline positions close to zero. However, with respect to binding strength and reversibility of dimer formation, the two compounds significantly differ from each other. With compound 1, which uses Fe(II)–ligand interactions for dimerization, dimer formation is virtually complete. On the other hand, with compound 2, capable of dimerization via hydrogen bonds, at maximum around 50% of the molecules were actually involved in dimer formation under the conditions used. This confirms the relative weakness of this type of bond which is well known for biological macromolecules, in particular proteins [15, 16]. In addition, dimer formation showed two peculiarities:

1. It was significant even in water, despite the strong competition expected from water–ligand hydrogen bonds. Possibly, steric conditions in the dimer lead to an exclusion of water from the binding site(s) (a case well known from proteins and nucleic acids [15, 16]).
2. The percentage of dimer present in the sample did not significantly decrease when the sample was diluted several fold, indicating that dimer formation was largely irreversible. At present, we are unable to explain this behaviour.

Concerning objective 2, our results again demonstrate the long-known ability of sedimentation equilibrium analysis to detect and quantify associations between macromolecules involving noncovalent bonds, an additional advantage being the large molar mass range which can be covered. This contrasts with the experience of many groups using MALDI-TOF-MS, a technique allowing very accurate molar mass determinations but suffering from the danger of breaking weak associations [17, 18]. In addition, the present study confirms that the buoyant density method, for determining $\bar{v}$ at compound concentrations used in molar mass determination [8, 11, 12], is a valuable auxiliary tool in sedimentation analysis. The THF/propylene carbonate solvent pair [8] again was shown to be ideally suited for studying PEO derivatives, owing to the good solubility of the compounds in both solvents and in mixtures thereof (with one exception where aggregation occurred). Thus, all tools required for characterizing molar mass and association behaviour of the compounds of the present study and of many related ones are available.

Acknowledgements The authors express their sincere thanks to Michael A. R. Meier for GPC and MALDI-TOF-MS measurements and the Dutch Polymer Institute, NWO, as well as the Fonds der Chemischen Industrie for financial support.

References

1. Lehn J-M (1995) Supramolecular chemistry – concepts and perspectives. VCH, Weinheim
2. Lohmeijer BGG, Schubert US (2003) J Polym Sci Part A Polym Chem 41:1413
3. Schubert US, Eschbaumer C (2002) Angew Chem Int Ed Engl 41:2892
4. Brunsveld L, Folmer BJB, Sijbesma RP, Meijer EW (2001) Chem Rev 101:4071
5. Lohmeijer BGG, Schubert US (2003) Macromol Chem Phys 204: 1072
6. Bosman AW, Folmer BJB, Hirschberg JHKK, Keizer HM, Sijbesma RP, Meijer EW (2002) Polym Prepr Am Chem Soc Div Polym Chem 43:322
7. Keizer HM, Ramzi A, Sijbesma RP, Meijer EW (2003) Polym Prepr Am Chem Soc Div Polym Chem 44:596
8. Tziatzios C, Precup AA, Weidl CH, Schubert US, Schuck P, Durchschlag H, Maechtle W, van den Broek JA, Schubert D (2002) Prog Colloid Polym Sci 119:24
9. Schuck P (1994) Prog Colloid Polym Sci 94:1
10. Schuck P, Legrum B, Passow H, Schubert D (1995) Eur J Biochem 230:806
11. Tziatzios C, Durchschlag H, Sell B, van den Broek JA, Maechtle W, Haase W, Lehn J-M, Weidl CH, Eschbaumer C, Schubert D, Schubert US (1999) Prog Colloid Polym Sci 113:114
12. Tziatzios C, Durchschlag H, Weidl CH, Eschbaumer C, Maechtle W, Schuck P, Schubert US, Schubert D (2002) ACS Symp Ser 812:185
13. Hogg R, Wilkins RG (1962) J Chem Soc 1962:341
14. Sullivan BP, Calvert JM, Meyer T (1980) J Inorg Chem 19:1404
15. Kauzmann W (1959) Adv Protein Chem 14:1
16. Dill KA (1990) Biochemistry 29:7133
17. Nielen MWF (1999) Mass Spectrom Rev 18:309
18. Karas M, Glückmann M, Schäfer J (2000) J Mass Spectrom 35:1

Progr Colloid Polym Sci (2004) 127: 61–66
DOI 10.1007/b98014

P. Lavrenko
N. Yevlampieva
B. Dardel
R. Deschenaux

Hydrodynamic investigations of the C_{60} fullerene effects on liquid-crystalline malonate-based dendrimers and hemidendrimers in solution

P. Lavrenko (✉)
Institute of Macromolecular Compounds,
Russian Academy of Sciences,
Bolshoy pr. 31, 199004
St. Petersburg, Russia
e-mail: lavrenko@mail.macro.ru

N. Yevlampieva
Institute for Physics,
Saint Petersburg State University,
Ulyanovskaya 1, 198504
St. Petersburg, Russia

B. Dardel · R. Deschenaux
Institut de Chimie,
Université de Neuchâtel, Av. de
Bellevaux 51, Case postale 2,
2007 Neuchâtel, Switzerland

Abstract The [60]fullerene (C_{60}) effects on behavior of hybrid liquid-crystalline malonate-based dendrimers and hemidendrimers, where C_{60} is attached directly to the start center, are investigated in dilute solutions using hydrodynamic methods. Differences were observed in diffusion–viscosity properties of hemidendrimers with and without C_{60}. Nonmolecular dispersity of the dendrimer solutions was found. It was established that the C_{60} effects on the hydrodynamic properties of dendrimers is weaker than the influence of the solvent. Additionally, nonmolecularity of the solutions is primarily caused by intramolecular and intermolecular interactions of the mesogenic groups. The origins of these differences are discussed in comparison with the effects of the solvent thermodynamic quality.

Keywords Malonate-based dendrimer · Hemidendrimer · Fullerene effect · Hydrodynamic dimensions · Diffusion

Introduction

Insertion of fullerene molecules, which influence the intensity of intramolecular interactions, is one of the new ways for modification of the thermotropic liquid-crystalline (LC) dendrimers used in production of materials with desired thermotropic and mesomorphic properties. This is particularly true for hybrid malonate-based dendrimers functionalizied with cyanobiphenylene substituents as the terminal groups [1]. Specifically strong intramolecular and intermolecular dipole–dipole interactions of such groups primarily determine mesomorphism of a dendrimer, the types of the mesophases, and their temperature interval.

To characterize the intensity of these interactions, particularly intramolecular ones that define the external form and size of the individual macromolecules, one can incorporate into the polymer matrix the fullerene, as a unique polyconjugated system. Interestingly the fullerene effect is observed in a melt as well as in dilute solution when intermolecular interactions are very weak. So, recently, using molecular hydrodynamic and electro-optical methods, we have registered the structuring effect of [60]fullerene (C_{60}) on molecular properties of poly(phenylene oxide) [2] and poly(methyl methacrylate) [3].

In the present work, hydrodynamic properties of fullerene-containing malonate-based dendrimers (DF) and hemidendrimers (hDF) of the second generation are investigated with the aim to characterize the possible influence of C_{60} on molecular properties of the LC dendrimers. These compounds were considered as branched structures with an internal or an external center of branching, the center being covalently bonded with C_{60}. The data obtained are compared with those found for the corresponding dendrimer (D) and hemidendrimer (hD) without fullerene. The chemical formulae of the dendrimers investigated are as follows:

$$\text{DF} \quad (((R-OCO)_2-C_6H_3-OCO)_2\text{–}C_6H_3-O_2C-C_6H_4-O-(CH_2)_6-O_2C)_2\text{–}C60$$
$$\text{D} \quad (((R-OCO)_2-C_6H_3-OCO)_2\text{–}C_6H_3-O_2C-C_6H_4-O-(CH_2)_6-O_2C)_2\text{–}CH_2$$
$$\text{hDF} \quad ((R-OCO)_2-C_6H_3-OCO)_2\text{–}C_6H_3-O_2C-C_6H_4-O-(CH_2)_6-O_2C-C60-CO_2-C_2H_5$$
$$\text{hD} \quad ((R-OCO)_2-C_6H_3-OCO)_2\text{–}C_6H_3-O_2C-C_6H_4-O-(CH_2)_6-O_2C-CH_2-CO_2-C_2H_5$$

where $R = \text{–}(CH_2)_{10}\text{–}O\text{–}C_6H_4\text{–}CO_2\text{–}C_6H_4\text{–}C_6H_4\text{–}C{\equiv}N$

The hDF and hD structures are represented in Schemes 1 and 2 in accord with the hydrodynamic dimensions given later.

Experimental

LC dendrimer derivatives were obtained as described earlier [1] by the reaction of C_{60} bonding to malonate-based hybrid dendrimers of the second generation with the terminal cyanobiphenylene mesogenic groups in the proportion of one fullerene molecule per dendritic particle. The compounds were characterized by the following phase transition temperatures (in degrees centigrade) (where S_A, N, and I are the smectic A, nematic, and isotropic phases, respectively) and the glass-transition temperature, T_g:

	T_g	$T(S_A \to N)$	$T(S_A \to I)$	$T(N \to I)$
DF	–	183	–	184
D	46	196	–	202
hDF	50	–	161	–
hD	34	–	182	–

One can see here the fullerene effects on the mesomorphic properties of the compounds under consideration in lowering the phase-transition temperatures and in increasing the glass-transition temperature.

The molecular properties were investigated at 26 °C in dilute solutions in pure benzene of analysis grade (with density $\rho_0 = 0.873$ g/ml, viscosity $\eta_0 = 0.590$ cP, and refractive index $n_D = 1.497$) and in chloroform of reagent grade ($\rho_0 = 1.477$ g/ml, $\eta_0 = 0.536$ cP, and $n_D = 1.444$). The solutions were prepared 1 h before the investigations. Lyophilically dried samples were used (the samples without fullerene appeared to be of higher crystallinity than those containing fullerene).

Hydrodynamic investigations were carried out as described earlier [2, 3]. Low-speed synthetic boundary experiments were performed in a 3180 analytical ultracentrifuge from the Hungarian Optical Works MOM to determine the diffusion coefficients. The rotor rotation frequency was $10{\times}10^3$ rpm. Fullerene-containing dendrimers absorbed considerably within the visual spectrum range; therefore, a light filter with a transmission maximum at $\lambda = 609$ nm was used in the optical recording system. The light absorption and the very low molecular weights of the samples were the reasons why dynamic light scattering and other types of analytical ultracentrifugation experiments were not successful in the study.

The apparent value of the refractive index increment, dn/dc, was used to control the complete dissolution of a sample. It was evaluated from the interference curve area [4] using the equation $dn/dc = (\lambda/abh)Q/c$, where a is a spar twinning, b is the interference fringe interval, h is the optical path length in a cell, and c is the solute concentration.

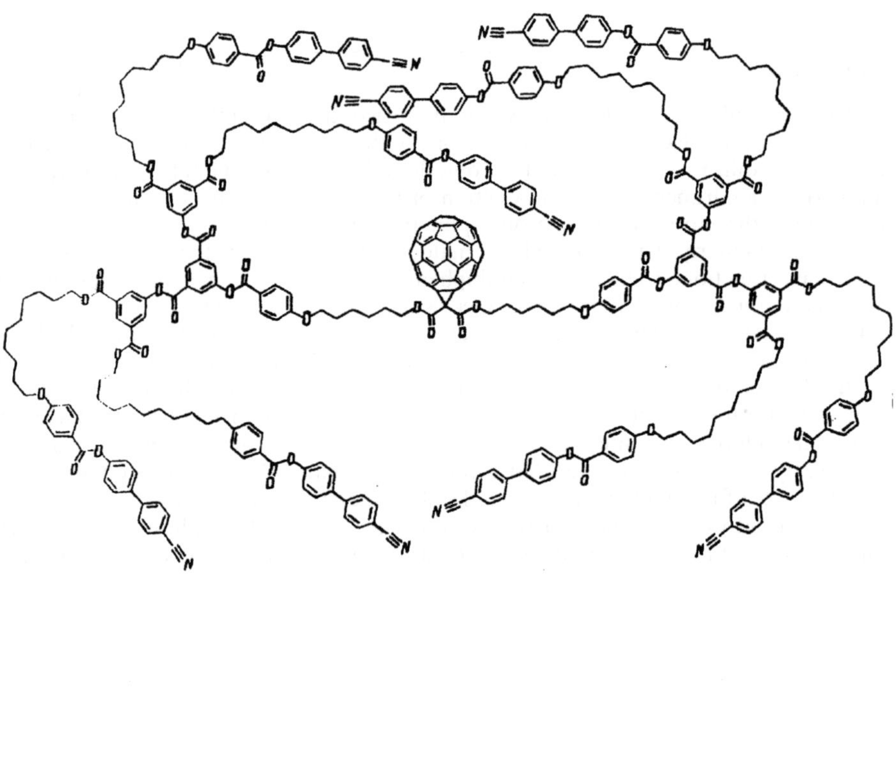

Scheme 1 Schematic plane representation of the fullerene-containing malonate-based dendrimer structure

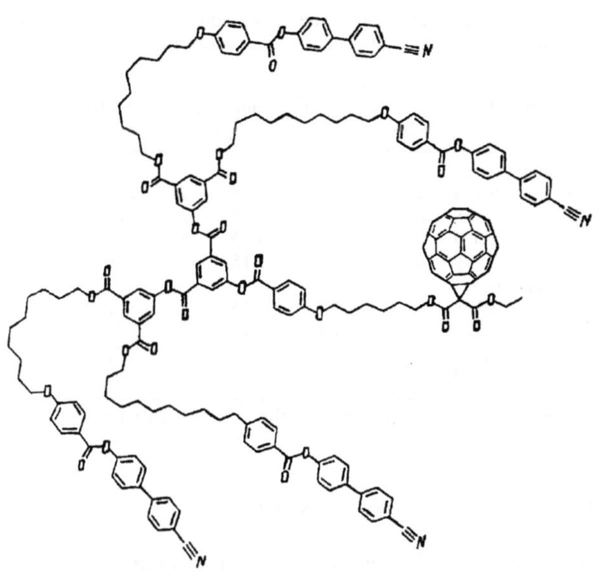

Scheme 2 Schematic plane representation of the fullerene-containing malonate-based hemidendrimer structure

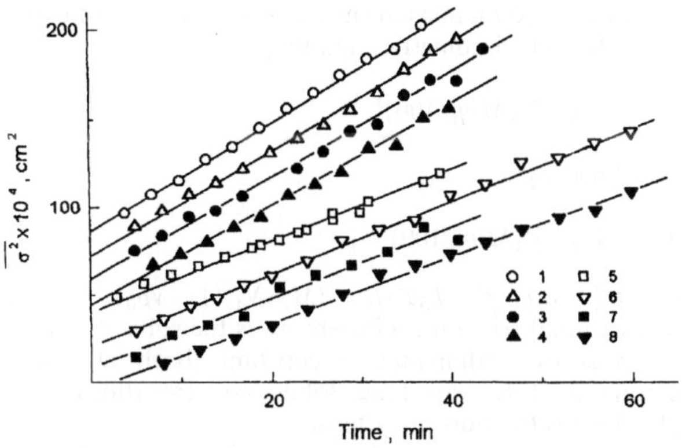

Fig. 1 Dispersion of the concentration boundary, $\overline{\sigma^2}$, versus time, t, for hemidendrimers (*hD*) (*1, 2*), fullerene-containing malonate-based hemidendrimers (*hDF*) (*3, 4*), dendrimers (*D*) (*5, 6*), and fullerene-containing malonate-based dendrimers (*DF*) (*7, 8*) in benzene (*1, 3, 5, 7*) and chloroform (*2, 4, 6, 8*) solutions at 26 °C. Every second curve is shifted up by 10 units

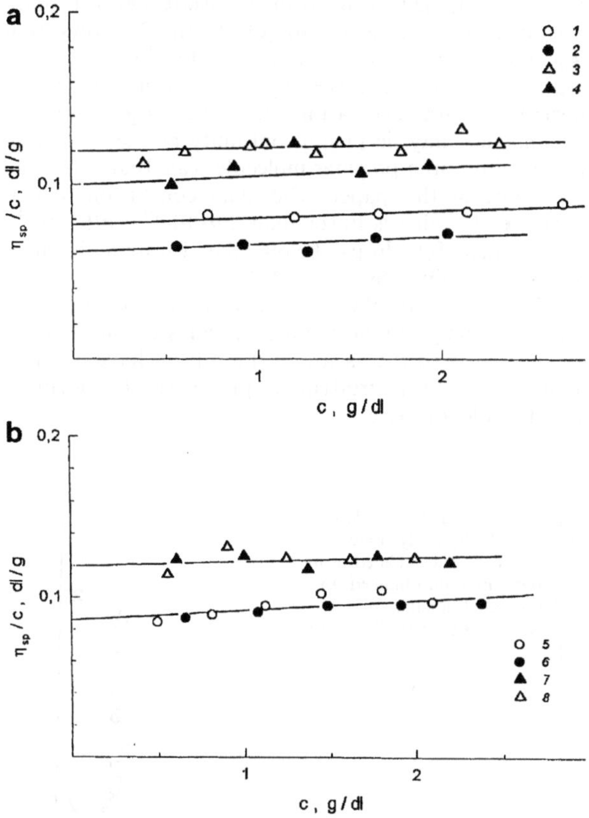

Fig. 2 Concentration dependence of the solution relative viscosity increment, η_{sp}/c, as measured at 26 °C for hD (*1, 3*), hDF (*2, 4*), D (*5, 7*), and DF (*6–8*) in benzene (*1, 2, 5, 6*) and chloroform (*3, 4, 7, 8*) solutions. The *solid lines* correspond to linear approximations

Results and discussion

Investigations were carried out under conditions of real solutions in chloroform and benzene as solvents of different polarity. One could expect markedly different (in intensity) solute–solvent interactions in these polymer–solvent systems.

The diffusion data obtained for dendrimers in benzene and chloroform solutions are shown in Fig. 1 as plots of the diffusion curve dispersion versus time, $\overline{\sigma^2}(t)$. The experimental points could be well approximated by linear functions, and the diffusion coefficients were calculated from the linear slope of these curves by $D = (1/2)\partial\overline{\sigma^2}/\partial t$. Note, that for the hemidendrimers (curves 1–4) the slope of the curves is greater than that for the dendrimers (curves 5–8), in qualitative accordance with the lower molecular weights of the hemidendrimers. The concentration dependence of D at the concentrations used (from 0.25 to 0.50 g/dl) was absent.

The particle size in solution was characterized by the hydrodynamic radius (radius of hydrodynamically equivalent sphere), R_H, calculated by

$$R_H = kT/6\pi\eta_0 D, \qquad (1)$$

where k is the Boltzmann constant.

The concentration dependencies of the reduced viscosity of the solutions η_{sp}/c are shown in Fig. 2. The intrinsic viscosities, $[\eta]$, were calculated using the Huggins equation $\eta_{sp}/c = [\eta] + k_x[\eta]^2 c$. The average value of the Huggins constant, k_x, was 0.60.

The hydrodynamic parameter A_0 was calculated using the diffusion–viscometric data by [5]

$$A_0 = (D\eta_0/T)(M[\eta]/100)^{1/3} \quad (2)$$

as defined by

$$A_0 = (k_B/[f])(M[\eta]/100)^{1/3},$$

where $[f]{\equiv}f/\eta_0$, $f = k_BT/D = (M/sN_A)(1 - \bar{v}\rho_0)$, f is the translational friction coefficient, η_0 is the solvent viscosity, s is the sedimentation constant (both D and s extrapolated to vanishing solute concentration), and $(1 - \bar{v}\rho_0)$ is the buoyancy term.

As shown earlier, A_0 is invariant to molecular weight, at least in the Gaussian range, and is equal to $(3.2 \pm 0.2){\times}10^{-10}$ erg/K mol$^{1/3}$ for all the flexible chain polymers in solutions in both θ and non-θ solvents [5]. That is why this parameter is often used in the M evaluation. It is important to note that within a low molecular weight range, A_0 is a more complicated function of M, as can be analyzed, for example, using the data obtained for poly(dimethylsiloxane) in bromocyclohexane at 29.5 °C (θ) [6]. Figure 3 shows that A_0 has really a constant value close to $3.2{\times}10^{-10}$ erg/K mol$^{1/3}$ at M exceeding 10^4, but its magnitude decreases with decreasing M. The origins of the dependence may lie in noncontinuity of the surrounding medium on a polymer molecule scale that is beyond the scope of this paper. The main conclusion here is that the A_0 value is lower than $2.9{\times}10^{-10}$ erg/K mol$^{1/3}$ only when $[\eta]$ drops below the Einstein intrinsic viscosity as given by $[\eta]_E = 2.5\bar{v}$.

The results of the present work are collected in Table 1. Analyzing the results, one can see that insertion of fullerene and the nature of the solvent have different influences on the hydrodynamic properties of dendrimers and hemidendrimers.

Table 1 Hydrodynamic properties of liquid-crystalline dendrimers and hemidendrimers with and without fullerene in benzene and chloroform at 26 °C

Polymer	M^a	$[\eta]$ (dl/g)	$D{\times}10^7$ (cm^2/s)	dn/dc_{609} (ml/g)	R_H (Å)	$A_0{\times}10^{10}$ (erg/K mol$^{1/3}$)
Benzene						
D	5,158	0.085	17.0	0.13	22	2.55
DF	5,878	0.085	17.0	0.15	22	2.66
hD	2,579	0.078	25.0	≤0.16	15	2.90
hDF	3,299	0.062	25.0	0.15	15	2.90
Chloroform						
D	5,158	0.12	17.4	0.15	24	2.66
DF	5,878	0.12	17.4	0.15	24	2.78
hD	2,579	0.12	24.0	0.18	17	2.91
hDF	3,299	0.10	24.0	0.18	17	2.97

[a] The values were calculated from the chemical structure

Effect of the nature of the solvent

It is easy to see that $[\eta]$ for any compound in chloroform solutions is higher than that in benzene. This can be considered as a consequence of the more intense thermodynamic polymer–solvent interactions (the better thermodynamic quality of chloroform) if the apparent value of the molecular weight remains invariable.

The comparison of curves 1–4 and 5–8 in Fig. 2 shows that the solvent effect is considerable greater in the case of hemidendrimers (60–70%) than in the case of dendrimers (around 30%). It also follows from the data that the geometrical form of the dendrimer molecules in solution differs from spherical because the uniform swelling of the spherical dendrimer molecule would not lead to a change in $[\eta]$.

The change in R_H of a particle on passing from benzene to chloroform solution, as follows from the Table 1, is also more noticeable for hemidendrimers (15%) than for dendrimers (8%).

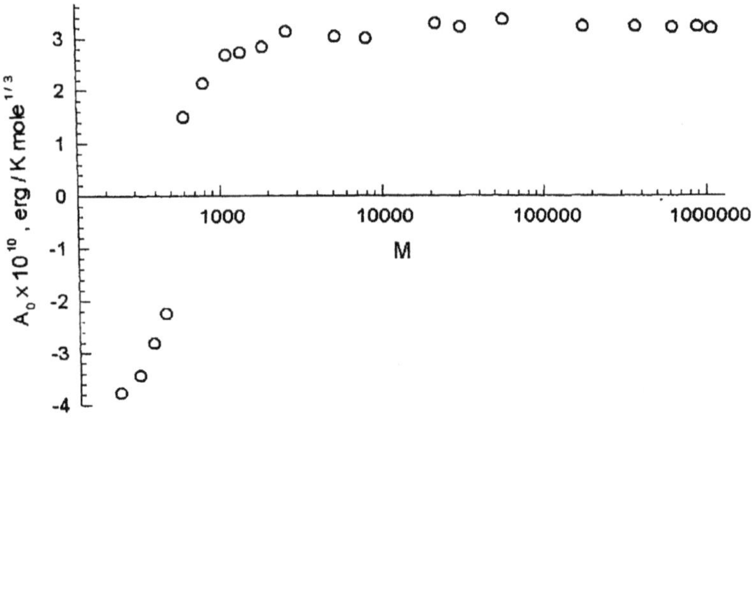

Fig. 3 Molecular weight dependence of the hydrodynamic parameter A_0 as calculated using the literature data obtained for poly(dimethylsiloxane) in bromocyclohexane at 29.5 °C (θ) [6]

The fullerene effect

Incorporation of C_{60} into the hemidendrimer molecule, as follows from comparison of curves 1 and 2 and 3, and 4 in Fig. 2, is accompanied by a decrease in $[\eta]$ in both solvents. On one hand, this lowering is insignificant (around 20%) but, nevertheless, it exceeds the experimental error limits. On the other hand, the effect of fullerene is here weaker than the effect of the solvent (60–70%) as follows from comparison of curves 1 and 2 and 1 and 3.

Incorporation of C_{60} into dendrimers is not detected in viscometric measurements. The scattering of the points does not exceed experimental error, as follows from comparison of points 5 and 7 and 6 and 8, whereas the solvent effect is more pronounced and is close to 30%.

R_H of a particle as calculated using Eq. (1) and given in Table 1 is not affected markedly by incorporation of C_{60} into the dendrimer and hemidendrimer molecules. To explain the result we analyze now all the diffusion–viscometric data associated with Eq. (2).

Hydrodynamic parameter and dispersity of solutions

Substituting into Eq. (2) the experimental values of $[\eta]$, D, and the molecular weight corresponding to the chemical structure, we obtain, for hD and hDF in both solvents, $A_0 = 2.9 \times 10^{-10}$ erg/K mol$^{1/3}$, which quantitatively coincides with the low limit of 2.9143×10^{-10} erg/K mol$^{1/3}$ predicted for the invariant theoretically in the frame of a hard-sphere model [5]. This result may confirm the molecular dispersity of the solutions of hD and hDF and the high symmetry of the geometric form of the hemidendrimer molecules in solution.

Analogous calculations for D and DF in both the solvents yield unreasonably low values of A_0, from 2.5×10^{-10} to 2.7×10^{-10} erg/K mol$^{1/3}$. We approach the experimentally reasonable values of A_0, from 2.9×10^{-10} to 3.2×10^{-10} erg/K mol$^{1/3}$ [5], if use the higher (1.4–1.9 times) M values. In other words, the experimentally determined hydrodynamic data are in agreement with the present theories if we assume that a significant part of the dendrimer molecules in solutions of D and DF is aggregated to dimer and trimer forms.

Some of the aggregated molecules are smaller in a solution of DF, and, for both dendrimers, in chloroform solutions.

In the assumption, solutions of D and DF may differ in inhomogeneity. We tried to evaluate the inhomogeneity by differentiating the diffusion curve from the Gaussian form [7]. The typical experimental traces obtained in benzene solutions of D and DF are presented in Fig. 4. They were treated with the method

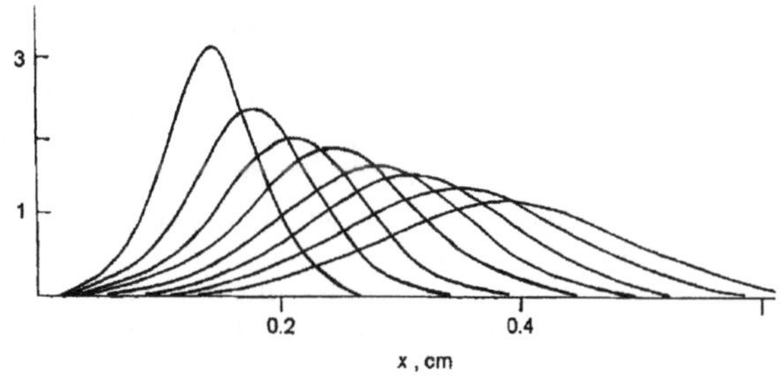

Fig. 4 Solvent (*left*)-solution(*right*) boundary profile (in relative units, b is the interference fringe interval) and its change with time as observed in a benzene solution of DF at 26 °C. Traces are recorded at times (from *left* to *right*) of 5, 9, 13, 17, 21, 25, 33, and 41 min, respectively, after the start. The solute concentration is 0.50 g/dl. Every second curve is shifted to the right by 0.035 units

of moments and with a Gaussian approximation. The corresponding $\overline{\sigma_m^2}$ and $\overline{\sigma_A^2}$ values were thus obtained. $\overline{\sigma_m^2}$ did not exceed $\overline{\sigma_A^2}$ for D in benzene solution. In turn, $\overline{\sigma_m^2}$ was slightly larger than $\overline{\sigma_A^2}$ for DF. So, inhomogeneity of the solution of DF is not significant but is greater than that of the solution of D. Nevertheless, we cannot quantitatively characterize the inhomogeneity with the method applied because of the low sensitivity of the diffusion coefficient to the molecular weight of the dendritic particles [8].

The result obtained may be qualitatively related to the presence of fullerene, which reduces the intermolecular interactions in a dendrimer solution and, hence, aggregation of the macromolecule.

Conclusions

We may conclude that the diffusion–viscometric investigations have indicated significant differences in the hydrodynamic behavior of LC hemidendrimers with and without C_{60} similar to those observed for polymers investigated earlier [2, 3]. Namely, insertion of fullerene is followed by decreasing intrinsic viscosity, which reflects (for invariable apparent molecular weight and macromolecule permeability) a decrease in the asymmetry of the geometry of the hDF molecule in solution. This fullerene effect is not greater than the effect of the nature of the solvent (polar/apolar solvent). However, the physical mechanism of the C_{60} effect remains unclear, if we take into account that C_{60} does not have a dipole moment, and this may be the subject of future investigations.

Note that the restructuring function of fullerene in the substances under discussion is formally equivalent to an increase in the generation number of the dendritic matrix [8].

In contrast to hemidendrimers, we have not found noticeable differences in the hydrodynamic behavior of the dendrimers with and without fullerene. These substances do not form real solutions in either solvent as this may be concluded from the underrated values of A_0 that are invariable to molecular weight. Similar results were obtained earlier for other hybrid dendrimers [9, 10] and can also be explained by partial aggregation of the macromolecules in solution.

Hence, the fullerene–dendrimer interactions are not intense enough to significantly disturb the intermolecular interactions of the dendrimers (functionalized groups) responsible for their aggregate state. Strong interactions of the mesogenic groups result obviously in close asymmetry of the dendrimer and hemidendrimer particles in a chloroform solution (coinciding values of $[\eta]$ as obtained for D and hD with molecular weights differing by a factor of 2). At the same time, they have noticeably different hydrodynamic sizes (different values of R_H as given in Table 1).

References

1. Dardel B, Guillon D, Heinrich B, Deschenaux R (2001) J Mater Chem 11:2814
2. Lavrenko PN, Yevlampieva NP, Volokhova DM, Vonogradova LV, Melenevskaya EY, Zgonnik VN (2002) Polym Sci 44:173
3. Yevlampieva NP, Lopatin MA, Lavrenko PN (2003) Polym Sci 45:46
4. Tsvetkov VN (1989) Rigid-chain polymers. Consultants Bureau, New York
5. Tsvetkov VN, Lavrenko PN, Bushin SV (1984) J Polym Sci Polym Chem Ed 22:3447
6. Yamada T, Koyama H, Yoshizaki T, Einaga Y, Yamakawa H (1993) Macromolecules 26:2566
7. Lavrenko P, Lavrenko V, Tsvetkov V (1999) Prog Colloid Polym Sci 113:14
8. Lavrenko PN (2001) Colloid Polym Sci 279:200
9. Lezov AV, Melnikov AB, Polushina GE, Ponomarenko SA, Boiko NI, Kossmel E, Ryumtsev EI, Shibaev VP (1998) Dokl RAN 362:638
10. Pavlov GM, Korneeva EV, Roy R, Michailova NA, Ortega PC, Perez MA (1999) Prog Colloid Polym Sci 113:150

Progr Colloid Polym Sci (2004) 127: 67–72
DOI 10.1007/b98015

P. Lavrenko
A. Cornell
V. Lavrenko
N. Pogodina

Host–guest interactions between poly(methylhydro-dimethylsiloxane) and solvent molecules studied by low-speed ultracentrifugation and viscometry

P. Lavrenko (✉)
Institute of Macromolecular Compounds, Russian Academy of Sciences, 199004 St. Petersburg, Russia
e-mail: lavrenko@mail.macro.ru

A. Cornell · V. Lavrenko · N. Pogodina
University of Massachusetts, Department of Chemistry, Amherst, MA 01003, USA

Abstract Low-speed ultracentrifugation data combined with viscometry results have been used in the characterization of intermolecular interactions in low molecular weight poly(methylhydro-dimethylsiloxane) (PMS) under ambient conditions. The same organic liquids of different nature were employed, firstly, as solvents and, secondly, as plasticizers of different polarity. The shape and the hydrodynamic size of the PMS molecules in dilute solutions are characterized and compared with the literature values known for poly(dimethylsiloxane). Translational and rotation mobilities of the PMS molecules in solutions are compared with the diluent effects observed in bulky PMS. Correlation is found between solvent–polymer interactions in organic solutions of PMS and intermolecular interactions in bulky PMS. The hydrodynamic problem of the local solvent structure (solvent continuity on a PMS molecular scale) is briefly mentioned.

Keywords Intermolecular interactions · Hydrodynamic behavior · Translational and rotational mobility · Poly(methylhydro-dimethylsiloxane) · Plasticizing effect

Introduction

Solubility of a polymer and many material and rheological properties of a polymeric solid and concentrated solutions are primarily determined by intermolecular coupling and interactions. In semidilute and dilute solutions, intense interactions between the macromolecules lead to the association and aggregation phenomena, which are often sensitive to temperature and the nature of the solvent.

In bulky polymers, the intermolecular attractions restrict the thermal motions of segments, providing the high viscosity of a polymeric solid. The force to deform a polymer depends upon the ease with which the polymer molecules change their configuration to accommodate themselves to a new strained state, and the solid becomes more feasible for deformation when swollen by penetrant species. The mechanisms were considered concerning the molecular structural relaxation caused by penetration of small molecules [1]. One is the scission of interchain bonds by the chemical action of the penetrant. The other is the plasticizing effect when small molecules penetrate into the interchain space of the polymer network, and as a result the segment gains more free volume for thermal fluctuations.

Chain–chain packing in the melt is also primarily determined by intermolecular interactions [2]. In low molecular weight polymers when the chain has only a limited number of Kuhn step lengths, in the absence of entanglement interaction, intermolecular coupling and intermolecular interactions are a response for the still unclear the viscoelastic anomalies, which are significantly reduced when the polymer is dissolved in a specific solvent [3]. Here, the coupling model successfully describes the crowded local segmental motions.

That is why characterization of the interactions is one of the fundamental aims of polymer physics. Silicone polymers have several characteristics that make them very suitable for these purposes. For example, poly(dimethylsiloxane) (PDMS), because of the high degree of flexibility of the polymer backbone, has the lowest glass-transition temperature of any known polymer. High chemical stability and a relatively low density are the reasons why materials based on PDMS are also ideal for many applications where PDMS exists in the bulk amorphous state as a liquid or as a cross-linked network.

Recently, we described the intermolecular effects observed in dilute solutions of silicone polymers [4]. The present work aims to evaluate the host–guest interactions in a plasticized bulky polymer via those in dilute solutions using hydrodynamic methods as applied to polar low molecular weight poly(methylhydrodimethylsiloxane) (PMS). The chemical formulas of the polymers under consideration (with x the degree of polymerization) are given by

PMS $CH_3[Si(CH_3)_2OSi(H)(CH_3)O]_{x/2}Si(CH_3)_2CH_3$

PDMS $CH_3[Si(CH_3)_2O]_xSi(CH_3)_2CH_3$

Benzene, methyl ethyl ketone (MEK), chloroform, and *N*-methylformamide (N-MF) were employed either as solvents or as diluents of very different polarity (with dipole moment, μ, of 0–3.86 D). Particularly, for chloroform, μ (1.18 D) is comparable with that of the Si–O bond, $\mu_{Si\text{-}O} = 1$ D [5].

The polymer–solvent interactions were analyzed by studying the PMS behavior in dilute solutions, whereas the polymer–polymer interactions were characterized by the changes in the bulky PMS viscosity caused by insertion of the named liquids (in small amounts) as the diluent dopes [6, 7, 8]. Translational mobilities of the PMS molecules in solutions were investigated with the hydrodynamic methodology used earlier in our study of low molecular weight polymers and examined with naphthalene [4]. The transport behavior and viscosity properties thus determined are compared with the literature values known for PDMS [9, 10, 11, 12, 13].

Experimental

Sample and solutions

PMS is a clear viscous liquid with a specific gravity of 0.99 g/cm^3 and kinematic viscosity of 10–15 cS at 25 °C (obtained from Flourochem, Glossop, UK, as the manufactured product PS 122.5). It is 50–55% methylhydro and 45–50% dimethylsiloxane copolymer. The molecular weight given by the manufacturer is 950 ± 50 g/mol, and the refractive index is 1.394. The PMS was used as received.

The hydrodynamic properties of the polymer were studied by the methods described previously [4] at 26 °C in benzene of reagent grade with density $\rho_0 = 0.873$ g/cm^3, viscosity $\eta_0 = 0.617 \times 10^{-2}$ g/cm/s, and refractive index $n_D = 1.497$. For chloroform the values were $\rho_0 = 1.477$ g/cm^3, $\eta_0 = 0.548 \times 10^{-2}$ g/cm/s, and $n_D = 1.444$. The viscosity of neat PMS at 26 °C is 6.767×10^{-2} g/cm/s as measured in a capillary viscometer. Concentrated solutions were prepared at ambient temperature by stirring for few hours, and density corrections were made in the calculations of the concentrations, c.

Methods

Low-speed analytical ultracentrifugation was performed in a 3180 analytical ultracentrifuge from the Hungarian Optical Works MOM equipped with interference optics as described elsewhere [14]. The rotor rotation frequency, n, was 10×10^3 rpm with an accuracy of ± 20 rpm. The temperature of the rotor (26.0 °C) was maintained with an absolute accuracy of ± 0.1 °C and with a constancy of regulation of ± 0.05 °C. A polyamide synthetic-boundary double-sector cell with a light path, h, of 1.2 cm and quartz windows was used. The pictures were recorded with the Tsvetkov interference detector system [14] operating at a wavelength, λ, of 546.1 nm with spar twinning, a, of 0.0209 cm and compensator fringe interval, b, of 0.0966 cm. A concentration boundary was formed between two solutions with different solute concentrations, c_1 and c_2, under the condition $\Delta c \equiv c_1 - c_2 \ll \bar{c} \equiv (c_1 + c_2)/2$. The results were related to the average concentration, $\bar{c}$.

The refractive increment, dn/dc, was estimated from the interference curve area, Q, by $dn/dc = (\lambda/abh)Q/\Delta c$.

Viscosity measurements were performed in a dilute regime ($c < 1/[\eta]$, where $[\eta]$ is the intrinsic viscosity) using an Ostwald viscometer with an average velocity gradient of 520 1/s (benzene) in a water bath at 26.0 ± 0.01 °C. The kinetic energy correction was negligible.

The partial specific volume, $\bar{v}$, of the polymer in solution was estimated densimetrically. The density of the benzene solution of PMS at 21.5 °C and $c = 20.00$ wt% was measured to be 0.892 g/cm^3 and $\bar{v} = 1.042$ cm^3/g.

All the data were evaluated via nonweighted linear regression analysis and refined by least squares.

Results and discussion

Translational mobilities of the macromolecules in solutions were determined using the low-speed ultracentrifugation method. Typical pictures obtained for the solution of PMS in benzene and in chloroform are illustrated in Fig. 1. Down-type traces are here due to the negative sign of the refractive index increment, and the greater area of the curves for benzene reflects a higher absolute value of dn/dc in benzene solution (−0.10 ml/g) relative to that in chloroform (−0.04 ml/g). The dn/dc values thus obtained are in good agreement with the refractive indices of PMS and the solvents [15], reflecting the complete dissolution of the polymer.

The interference curves in Fig. 1 are sufficiently symmetric in form. They were treated, therefore, in a Gaussian approximation with the "height–area" method [14]. The second central moment (dispersion), $\overline{\sigma^2}$, of the $\partial c/\partial x(x)$ distribution was evaluated by $\overline{\sigma^2} = (a^2/8)[\mathrm{argerf}(aH/Q)]^{-2}$, where H is the maximum ordinate of the curve, x the radial distance, and argerf means an argument of the probability integral. The diffusion coefficient, D, was calculated from the $\overline{\sigma^2}$ dependence on time, t, by

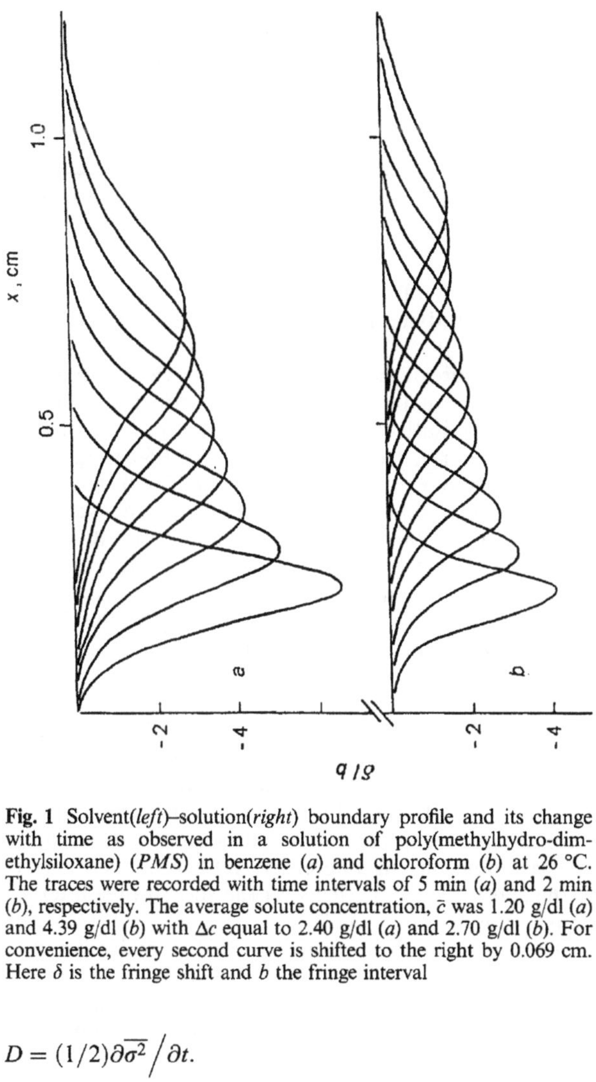

Fig. 1 Solvent(*left*)–solution(*right*) boundary profile and its change with time as observed in a solution of poly(methylhydro-dimethylsiloxane) (*PMS*) in benzene (*a*) and chloroform (*b*) at 26 °C. The traces were recorded with time intervals of 5 min (*a*) and 2 min (*b*), respectively. The average solute concentration, $\bar{c}$ was 1.20 g/dl (*a*) and 4.39 g/dl (*b*) with Δc equal to 2.40 g/dl (*a*) and 2.70 g/dl (*b*). For convenience, every second curve is shifted to the right by 0.069 cm. Here δ is the fringe shift and b the fringe interval

$$D = (1/2)\partial\overline{\sigma^2}/\partial t.$$

The Lamm correction for the field effect [16] was negligible (0.007% or less) as estimated using $D = D_{sed}(1 - s\omega^2 t)$, where D_{sed} is the value calculated from the sedimentation patterns, s the sedimentation coefficient (around 0.176 Sv as estimated using Svedberg equation), and $\omega^2 t$ the magnitude of the centrifugal field.

The results are represented in Fig. 2, where the $\overline{\sigma^2}$ values are plotted against t. One can see that the experimental points as obtained here at different solute concentrations form the common functions for PMS in both chloroform (points 1, 2) and benzene (points 3–5) solutions. Hence, the diffusion coefficient, in error limits, is independent of the solute concentration over the $\bar{c}$ range investigated, from 1.20 to 17.49 g/dl in benzene and from 0.95 to 4.39 g/dl in chloroform solutions. The $D_0\times 10^7$ values (in centimeters squared per second) of

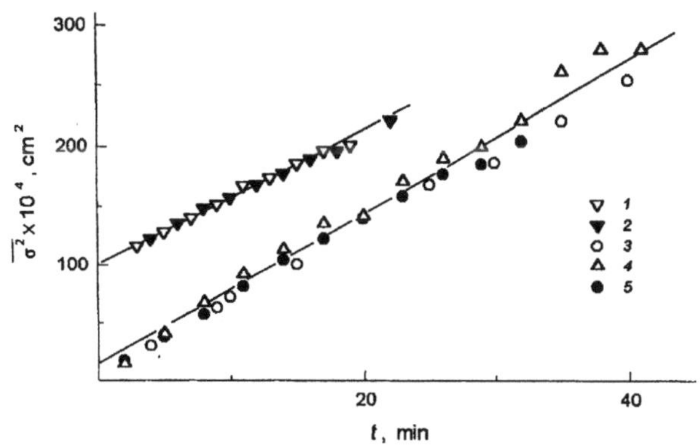

Fig. 2 Dispersion of the concentration boundary, $\overline{\sigma^2}$, versus time for PMS in chloroform (*1*, *2*) and benzene (*3*–*5*) at 26 °C. $\bar{c}$ (g/dl) was 0.95 (*1*), 4.39 (*2*), 1.20 (*3*), 8.28 (*4*), and 17.49 (*5*). Points *1* and *2* are shifted up by 100 units

53 ± 2 and 54 ± 2 were thus obtained for PMS in benzene and chloroform, respectively.

The hydrodynamic radii of the polymer particles at infinite dilution, R_H, were calculated with aid of the following equation:

$$R_H = kT/6\pi\eta_0 D,$$

where k is Boltzmann's constant and T the temperature in Kelvin. The numerical coefficient 6 corresponds here to strong engagement of the particle surface with a continuous surrounding medium and, respectively, to transport of the neighboring solvent layer together with the carrier.

The viscosities of the solutions were measured in a routine way. The flow time was measured to a precision of 0.1 s. Density corrections were made in the calculation of the relative viscosity, η_r, from the flow times of the solution, τ, and the solvent, τ_0, by $\eta_r = (\tau/\tau_0)(\rho/\rho_0)$. The data obtained for the specific viscosity, η_{sp}, were treated as usual by the Huggins plot to determine the intrinsic viscosity, $[\eta]$, by

$$\eta_{sp}/c = [\eta] + k_H[\eta]^2 c,$$

where k_H is the Huggins coefficient. The results are represented in Fig. 3. Linear Huggins plots were obtained, reflecting the absence of any possible deformation or dissociation due to the shear forces employed in the viscometer capillary, at least for the concentration and shear rate ranges used for the measurements. For PMS in chloroform solution (curve 3) we obtained $[\eta] = (0.027 \pm 0.002)$ dl/g and $k_H = 0.33$. In benzene solutions of PMS (curve 2), the $[\eta]$ and k_H values were (0.019 ± 0.001) dl/g and 0.40, respectively. The last $[\eta]$ value agrees well with the $[\eta](M)$ dependence known for PDMS in toluene at 25 °C [9]. For PMS in MEK

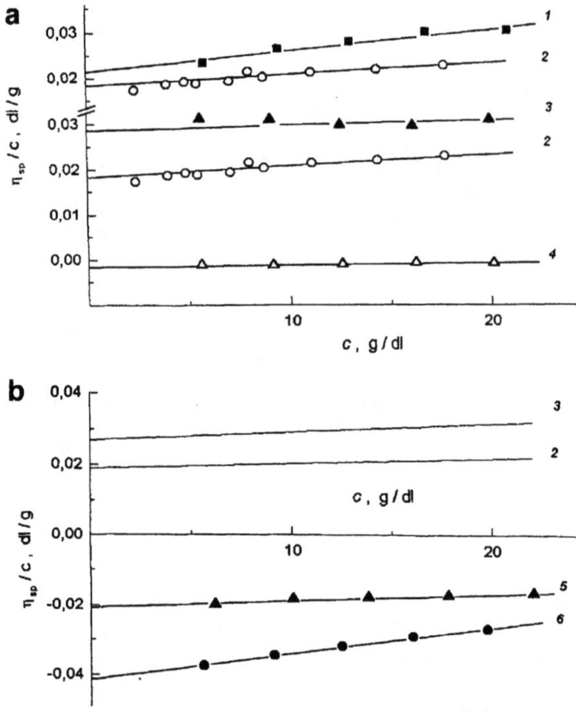

Fig. 3a, b Concentration dependence of the solution relative viscosity increment, η_{sp}/c, as measured at 26 °C for PMS in chloroform (*3*), benzene (*2*), methyl ethyl ketone (*1*), and *N*-methylformamide (*4*) solutions, respectively, and for chloroform (*5*) and benzene (*6*) in PMS solutions. The *lines* correspond to linear approximations

solution (curve 1) at 20 °C (a θ solvent [10]), we obtained $[\eta] = 0.021 \pm 0.001$) dl/g.

The viscometric radii, R_η, were calculated using $R_\eta = (0.3M[\eta]/\pi N_A)^{1/3}$, where N_A is the Avogadro number.

The hydrodynamic invariant, A_0, was calculated by [17]

$$A_0 = (D\eta_0/T)(M[\eta]/100)^{1/3}.$$

The results are reported in Table 1. The A_0 values of $(2.9 \pm 0.1)\times10^{-10}$ erg/K mol$^{1/3}$ thus obtained are somewhat lower than that typical for the coiled macromolecules (3.2×10^{-10} erg/K mol$^{1/3}$) but are in a good agreement with those available for PDMS with equivalent low molecular weight [10] and are very close to $A_0 = 2.9143\times10^{-10}$ erg/K mol$^{1/3}$ predicted for hard spheres [17].

PMS in solution

The behavior typical of flexible (nonaggregated) polymers in dilute solution is indicated by the data obtained. The unperturbed dimensions of the PMS molecules in

Table 1 Hydrodynamic and optical parameters of poly (methylhydro-dimethylsiloxane) (the manufacturer's value of the molecular weight is 950 ± 50) in benzene and chloroform solutions at 26 °C

Solvent	R_η (Å)	R_H (Å)	$A_0 \times 10^{10}$ (erg/K mol$^{1/3}$)	$dn/dc_{546.1}$ (ml/g)
Benzene	6.6	6.7	2.9 ± 0.1	−(0.10 ± 0.01)
Chloroform	7.4	7.4	2.9 ± 0.1	−(0.04 ± 0.01)

solution are characterized by $R_\eta = 6.9$ Å as calculated from the hydrodynamic data obtained in MEK at 20 °C (a θ solvent [9]).

The structure of the PMS chain is very close to that for PDMS studied in detail; therefore, the properties of PMS may be quantitatively compared with those of PDMS [10]. Particularly, for PDMS with $M = 950$ (degree of polymerization $x = 12$) in MEK at 20 °C (θ) we found $\mu \approx 1.56$ D, $[\eta]_\theta = 0.0175$ dl/g, $k_H \approx 0.9$; in bromocyclohexane at 28 °C (θ) $D \approx 20.5\times10^{-7}$ cm^2/s and $D\eta_0 = 40.4\times10^{-9}$ g cm. The following conformational parameters are also available from the cited paper, $M_L \approx 21$ 1/Å, $A \approx 28$ Å, and $d \approx 2$ Å as the mass per unit chain length, the Kuhn segment length, and the hydrodynamic diameter, respectively.

It is easy to see that PMS molecule ($M = 950$, $x = 14$, $M_L = 19$ 1/Å) consists of approximately 1.5 Kuhn lengths, and no significant excluded-volume effects may be expected in the hydrodynamic behavior of PMS. This is strictly confirmed by our measurements under θ conditions. Hence, different hydrodynamic dimensions of PMS molecules in different solvents must be referred to the specific solvent effects on the short-range interactions in the macromolecule.

The ratios of the viscometric to the hydrodynamic radii, R_η/R_H, are close to unity, the limit predicted for hard spheres [18]. They are 0.98 and 1.0 in benzene and chloroform, respectively. Small deviations from unity can be explained either by random experimental errors or by polydispersity of the sample.

For PMS in benzene solution the $[\eta]$ value (0.019 dl/g) is lower than the Einstein intrinsic viscosity (0.026 dl/g) as predicted in the hard-sphere approximation given by $2.5\bar{v}$[19]. Similar effects were observed earlier for oligo(dimethylsiloxane)s [10] and for individual compounds such as naphthalene ($[\eta] = 0.0125$ dl/g in benzene solution [4]). Such anomalous behavior was regarded as "arising from specific interactions between polymer and solvent molecules and can never be treated molecular-theoretically within the framework of classical hydrodynamics" [10].

We can take into account that the numerical coefficient 2.5 corresponds to strong engagement of the hard-sphere surface with a surrounding medium and, respectively, to movement of the neighboring solvent layer together with the rotating sphere without slipping.

In this way, the lower experimental $[\eta]$ value means that this assumption is not valid for PMS molecules in benzene solution, and there is particle slipping on the medium at its rotation in the capillary shear field (slip boundary conditions [20]). The other possible explanation lies in the permeability of the spheres then, $\bar{v} \leq [\eta] \leq 2.5\bar{v}$ [21]) or in an interrupted structure of the solvent on the PMS molecular scale.

A higher value of $[\eta]$ was obtained for PMS in chloroform solution (0.027 dl/g), meaning an increase in R_η from 6.6 to 7.4 Å along with a simultaneous increase in R_H from 6.7 to 7.4 Å. To analyze the effect, we calculated the relation $D\eta_0[\eta]^{1/3}$, which is sensitive to the molecular weight only [17]. Within experimental error, the same values of the invariant A_0 (which is proportional to $D\eta_0[\eta]^{1/3}$) were obtained in both solvents. Hence, the apparent value of the molecular weight is invariable on passing from benzene to chloroform solution of PMS, and there is the other reason for the difference in the polymer hydrodynamic behavior. The lower D and higher $[\eta]$ values determined for PMS in chloroform solution reflect the greater extension of the macromolecules in the solvent, which is in accord with the coil behavior of PMS in a dilute solution.

Figure 3 (curve 4) shows that $[\eta]$ drops drastically to zero in an N-MF solution of PMS. The plasticizing effect of N-MF as a diluent on PMS is clearly also close to zero. Taking into account the origins of the viscosity phenomenon, we have to relate the result to the interrupted structure of the medium surrounding the PMS molecules. Really, the zero $[\eta]$ value means the same rotation mobilities of both the guest and host molecules, and this may result from either specifically equivalent host–host and host–guest interactions or their close geometric form and sizes. The neat N-MF molecule is smaller in size than PMS. In size terms, the result $[\eta]=0$ would mean that N-MF molecules are strongly aggregated in neat N-MF at ambient temperature. The origins of the effect may lie in the very high dipole moment (3.86 D) of the N-MF molecule.

Benzene and chloroform in PMS solutions

As already mentioned, PMS is a liquid under ambient conditions; therefore, a mixture of a small amount of benzene (or chloroform) with PMS may be considered as a solution of the corresponding liquid in PMS. The viscosity of the solution is dependent on the solute content as illustrated by points 5 and 6 in Fig. 3. Formally treated, the data yield the $[\eta]$ values of −0.024 and −0.041 dl/g for chloroform and benzene in PMS solutions with the k_H values of 0.43 and 0.45, respectively. The intrinsic viscosity is here a measure of the plasticizing effect of a given diluent on a given polymer.

Negative (in sign) intrinsic viscosities were obtained earlier for oligo(dimethylsiloxane)s [10], the molecular dimensions of which are smaller than or comparable with that of the solvent. In that case, the negative $[\eta]$ value is a parameter which reflects how strongly the intermolecular interactions and ordering in a liquid medium might be affected by the small diluent dopes or by specific solute–solvent interactions. As is generally known, the viscosity of a polymer liquid is determined by the inner net of couplings with the fluctuating knots formed by any types of the physical interactions, which are localized periodically along the polymer chains [6]. From this point of view, the lower $[\eta]$ value, found for benzene in PMS solution, reflects a stronger disturbance of the net by this diluent. The disturbance, in turn, facilitates the macromolecule slipping relatively to the coupling points in a polymer liquid. An alternative explanation can be found in the decreasing dimensions of the kinetic chain units (segments) that provide the elementary acts of the solute movements in space.

Conclusions

Analyzing the hydrodynamic properties presented, we may conclude that PMS forms real solutions in all the solvents used. PMS molecules in a benzene solution are compact and close to spherical in form. In a chloroform solution, the polymer molecules are more extended owing to specific solute-solvent interactions (the chloroform–polymer contacts are preferable to the polymer–polymer ones), and the polarity of the solvent and polymer may play an important role. Chloroform is a much more active agent, relative to benzene, in the disturbance of the intramolecular polymer–polymer interactions in a dilute solution of PMS (for PMS, the R_H and R_η values in chloroform exceed those in benzene at the invariable apparent molecular weight).

There is the opposite situation in condensed PMS: benzene is the much more active agent, relative to chloroform, in the disturbance of the intermolecular polymer–polymer interactions (the absolute value of $[\eta]$ found for benzene in PMS solution exceeds that of chloroform).

The effect found here might reflect a more general correlation, the stronger are the physical interactions in the solvent–polymer system, the weaker are the disturbing influences of the probe molecules on the intermolecular interactions in the condensed polymer. In other words, intermolecular interactions are better kept with insertion of more active dopes (or they are better propagated via more active dopes). As a result, the inner net of couplings formed by these interactions in a liquid polymer is saved more perfectly via interactions with more active diluent dopes. In turn, the good plasticizer effect may be expected with a diluent dope, which is inactive as a polymer solvent.

The effect of the nature of the solvent might also be important in analyzing the aggregation (complex formation) phenomena in various binary polymer systems via hydrodynamic investigations.

Acknowledgement The authors express their gratitude to D. Lacey for providing the polymer sample.

References

1. Fujita H, Kishimoto A (1958) J Polym Sci 23:547
2. Sides SW, Curro J, Grest GS, Stevens MJ, Soddemann T, Habenschuss A, Londono JD (2002) Macromolecules 35:6455
3. Ngai KL, Plazek DJ, O'Rourke VM (1997) Macromolecules 30:5450, and references therein
4. Lavrenko P, Yevlampieva N, Okatova O, Kolbina G, Olbrich M (1997) Macromol Chem Phys 198:3581
5. Dasgupta D, Smyth CP (1967) J Chem Phys 47:2911
6. Vinogradov GV, Malkin AYa (1977) Rheology of polymers. Khimiya, Moscow (in Russian)
7. Vinogradov GV, Malkin AY, Blinova NK, Sergeyenkov SI, Zabugina MP, Titkova LV, Yanovsky YG, Shalganova VG (1973) Eur Polym J 9:1231
8. Andrianova GP (1975) J Polym Sci Polym Phys Ed 13:95
9. Bianchi U, Dalpiaz M, Patrone E (1964) Makromol Chem 80:112
10. Yamada T, Koyama H, Yoshizaki T, Einaga Y, Yamakawa H (1993) Macromolecules 26:2566
11. Haug A, Meyerhoff G (1962) Makromol Chem 53:91
12. Kuwahara N, Miyake Y, Kaneko M, Furuichi J (1962) Rep Prog Polym Phys Jpn 5:1
13. Yamada T, Yoshizaki T, Yamakawa H (1992) Macromolecules 25:1487
14. Lavrenko P, Lavrenko V, Tsvetkov V (1999) Prog Colloid Polym Sci 113:14
15. Giebel L, Benmouna M, Borsali R, Fischer EW (1993) Macromolecules 26:2433
16. Lamm O (1929) Z Phys Chem A143:177
17. Tsvetkov VN, Lavrenko PN, Bushin SV (1984) J Polym Sci Polym Chem Ed 22:3447
18. Roovers J, Martin JE (1989) J Polym Sci Polym Phys Ed 27:2513
19. Einstein A (1906) Ann Phys 19:2899
20. Subramanian G, Davis HT (1975) Phys Rev 11:1430
21. Taylor GI (1932) Proc R Soc Lond Ser A 138:41

Progr Colloid Polym Sci (2004) 127: 73–82
DOI 10.1007/b98016

Christine Ebel

Analytical ultracentrifugation for the study of biological macromolecules

This contribution was presented at the 13th International Symposium on Analytical Ultracentrifugation, Osnabrück, 6–7 March 2003

C. Ebel
Laboratoire de Biophysique Moléculaire
Institut de Biologie Structurale
UMR 5075 CEA-CNRS-UJF
41 rue Jules Horowitz
38027 Grenoble Cedex 1, France
e-mail: christine.ebel@ibs.fr
Tel.: +33-4-38785494
Fax: +33-4-38789638

Abstract The main aim of the review is to present the potential of analytical ultracentrifugation (AUC) in biology to physicists. It is based on the presentations made in *Advances in AUC and hydrodynamics*, a practical and lecture course (61–8 June 2002) and the following Euro Conference (8–11 June 2002) in Autrans, France. The specificities of biological materials compared with those of polymers or colloid physics and the reason why AUC is so powerful in routine studies and how it can address more complex systems are presented.

Keywords Analytical ultracentrifugation · Hydrodynamics · Proteins · Biological macromolecules · Complexes

Introduction

Analytical ultracentrifugation (AUC) is widely used for the characterization of the interactions of biological macromolecules and assemblies. The main aim of this overview is to present the potential of AUC in biology to physicists. I shall present – very generally – the specificities of biological materials compared with those of polymers or colloid physics and why AUC is so powerful in routine studies and how it can address more complex systems. This review and the cited references are based on the presentations made in *Advances in AUC and hydrodynamics*, a practical and lecture course (6–8 June 2002) and the Euro Conference (8–11 June 2002) in Autrans, France. The interested reader is advised to consult other reviews on AUC, such as the recent ones in Refs. [1, 2].

Biological macromolecules

"All living creatures are made of cells". Thus begins one of the most famous textbooks on molecular biology: *Molecular biology of the cell* [3]. Living organisms are monocellular or multicellular. All can be classified into prokaryotic (comprising Archaea and bacteria), and eukaryotic. While the latter are more compartmentalized and contain distinct nuclei, the general organization of all cells presents strong similarities. Four types of constituents can be distinguished: proteins, nucleic acids, polysaccharides and lipids. A cell, as nicely represented by Goodsell [4], shows two main features. First, the macromolecular crowding is obvious within the cytosol, the macromolecular concentration being above 0.1 g/ml. The local environment will control macromolecular interactions. Secondly, molecular machines can be defined, in which different macromolecules interact to accomplish a precise function that is needed for the life of the cell. The identification of the partners that are involved in a function, their interactions and general organization, the factors that control the associations, are the objects of most of the AUC studies in biology.

For the purpose of AUC, a specificity of biological material compared with polymer or synthetic colloids is that, in most of the cases, it is highly defined. Proteins are extremely precisely defined in their amino acid sequences and posttranslational modifications (e.g.

specific addition of a sugar). Nucleic acids (DNA, RNA) also are precisely defined in terms of their sequences and modifications. Thus, the atomic composition and molar mass of biological macromolecules are usually either known or can be measured with extreme precision, using, for example, mass spectrometry (MS). From the progress in molecular biology and in purification techniques, precisely defined macromolecules can be obtained in rather large amounts (a few milligrams) and with a very high level of purity. A second specificity is that, most often, biological macromolecules can be defined as compact particles of defined shape. This is known from high-resolution structures obtained by crystallography, which allow us to assign a spatial position to about all the atoms, as can be seen in Fig. 1 in the example of a globular protein [5]. This can be true even for extremely large complexes, such as a ribosome, whose size is of the order of 30 nm [6]. However, crystallography detects only ordered particles. The uniqueness of the shape is, however, not always valid. NMR studies allow us to detect flexible loops or domains, as can be seen in Fig. 2 [7]. Also, the relative arrangement of the domains in multidomain proteins could be quite flexible.

AUC in a routine mode for the study of biological macromolecules

Basic equations

AUC in a routine mode will address a very limited number of different precisely defined macromolecules, which are diluted in an aqueous buffered solvent containing typically 150 mM salt. Thus, the sedimentation of ideal particles in a homogeneous solvent is considered. The Lamm equation expresses the spatial and temporal evolution of the concentration, c, of the solute in the ultracentrifuge:

$$(\partial c/\partial t) = -1/r \cdot \partial/\left[r\left(cs\omega^2 r - D \cdot \partial c/\partial r\right)\right], \qquad (1)$$

where t is time, r the distance to the center of rotation and ω the angular velocity.

The transport is a function of the sedimentation and diffusion coefficients of the particle, s and D, respectively, which, in the case of ideal solutes, are constant and well defined. They depend on the buoyancy and hydrodynamic properties of each particle and are related to both solvent and macromolecule characteristics. Each particle is characterized by its molar mass, M, partial specific volume, $\bar{v}$, hydrodynamic or Stokes radius, R_H, and the solvent by its density, ρ, and viscosity, η. With Avogadro's number, N_A, the Svedberg and the Stokes–Einstein equations can be expressed as

$$s = M(1 - \rho\bar{v})/(N_A 6\pi\eta R_H) \qquad (2)$$

and

$$D = RT/(N_A 6\pi\eta R_H). \qquad (3)$$

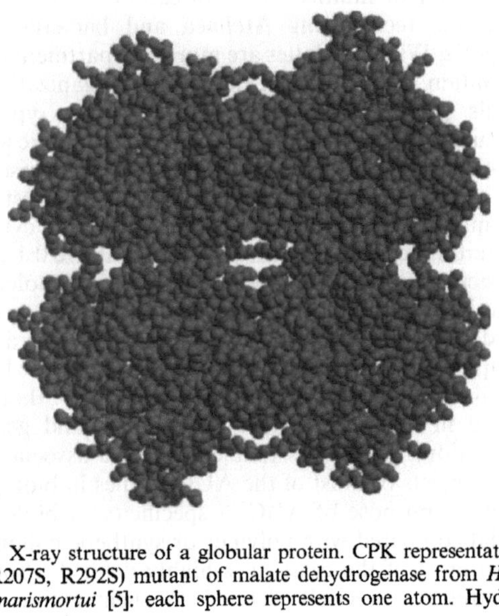

Fig. 1 X-ray structure of a globular protein. CPK representation of the (R207S, R292S) mutant of malate dehydrogenase from *Haloarcula marismortui* [5]: each sphere represents one atom. Hydrogen atoms are not seen

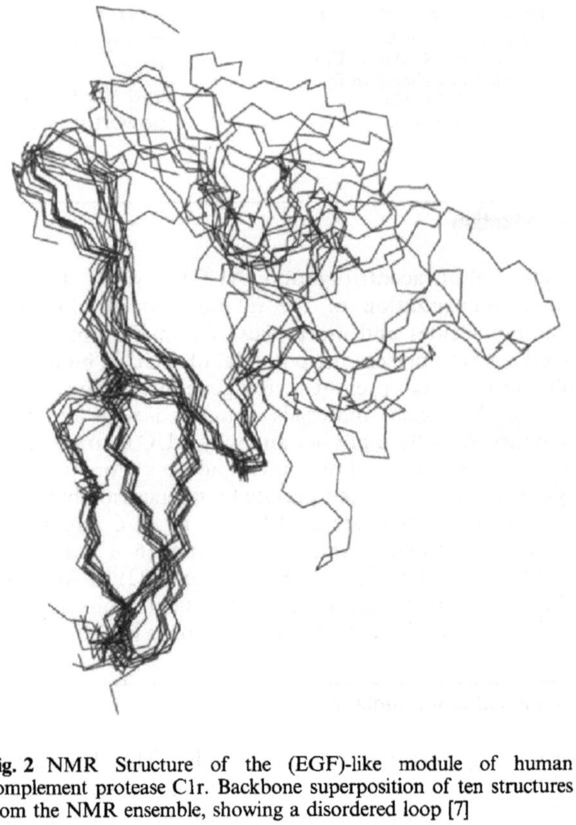

Fig. 2 NMR Structure of the (EGF)-like module of human complement protease C1r. Backbone superposition of ten structures from the NMR ensemble, showing a disordered loop [7]

In most cases, the system is highly defined. The solvent density and viscosity can be measured, or calculated from tabulated data using the program SEDNTERP [8]. The molar mass, M, of the particle is a multiple or combination of precise numbers. The partial specific volume can be estimated for proteins from the amino acid composition with the program SEDNTERP with a good precision [9]. Very often the hydrodynamic radius can also be estimated from independent measurements, from quasielastic light scattering or calibrated (with a set of proteins of known R_H) gel filtration experiments. R_H is related to R_0, the minimum theoretical hydrodynamic radius compatible with the volume of the particle – corresponding to a sphere –, by the frictional ratio f/f^o:

$$R_H = (f/f^o)R_0. \tag{4}$$

The values of the frictional ratio depend on the hydration and shape. Globular compact macromolecules are characterized by f/f^o values of 1.25 ± 0.05, as can be seen from tabulated frictional ratio of proteins and viruses [10] and the pictures of the related three-dimensional structures [11].

Generally, two types of AUC protocols, schematized in Fig. 3, are used for the study of biological macromolecules: sedimentation velocity and equilibrium sedimentation. The experimental profiles are related in the former to s and D and in the latter to the ratio s/D, i.e. the buoyancy mass.

Homogeneity of the macromolecules and analysis model

The Lamm equation (Eq. 1) is used to analyze the sedimentation profiles to determine constant values of s and D for a noninteracting ideal system. But first the qualitative description of the system has to be obtained. Is it homogenous or not? How many different species are there? Do these species interact or not? These are key pieces of information required to define a model for a quantitative analysis. The easier way to proceed is to undertake sedimentation velocity experiments at different dilutions and to characterize each experiment by a distribution of sedimentation coefficients [12, 13]. Different software can be used:

- The time-derivative method DCDT developed by Stafford [14]. An improved commercial version is available (DCDTPLUS from Philo [15]).
- The Van Holde and Weichchet method introduced by Demeler in the software platform ULTRASCAN (which exists in free or commercial versions, depending on the operating systems) [16].
- The continuous size distribution analysis in the program SEDFIT of Schuck [17].

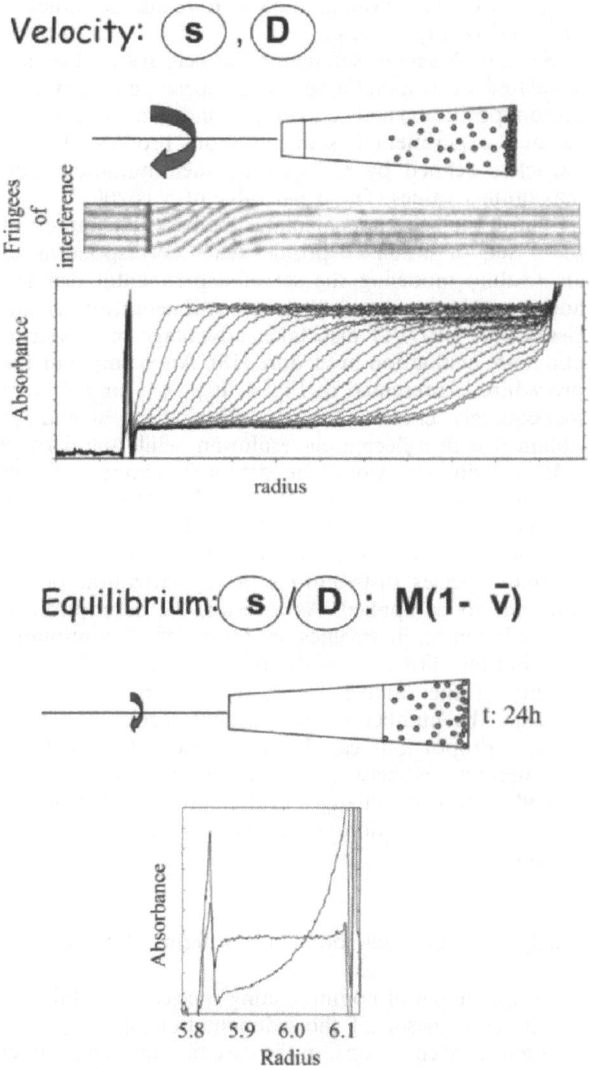

Fig. 3 The two usual types of analytical ultracentrifugation experiment. Sedimentation velocity uses a long column (typically 1-cm long) and high rotor speed. Sedimentation profiles obtained as a function of time, and measured by interference or absorbance, give insight into the sedimentation and diffusion coefficients, s and D. The transport is dominated by the sedimentation process. Equilibrium sedimentation uses a short column (typically 3-mm long) and rather low rotor speed. One equilibrium profile is obtained after typically 24 h and is related to the ratio s/D, i.e. the buoyancy molar mass. Users often use three equilibrium profiles obtained at three different rotor speeds to characterize their system

Because SEDFIT is the most recent one, I will briefly present its principles. SEDFIT, in the subprogram c(s), uses simulated sedimentation profiles, based on the Lamm equation. First, the user is asked to give the values of ρ and η of the solvent, and $\bar{v}$ and f/f^o of the particles. These values allow us to relate the values of M, s and D (Eqs. 2, 3, 4).

Second, N sets of sedimentation velocity profiles are simulated, corresponding to the geometry (meniscus and bottom positions) and centrifugal conditions ($\omega^2 t$) of the various experimental sedimentation profiles, for N particles, defined by the user by their minimum and maximum s values. The usual value of N is 200.

Third, the program calculates the best distribution of the N sets of simulated profiles (each corresponding to an s value) modeling the set of experimental profiles, and provides a distribution of sedimentation coefficients, $c(s)$. SEDFIT also takes advantage of a systematic noise evaluation procedure [18]. An example of the procedure is presented in Fig. 4. If ρ, η, $\bar{v}$ or f/f^0 were inadequately chosen, the $c(s)$ distribution would be obtained with a decreased resolution, while the derived $c(M)$ distribution would be definitely wrong (s is an experimental parameter, M is calculated from the s values and an estimation of the shape and $\bar{v}$). For heterogeneous systems, and particularly for broad or complex species distributions, a regularization procedure has to be applied. As a result, the description of the sedimentation profiles in terms of a continuous distribution allows us to determine the level of homogeneity of the sample. In the case of noninteracting systems, the number of independent species and their relative proportions can be determined. The evidence for interactions between the macromolecules can be provided from the changes in the apparent distribution of the sedimentation coefficients with the sample concentration.

Analysis of the whole boundary in terms of s and D

Once the model of noninteracting species is established, by the comparison of the sedimentation of samples at different concentrations, in the case of a limited number of one, two or three species, the analysis of the boundary can be done to extract s and D. The model has to be chosen properly: the analysis of a slightly heterogeneous system in terms of one particle will lead to an overestimation of D, and thus an underestimation of M. Different programs can be used:

- DCDT, from Stafford [14], or DCDTPLUS, from Philo [15].
- SVEDBERG, from Philo [19].
- LAMM, from Behlke and Ristau [20, 21].
- ULTRASCAN, mentioned earlier, from Demeler [22, 23].
- SEDFIT, mentioned earlier, from Schuck, in the noninteracting species model [24].

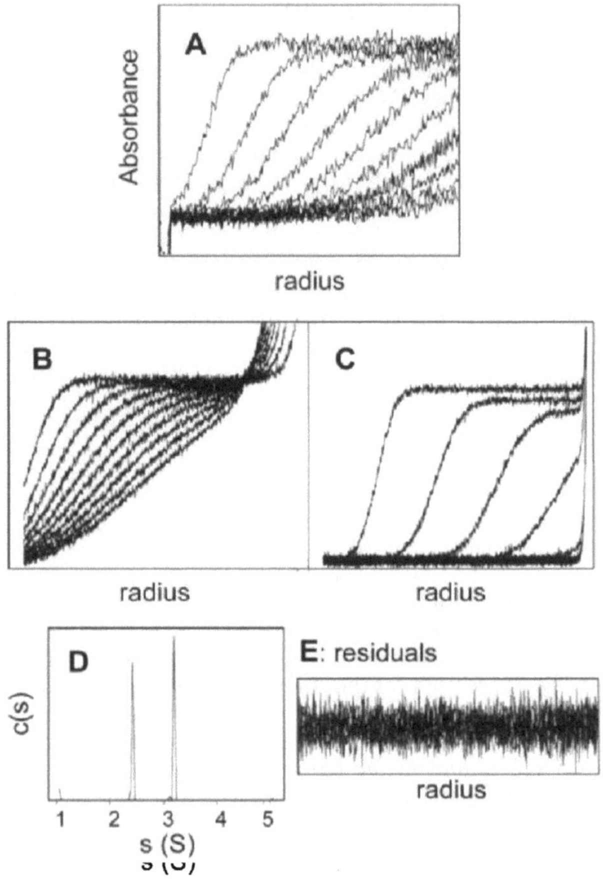

Fig. 4A–E Obtaining a continuous distribution of sedimentation coefficients, $c(s)$, by SEDFIT. **A** Selection of the experimental files, for a solution comprising a fragment of heparin of 14 kDa and a protein of 34 kDa (interferon-γ) in the same proportion. Ten profiles obtained every hour at 42000 rpm and 10 °C are shown. **B,C** Simulated profiles with $\rho = 1.0105$ g/ml, $\eta = 1.429$ cP, $\bar{v} = 0.64$ml/g, $f/f^0 = 1.25$. The values of ρ and η are those of the solvent, those of $\bar{v}$ and f/f^0 are reasonable assumptions corresponding to globular, usually hydrated, complexes with about 1 g heparin per gram protein. The geometry of the cell, the rotor speed and the times of sedimentation correspond to the experimental profiles shown in **A**. **B** corresponds to the smallest particle considered, with $s = 1$ S and a corresponding calculated D value of 10×10^{-7} cm^2/s. **C** corresponds to the largest particle considered, with $s = 5$ S and a corresponding calculated D value of 4.5×10^{-7} cm^2/s (see the text). The comparison of **B** and **C** shows clearly how broadening of the sedimentation profiles is taken into account by the program. SEDFIT was used to simulate 100 sets of sedimentation profiles corresponding to s values between 1 and 5 S. The best combination of these profiles for the description of the experimental profiles corresponds to the continuous distribution $c(s)$ in **D**. **E** shows the superposition of the differences between the profiles calculated with the distribution of particles in **D** and the experimental profiles. The $c(s)$ distribution in **C** suggests (but does not certify) two types of species in solution. For this study, this result was confirmed by other sedimentation experiments performed with various proportions of the partners

DCDT, SVEDBERG and LAMM use approximate analytical solutions of the Lamm equation, while ULTRASCAN and SEDFIT use numerical solutions. Precise values of s (to percent accuracy) and D (at typically 10%, reaching some percent for really homogeneous samples and with data of quality) can very easily be obtained from these procedures, from which a quite precise value of the buoyancy molar mass, and then, knowing ρ and $\bar{v}$, the molar mass of the macromolecule can be calculated (Eq. 2). As an example, we have studied the relationship between the assembly and enzymatic activity properties of fragments of human complement protease C1r [24]. The molar masses of different constructions from MS and from sedimentation velocity experiments are compared in Table 1. MS provides the molar mass of the subunit and sedimentation velocity that of the assembly in solution. The results indicate clearly that the CCP1/2-SP constructions are dimers (molar mass from sedimentation velocity is twice that from MS), and the CCP2-SP are monomers. In that example, the molar masses (per subunit) from AUC are slightly smaller than the values calculated from MS, but the conclusions are unambiguous, given the fact that sedimentation velocity gave the same results at three concentrations.

Distribution of sedimentation coefficients in the case of interacting systems

In the case of interacting systems, the shape of the sedimentation velocity profiles depends on the stoichiometry and kinetics of the association process, and is rather complex, even in the case of infinitely rapid association, for processes more complicate than dimerization [14, 25]. The $c(s)$ distribution profiles obtained for a protein solubilized in the presence of detergent [26] are shown in Fig. 5. The "continuous distributions" show a single peak or a more complex pattern, with a displacement to higher s values with a higher protein concentration or a decreased detergent concentration. This indicates that the protein is subjected to an autoassociation equilibrium modulated by the detergent. The appearance of the sedimentation velocity profiles depends on the kinetics of the equilibrium: for instantaneous monomer–dimer equilibrium (which is a simple case), we would expect symmetrical peaks characterized by s values between those of the monomer and the dimer. For an infinitely slow process, we would expect two narrow peaks in variable proportions corresponding to the individual species. We were able to slow the kinetics of association in a monomer–dimer system, the Epstein–Barr virus protease, by adding 20% glycerol to the solvent without affecting the monomer–dimer association constant [27], but the use of a cosolvent in general can change the macromolecular interactions. The evolution of the mean sedimentation coefficients of the sample with the sample concentration can be used to determine association constants [28].

Table 1 Comparison of molecular masses M, of fragments of human complement protease C1r from sedimentation velocity (*SV*) and mass spectrometry (*MS*) experiments. See the details in Ref. [24]

Fragment	M from SV (kDa ± 10%	M from MS (kDa ± 0.025)
CCP1/2-SP, wild type	86	49.657
CCP1/2-SP, R444Q mutant	93	50.079
CCP1/2-SP, S637A mutant	87	49.974
CCP2-SP, wild type	38	40.490
CCP2-SP, R444Q mutant	36	40.405
CCP2-SP, S637A mutant	39	40.362

Sedimentation equilibrium

Equilibrium sedimentation is more widely used than sedimentation velocity for the study of autoassociating systems [29]. Derived from the Lamm equation, the equilibrium condition of a heterogeneous dilute system can be described by

$$a(r) = \sum c_{n,0}\varepsilon_n l \exp\left[\omega^2 M_n(1-\bar{v}_n\rho)/2RT \cdot (r^2 - r_0^2)\right] + \delta, \quad (5)$$

where $a(r)$ is the total signal, absorbance or fringe of interference, at radial position r; the summation addresses all the species that sediment, the indices n and 0 the different species and the reference radial position, respectively; $c_{n,0}\varepsilon_n l = a_n(r_0)$ is the signal at the

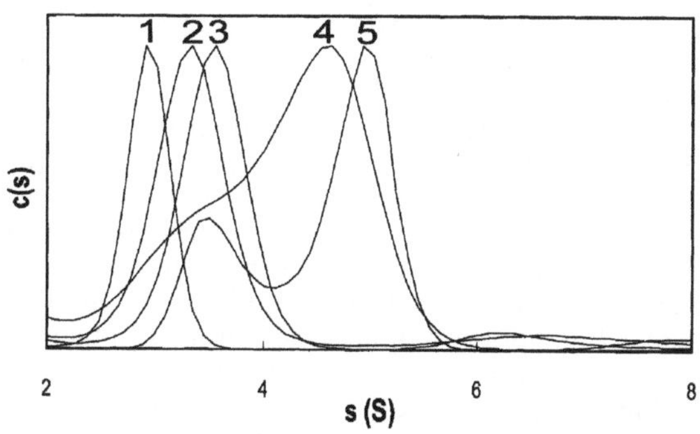

Fig. 5 Distribution of sedimentation coefficients for human paroxonase (*HUPON1*). Different distributions were obtained at various protein, salt and octaethylene glycol monododecyl ether ($C_{12}E_8$) detergent concentrations: 6 μM HUPON1, 7.29 mM $C_{12}E_8$, 140 mM NaCl (*1*); 6 μM HUPON1, 1.54 mM $C_{12}E_8$, 280 mM NaCl (*2*); 6 μM HUPON1, 0.81 mM $C_{12}E_8$, 280 mM NaCl (*3*); 18 μM HUPON1, 0.36 mM $C_{12}E_8$ (*4*); 6 μM HUPON1, 0,36 mM $C_{12}E_8$ (*5*). The details of the experiments are described in Ref. [26]. The set of experiments indicates that the protein is subjected to an autoassociation equilibrium modulated by the detergent

reference position r_0 for species n, with $c_{n,0}$ its concentration in weight units at r_0, ε_n the proportionality factor and l the path length (ε is the extinction coefficient in weight units at the wavelength of measurement for absorbance measurements, ε is the ratio J/c when the signal is a number of fringes, J, i.e. in interference measurement; J/c is related to the refractive index increment, $\Delta n/\Delta c$, by the relation $\Delta n/\Delta c = (J/c)(\lambda/Kl)$, λ being the wavelength of the laser light (675 nm) and K the magnifying coefficient supposed to be 1; we measured a value of 0.975 in our centrifuge [30]; T is the temperature (in Kelvin); δ is a signal offset, relating the sum of the contribution of the macromolecules considered and the measured signal.

We should bear in mind that

1. The equilibrium condition has to be obtained: it excludes unstable systems.
2. The analysis assumes a signal proportional to the concentration of the macromolecular species considered: absorbing and unstable solutes (cofactors such as reduced nicotinamide adenine dinucleotide or guanosine-5′-triphosphate, reducing agents such as dithiothreitol) which can also redistribute in the centrifuge must be avoided.

In routine treatments, the δ (baseline offset) factor is constant over the whole radial range, an assumption that has to be critically considered.

Routine studies of equilibrium sedimentation address autoassociating systems with one type of autoassociating macromolecule. The program NONLIN developed by Yphantis and Johnson [21], a version of which being including in the software based on ORIGIN (Microcal) supplied by Beckman, is widely used for this purpose. In the context of the model, Eq. (5) can be greatly simplified: all ε_n and $\bar{v}_n$ values are constant, c_n values are related to c_1 by the association constants K_n (in weight units): $K_n = c_n/c_1^n$, and $M_n = nM_1$. In general, only one or a maximum of two reactions of equilibrium are considered, corresponding to two or three different species in solution, so the equation is greatly constrained, the unknowns being δ, the K_n and/or n values. The residuals are the differences between the experimental and calculated signal values according to the results of the modeling process. Equilibrium sedimentation profiles obtained at different concentrations and rotor speeds are analyzed simultaneously in a global analysis, and the careful inspection of the residuals, which have to be randomly distributed around a null value, allows us to consider the model and the results of the analysis as relevant.

AUC for more complex systems

The present and future of AUC addresses more complex systems. What I will evoke briefly is the potential of AUC in order to address (1) the general shape of a macromolecule, (2) macromolecules that are heterogeneous in their composition, (3) association processes involving different partners, (4) systems that are complex in their size or (5) nonideal systems and (6) what insights can AUC give on the kinetics of the association and dissociation processes.

Some general recommendations should be made first. For complex systems, the design of the experiments is of particular importance: best use should be made of the capabilities of the centrifuge: the different systems of acquisition (interference, absorbance), the different cells, the choice of rotor speeds. The properties of the constituents or complexes have to be considered: mass, buoyancy, hydrodynamic radius, optical properties (including labeling), concentration dependence. We can modify the solvent properties, the composition (density) and use density gradients. To what extent AUC will be able to provide the appropriate response can be estimated in advance, by exploring the analysis of simulated conditions, which are easily obtained with the software using numerical solutions of the Lamm equation. This procedure allows us to determine which information can be reliably retrieved for each experimental condition.

Most probably, as described in the routine procedure, the analysis of the experimental data will follow the following steps: from a model of analysis, use solutions of the Lamm equation in order to obtain synthetic data, (for complex systems, heterogeneous or interacting systems, combinations of Eq. 1 with nonconstant s and D and/or taking into account the reactions have to be considered) compare them with experimental data, evaluate the systematic noise and the deviations, and come back if needed to a new, more appropriate model of analysis.

Also, the information given by AUC tends to be integrated in a global analysis. This is routinely done only for sedimentation equilibrium profiles. The utility of global analysis of AUC experiments at high and low speeds was theoretically predicted some years ago [23]. The global analysis of three sets of sedimentation velocity profiles, obtained at three protein concentrations, for the characterization of sample nonideality and protein characteristics, was, to my knowledge, the first publication concerning global analysis of different samples in sedimentation velocity [31]. The very recent program SEDPHAT from Schuck is designed to allow the combined analysis of data from various experiments: sedimentation equilibrium, velocity experiments and quasielastic light scattering.

The analysis of complex systems benefits from the use of complementary techniques providing R_H or size distribution analysis, such as gel filtration, quasielastic light scattering or flow-field flow fractionation [32]. AUC is, of course, used together with small-angle X-ray or neutron scattering, electron microscopy, or hydrodynamic modeling, amongst others (see Ref. [33] as an example of complementary AUC and small-angle X-ray scattering) to obtain structural information at low or high resolution. Future data analysis will tend to use the raw information provided by these techniques in an integrated way.

Advances in hydrodynamics

Can we propose a three dimensional arrangement compatible with solution properties? The fascination of hydrodynamic modeling is related to this question. It is clear that the translational friction measured by AUC is not the parameter that is the most sensitive to the shape, different shapes corresponding to the same experimental value [34]: this is why there is the coexistence of hydrodynamic modeling approaches at high and low resolution, in order to select improbable structures. The bead modeling approach has been developed by Garcia de la Torre et al. [35]; for a recent review, see Ref. [36]. The suite of computer programs HYDRO/SOLPRO from the laboratory of Garcia de la Torre are described at and can be downloaded from http://leonardo.fcu.um.es/macromol: they calculate solution parameters, including hydrodynamic coefficients but also NMR or scattering parameters (amongst others) for particles defined at low-level or high-level resolution (from atomic coordinates, cryogenic electron microscopy, or arbitrarily shaped particles), using bead modeling (the particle is an assembly of spheres) or shell modeling (the surface of the particle is represented by a shell of minibeads). For example, from high resolution description (Protein Data Bank files), HYDROPRO computes the hydrodynamic properties of rigid macromolecules using the later approach with a good precision [37]. The SOMO program of Rai et al. [38] uses the former approach, which is potentially more powerful to describe the scattering properties of the macromolecules. Protein hydration was critically examined by Durchschlag and Zipper [39, 40]. "Crystallohydrodynamics" uses the high resolution information of crystallography for the estimation (at low resolution) of domain orientation of antibodies [41, 42]. The program MacBEADS [21] can also be used to set up bead models. The progress in hydrodynamic modeling also concerns calculations from flexible macromolecules [43].

Systems heterogeneous in their composition

The knowledge of the structure and interactions of macromolecular systems consisting of chemically different partners is very important for the understanding of the cell function. Membrane proteins that constitute about 30% of the proteins present at a given time in the cell, require, in order to be obtained in a pure and soluble folded state, the use of amphiphilic compounds, generally detergents but also polymers, that bind the proteins in variable proportions of the order of some grams per gram [44]. Examples of their studies by AUC are given in Refs. [1, 26, 45, 46]. The evolution of the genomes is related to global variations in the ATGC content in the DNA, that can precisely be measured with AUC [47]. The transduction (regulation of cell life from an external signal), transcription (from DNA to m-RNA) and translation (synthesis of proteins from mRNA using tRNA and various other factors on the ribosome, an enormous machinery made of RNA and proteins) events all require protein–nucleic acid interactions. The sugar part of glycosylated proteins frequently reaches 30% of the total mass [48, 49]. Sugars at the exterior of the cells have important functions based on their interactions with proteins. Serum lipoproteins are edifices of lipids and proteins in variable proportions [50, 51]. Ferritin is an iron storage protein, with variable relative content of inorganic material that can reach 50% [32]. In general, the question of ligand binding regulation is associated with modulation of the stoichiometry [33, 52]. For each type of heterogeneous system, particular strategies have been developed based, in general, on the comparison of systems of detection in AUC and/or on the difference in the buoyancies of the partners, often exploited through contrast variation studies in solvents of various densities.

Sedimentation equilibrium of heterogeneous complexes

Sedimentation equilibrium of heterogeneous complexes generally uses Eq. (5). These studies are, however, significantly more difficult than in the case of autoassociation because there are fewer constraints in the analysis. Excellent samples and AUC instrumentation (optics) are required. Acquisition of intensities instead of absorbance values is recommended. Different programs are available: ULTRASCAN from Demeler can fit noninteracting components or heteroassociating systems (A + B< = >AB); ULTRASPIN from Veprintsev [53] allows the easy consideration of various schemes of association; POLYMOLE from Behlke [54] allows multiwavelength analysis, and uses mass conservation and optionally statistically scaled binding constants. All these analyses compare the details of experimental and calculated profiles in the framework of a model of association, which can be difficult to define. To help in this purpose, SEGAL, developed by Machaidze and Lustig [55], allows easy determination of apparent mean molar masses from parts of the experimental profiles. The purpose of MSTAR [56] is to determine the mean molar masses along the sedimentation profiles, in a model-independent way, from interference data. Minton

[57, 58] developed an original strategy for the evaluation of the association scheme based on the measurement of signal average molar masses obtained from low-speed experiments on a number of samples of various composition.

Systems heterogeneous in size

AUC contributed significantly to the characterization of complex self-associating systems. Tubulin [59] or Fts-Z polymerization [60] and amyloid fibril self-association [61, 62] are famous examples. Another kind of size heterogeneity is found in the study of polysaccharides. In contrast to proteins and nucleic acids, polysaccharides are very heterogeneous materials, in composition and size. Furthermore, they have very anisotropic shapes and are often highly charged, making their characterization, in molar mass and size, particularly difficult. Particular strategies, similar to those used in the studies of synthetic polymers, and based on different measurable hydrodynamic characteristics, on their evolution with concentration and size of the polysaccharides, have been developed for these materials [30, 63, 64, 65].

Nonideal systems

A nonideal system is a system in which weak interactions affect the parameters measured. More generally, any event that weakly affects the properties of the system can be considered as providing nonideality. Solvent redistribution around the particle has to be considered in the case of multicomponents solvents, when the solvent density differs significantly from the water density, or when solvent interactions are large, a case for which the macromolecule can alternatively be considered as a complex (as in membrane proteins), or in the case of polyelectrolyte macromolecules. The rigorous formalism to account for such interactions is clearly established and does not require the precise description of the solvation or hydration [66, 67, 68, 69, 70]. Another nonideal situation corresponds to highly concentrated solutions of macromolecules. Because of the weak interactions between macromolecules, their positions and movements are correlated. The consideration of nonideality in AUC was historically related to the necessary use of high macromolecule concentrations to obtain a good signal-to-noise ratio, but is now studied per se in the framework of the macromolecular crowding in the cell [71] or of macromolecule crystallization. Virial coefficients are related (through different approaches) to the macromolecule volume, charge and shape [72, 73] and to the interactions between macromolecules or between the solvent and the macromolecule [74, 75, 76, 77, 78, 79]. Now the increased capabilities of data treatment in AUC allows us to consider the effect of nonideality even in the case of rather dilute systems, for example, in sedimentation velocity experiments [31] or equilibrium ones (most of the programs have options to take into account nonideality at high concentration). The recent attempt to consider the solvent compressibility in the ultracentrifuge can also be considered as a kind of nonideality. It can be more appropriate to use a thermodynamic notation, consider $(\partial\Pi/\partial c)_{\mu s}$ instead of M, $(\partial\Pi/\partial C)_{\mu s}$ instead of D (C is the concentration in molar units) and $(\partial\rho/\partial c)_{\mu s}$ instead of $(1 - \rho\bar{v})$ (μs is for constant chemical potential of the solvent components). In this context, it is essential to promote the links between the scientific communities of biologists and physicists from synthetic polymer and colloid science who are using these thermodynamics approaches. One example can be found in the study of phase separation of calcium alginate gels [80].

Kinetics of the association–dissociation processes from sedimentation velocity

In sedimentation velocity experiments, association processes characterized by rate scales below 10^{-5} 1/s are slow: the species can be considered as moving independently. Rate scales larger than 0.1 1/s will characterize fast processes indistinguishable from instantaneous equilibrium. In between, the transport depends on the kinetics of association and of dissociation. Stafford [14] has developed the program SEDANAL for the global analysis of the transport in systems in autoassociations or heteroassociations (A_nB_m system) using kinetics equations. Even if such studies are not at all routine, the highly informative content of the sedimentation velocity profiles (concerning the sample) combined with the high precision and the large number of data analyzed mean that rapid progresses could be expected in that field.

Conclusion: AUC, past, present and future

In a routine mode, AUC is used first for the qualitative characterization of the homogeneity of biological samples and for evidence of interactions between macromolecules. If the sample is homogeneous, precise values of the molar mass, hydrodynamic radius and association constant can be obtained. The challenge consists in the reliable description of more complex systems in solution, in terms of the number and kind of components and/or interactions. To my mind, the tools that have to or will be used for this purpose are

1. Well-designed experiments in view of the particular properties of the biological material studied

in order to make the best use of the capability of AUC.

2. The analysis of AUC data through numerical solutions of the Lamm equation expressed considering the complexity of the samples.
3. Global analysis of the data from AUC and other related techniques.

Acknowledgements Thanks are due to Guiseppe Zaccai for a critical reading of the manuscript and to Beate Bersch and Adriana Irimia for providing Figs. 1 and 2.

References

1. Lebowitz J, Lewis MS, Schuck P (2002) Protein Sci 11:2067
2. Laue T (2001) Curr Opin Struct Biol 11:579
3. Alberts B (1994) Molecular biology of the cell. Garland, New York
4. Goodsell DS (1991) Trends Biochem Sci 16:203
5. Irimia A, Ebel C, Madern D, Richard SB, Cosenza LW, Zaccai G, Vellieux FM (2003) J Mol Biol 326:859
6. Yonath A (2002) Curr Protein Pept Sci 3:67
7. Bersch B, Hernandez JF, Marion D, Arlaud GJ (1998) Biochemistry 37:1204
8. Hayes DB, Laue T, Philo J http://www.bbri.org/RASMB/rasmb.html
9. Harpaz Y, Gerstein M, Chothia C (1994) Structure 2:641
10. Cantor CR, Schimmel PR (1980) Techniques for the study of biological structure and function. Freeman, San Francisco
11. Goodsell DS, Olson AJ (1993) Trends Biochem Sci 18:65
12. Stafford WF (1997) Curr Opin Biotechnol 8:14
13. Schuck P, Perugini MA, Gonzales NR, Howlett GJ, Schubert D (2002) Biophys J 82:1096
14. (a) Stafford WF (2000) Methods Enzymol 323:302; (b) ftp://alpha.bbri.org/rasmb/spin/ms_dos/sedanal-stafford/
15. (a) Philo JS (2000) Anal Biochem 279:151; (b) http://www.jphilo.mailway.com
16. (a) Demeler B, Saber H, Hansen JC (1997) Biophys J 72:397; (b) http://www.cauma.uthscsa.edu/
17. (a) Schuck P (2000) Biophys J 78:1606; (b) http://www.analyticalultracentrifugation.com/
18. Schuck P, Demeler B (1999) Biophys J 76:2288
19. (a) Philo JS (1997) Biophys J 72:435
20. Behlke J, Ristau O (2002) Biophys Chem 95:59
21. http://www.bbri.org/RASMB/rasmb.html
21. Demeler B, Saber H (1998) Biophys J 74:444
22. Demeler B, Behlke J, Ristau O (2000) Methods Enzymol 321:38
23. Schuck P (1998) Biophys J 75:1503
24. Lacroix M, Ebel C, Kardos J, Dobo J, Gal P, Zavodszky P, Arlaud GJ, Thielens NM (2001) J Biol Chem 276:36233
25. Fujita H (1975) Foundations of ultracentrifugal analysis. Wiley, New York
26. Josse D, Ebel C, Stroebel D, Fontaine A, Borges F, Echalier A, Baud D, Renault F, Le Maire M, Chabrieres E, Masson P (2002) J Biol Chem 277:33386
27. Buisson M, Valette E, Hernandez JF, Baudin F, Ebel C, Morand P, Seigneurin JM, Arlaud GJ, Ruigrok RW (2001) J Mol Biol 311:217
28. Correia JJ (2000) Methods Enzymol 321:81
29. Laue TM (1995) Methods Enzymol 259:427
30. Pavlov G, Finet S, Tatarenko K, Korneeva E, Ebel C (2003) Eur Biophys J 32:437
31. Solovyova A, Schuck P, Costenaro L, Ebel C (2001) Biophys J 81:1868
32. Cölfen H, Völkel A (2003) Eur Biophys J 32:432
33. Solovyova AS, Meenan N, McDermott L, Garofalo A, Bradley JE, Kennedy MW, Byron O (2003) Eur Biophys J 32:465
34. Errington N, Rowe AJ (2003) Eur Biophys J 32:511
35. García de la Torre J, Navarro S, Lopez Martinez MC, Diaz FG, Lopez Cascales JJ (1994) Biophys J 67:530
36. Byron O (2000) Methods Enzymol 321:278
37. García de la Torre J, Huertas ML, Carrasco B (2000) Biophys J 78:719
38. Rai N, Rocco M, Byron O http://www.chem.gla.ac.uk/˜olwyn/somo
39. Durchschlag H, Zipper P (2003) Eur Biophys J 32:487
40. Durchschlag H, Zipper P (2001) Biophys Chem 93:141
41. Carrasco B, Garcia de la Torre J, Davis KG, Jones S, Athwal D, Walters C, Burton DR, Harding SE (2001) Biophys Chem 93:181
42. Longman E, Kreusel K, Tendler SB, Fiebrig I, King K, Adair J, O'Shea P, Ortega A, Garcia de la Torre J, Harding SE (2003) Eur Biophys J 32:503
43. García de la Torre J, Pérez Sánchez HE, Ortega A, Hernández JG, Fernandez MX, Díaz FG, López Martinez MC (2003) Eur Biophys J 32:477
44. le Maire M, Champeil P, Moller JV (2000) Biochim Biophys Acta 1508:86
45. Fleming KG (2000) Methods Enzymol 323:63
46. Lustig A, Engel A, Tsiotis G, Landau EM, Baschong W (2000) Biochim Biophys Acta 1464:199
47. Clay O, Douady CJ, Carels N, Hugues S, Bucchiarelli G, Bernardi GB (2003) Eur Biophys J 32:418
48. Lewis MS, Junghans RP (2000) Methods Enzymol 321:136
49. Center RJ, Leapman RD, Lebowitz J, Arthur LO, Earl PL, Moss B (2002) J Virol 76:7863
50. Bozoky Z, Fulop L, Kohidai L (2001) Eur Biophys J 29:621
51. Perugini MA, Schuck P, Howlett GJ (2002) Eur J Biochem 269:5939
52. Chebotareva NA, Klinov SV, Kurganov BI (2001) Biotechnol Genet Eng Rev 18:265
53. Veprintsev D http://www.mrc-cpe.cam.ac.uk/~dbv/soft/download/ultraspin/
54. Behlke J, Ristau O (2003) Eur Biophys J 32:427
55. Machaidze G, Lustig A http://www.biozentrum.unibas.ch/personal/jseelig/AUC/software.html
56. ftp://alpha.bbri.org/rasmb/spin/ms_dos/mstar-coelfen/
57. Minton AP (1997) Prog Colloid Polymer Sci 107:11
58. Rivas G, Stafford W, Minton AP (1999) Methods 19:194
59. Lobert S, Boyd CA, Correia JJ (1997) Biophys J 72:416
60. Rivas G, Lopez A, Mingorance J, Ferrandiz MJ, Zorrilla S, Minton AP, Vicente M, Andreu JM (2000) J Biol Chem 275:11740
61. Perugini MA, Schuck P, Howlett GJ (2000) J Biol Chem 275:36758
62. MacRaild CA, Hatters DM, Lawrence LJ, Howlett GJ (2003) Biophys J 84:2562
63. Harding SE, Colfen H (1995) Anal Biochem 228:131
64. Hokputsa S, Jumel K, Alexander C, Harding SE (2003) Eur Biophys J 32:450

65. Fee M, Errington N, Jumel K, Illum L, Smith A, Harding SE (2003) Eur Biophys J 32:457
66. Eisenberg H (2000) Biophys Chem 88:1
67. Eisenberg H (2003) Eur Biophys J 32:406
68. Ebel C (1995) Prog Colloid Polym Sci 99:17
69. Ebel C, Costenaro L, Pascu M, Faou P, Kernel B, Proust-De Martin F, Zaccai G (2002) Biochemistry 41:13234
70. Kernel B, Zaccai G, Ebel C (1999) Prog Colloid Polym Sci 113:168
71. Rivas G, Fernandez JA, Minton AP (2001) Proc Natl Acad Sci USA 98:3150
72. Harding SE, Horton JC, Jones S, Thornton JM, Winzor DJ (1999) Biophys J 76:2432
73. Rowe AJ (1992) In: Harding SE, Rowe AJ, Horton JC (eds) Analytical ultracentrifugation in biochemistry and polymer science. The Royal Society of Chemistry, Cambridge, p 394
74. Pusey PN, Tough JA (1985) In: Pecora (ed) Dynamic light scattering. Applications of photon correlation spectroscopy. Plenum, New York, p 85
75. Behlke J, Ristau O (2000) Biophys Chem 87:1
76. Costenaro L, Ebel C (2002) Acta Crystallogr Sect D 58:1554
77. Costenaro L, Zaccai G, Ebel C (2002) Biochemistry 41:13245
78. Wills PR, Jacobsen MP, Winzor DJ (2000) Biophys J 79:2178
79. Minton AP (2000) Biophys J 78:101
80. Straatmann A, Borchard W (2003) Eur Biophys J

Progr Colloid Polym Sci (2004) 127: 83–88
DOI 10.1007/b98017

A. Seifert
H. M. Rawel
S. E. Harding
J. Kroll

Characterization of bovine serum albumin/chlorogenic acid solution mixtures by analytical ultracentrifugation

Abstract The aim of this work was to investigate a possible interaction between bovine serum albumin (BSA) and chlorogenic acid (CGA) in KH_2PO_4/K_2HPO_4 buffer (pH 7.0; ionic strength 0.1 mol/l) depending on the BSA/CGA molar ratio by analytical ultracentrifugation (sedimentation velocity) using an OPTIMA XL-A AUC equipped with a UV–VIS absorption optical system. The protein concentration of all the solution mixtures with BSA-to-CGA molar (mass) ratios of 1:19 (10:1) up to 1:95 (2:1) was 0.50 g/l. The investigations were carried out under experimental conditions where the possibility of derivatization of the BSA (covalent bonding) is minimized (pH, temperature, investigation of the solutions immediately after mixing). At 280 nm both sedimenting BSA and BSA with non-covalent-bound CGA can be detected. In the region of 320 nm BSA shows no absorption, and thus only BSA with bound CGA (absorption maximum at 325 nm) can be monitored. The results indicate weak interaction between BSA and CGA. At a scan wavelength of 280 nm sample solutions with a BSA-to-CGA molar ratio of 1:19 revealed a major component of 4.2 S similar to the major component of the BSA control sample. However a further component with a lower $s_{c,20°}$ value (2.3–2.5 S) appears and this is not present in the BSA control sample. At 320 nm the sedimentation of a major component with an $s_{c,20°}$ value of 1.8–2.0 S followed by a 3.1 S component could be demonstrated. This indicates that the binding of CGA to BSA results in considerably lower $s_{c,20°}$ value components. The cause could be a more asymmetric particle, a "swollen" structure and/or ligand-induced partial unfolding of the BSA molecule. Similar results have been obtained with BSA/CGA solutions up to a molar ratio of (1:95). UV–VIS absorption measurements and studies of intrinsic tryptophan fluorescence of BSA/CGA solution mixtures at various CGA concentrations also indicate such interactions between BSA and CGA. Some possible physiological consequences of the BSA/CGA interaction are considered.

Keywords Analytical ultracentrifugation · Sedimentation velocity · Interactions · Proteins · Chlorogenic acid

A. Seifert · H.M. Rawel · J. Kroll (✉)
Institute of Nutritional Science, University of Potsdam,
Arthur-Scheunert-Allee 114–116,
14558 Bergholz-Rehbrücke, Germany
e-mail: jkroll@rz.uni-potsdam.de
Tel.: +49-33200-88262
Fax: +49-33200-88306

S.E. Harding
National Centre for Macromolecular Hydrodynamics, School of Biosciences,
University of Nottingham,
Sutton Bonington LE12 5RD, UK

Introduction

Chlorogenic acid (CGA) is widely distributed in the plant kingdom. For example, green coffee beans contain 6–10% CGA, and a 200-ml cup of coffee may have 20–675 mg CGA, depending on the roasting and brewing conditions [1]. In potato tubers CGA constitutes up to 90% of the total phenolic content [2]. CGA plays an important role in many biological and metabolic processes, for example, antiviral activity and inhibition of tumour promotion. Because of its chemical structure CGA (Fig. 1) may interact with proteins by "weak" interactions (e.g., hydrogen, hydrophobic, ionic bonding) [3, 4, 5, 6, 7, 8]. Under defined reaction conditions (pH 9) nucleophilic binding to proteins can be attained, resulting in protein derivatives (covalent bonding) [8, 9, 10].

The aim of the work was to investigate the possible interaction between bovine serum albumin (BSA) and CGA depending on the BSA-to-CGA molar ratio by analytical ultracentrifugation (sedimentation velocity) under experimental conditions where the possibility of derivatization of the BSA (covalent bonding) is minimized.

Materials and Methods

Materials

BSA [Assay ≥98%; crystallized; lyophilized; $H_2O \leq 5\%$; ash ≤ 5%; $M_r \approx 67{,}000$ (lit)] and CGA hemihydrate (pure; ~97%; $M_r = 363{,}32$) were employed as supplied by Fluka Chemie, Buchs, Switzerland. Potassium dihydrogen phosphate and dipotassium hydrogen phosphate (p.a. Merck, Darmstadt, Germany) were dried overnight at 120 °C before use.

Solutions

The KH_2PO_4/K_2HPO_4 buffer solution (pH 7.0; ionic strength 0.1 mol/l) was prepared according to the method in Ref. [11] and gave a measured pH of 6.98.

Stock solutions of BSA (about 1.0 g/l) and CGA (exactly 1.00 g/l related to the weighed amount) were prepared by dissolving them in the buffer at room temperature. After determination of the protein content of the BSA stock solutions, the BSA control solutions and BSA/CGA solution mixtures of different BSA-to-CGA molar (mass) ratios of 1:19 (10:1), 1:38 (5:1) and 1:95 (2:1) were prepared. The protein concentration in all the solutions was 0.50 g/l. To minimize any possibility of derivatization of the BSA (covalent bonding) a moderate pH condition was applied and the solutions were investigated with the analytical ultracentrifuge (AUC) immediately after mixing (start of the runs within 2 h).

Methods

Analytical ultracentrifugation

The AUC employed was the OPTIMA XL-A (Beckman/Coulter) with a UV-VIS absorption optical system (190–800 nm). The solutions were investigated at 45,000 rpm (approximately

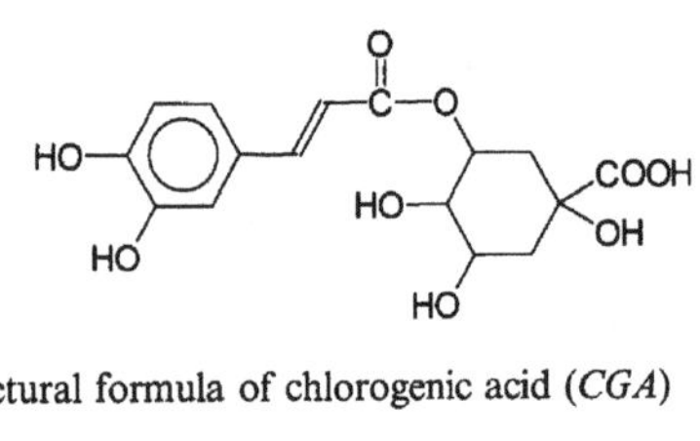

Fig. 1 Structural formula of chlorogenic acid (*CGA*)

160,000*g*) in double-sector cells which contained the solution and the reference solution (buffer solution or CGA solutions). CGA solutions in the reference sector were used because of the large basic CGA absorbance of more than 1.5 with increasing CGA content in the BSA/CGA solution mixtures. This experimental procedure gave absorbance values below 1.5 and also made it possible to scan the solutions in the region conventionally used for recording protein absorption (280 nm). During the runs "radial scans" were carried out in periods of 6–9 min, which provide figures of the absorbance as a function of radial displacement (distance from the centre of rotation) in the solution. The main aim of such high-speed sedimentation velocity investigations was to determine sedimentation coefficients and their distributions at the given concentration at 20.0 °C ($s_{c,20°}$; unit 10^{-13} s≡1 S [Svedberg]). The $s_{c,20°}$ values characterize the hydrodynamic behaviour of the BSA molecules in solution in the presence of CGA and in comparison to the reference molecule (BSA) dissolved in the same buffer. The sedimentation coefficient is a function of the mass, size and shape of the dissolved particles, of the properties of the solvent and of the interaction properties within the system. Additional runs were performed at 3,000 rpm prior to the high speed sedimentation velocity investigations to check if any possible higher molecular weight "impurities" could be formed in the preparations. Further "wavelength scans" (190–800 nm) were carried out at 3,000 rpm, giving figures of the absorbance depending on the wavelength at a fixed radius in the solution.

Two distinguishing features of the BSA and CGA molecules were exploited in the investigations. The first one is that the CGA molecules with a molar mass of about 350 are too small to sediment at the centrifugal fields generally realized with an AUC, whereas CGA bound to BSA by "weak" interactions can sediment. Secondly, CGA shows an absorption maximum at about 320 nm, but BSA does not absorb at this wavelength. By recording the absorbance in the region of 320 nm, it is thus possible to provide evidence for BSA with bound CGA. At 280 nm, both BSA and CGA absorb. A comparative evaluation of the sedimentation figures at wavelengths of 280 and 320 nm thus provides the basis of our assay procedure.

The evaluation of the results on the basis of the absorbance/radius figures was carried out with the program SEDFIT5 [12, 13] according to the least-squares $g^*(s)$ method, which facilitates the evaluation of a $g^*(s)$ distribution, based on direct boundary fitting and algebraic systematic noise deconvolution. The result of this analysis is an apparent sedimentation coefficient equivalent to the $g(s^*)$ analysis of dc/dt [14]. However, there is no need to differentiate the data and calculate dc/dt here, thus eliminating a source of broadening from finite time differences between the scans. Therefore, this method is, as a consequence, far more tolerant of large time steps, such as commonly occur in absorbance optical data. The model requires the following user input:

1. Resolution: This specifies the increments of the s values for the calculation. The recommended value of 100 was applied, with a value of 200 chosen as a check. Generally, with higher values, more noise is introduced into the distribution, but this can be efficiently reduced by Tikhonov–Phillips regularization [15].
2. s_{min} and s_{max}: These values should confine the range of s values expected for the sample. In all evaluations s_{min} and s_{max} were 1 and 20 S, respectively.
3. Confidence level (F ratio): This input is required for the statistical adjustment of the regularization constraint. The recommended values of 0.68 (one standard deviation) or 0.95 (two) were applied.

Further, a floating baseline was allowed.

UV–VIS absorption spectra

The UV–VIS absorption spectra of BSA, CGA and the solution mixtures were determined using a UV–VIS spectrophotometer

(Spekol, Carl Zeiss Jena, Germany) and additionally using the AUC as described earlier.

Protein content

The protein content of the BSA solutions was determined according to the method in Ref. [16] with the modification described in Ref. [17]. BSA was used to calibrate the regression curve after determining its protein content by semimicro Kjeldahl analysis (Kjeldatherm System KT40, Gerhardt Laboratory Instruments, Bonn, Germany).

Tryptophan fluorescence

Intrinsic fluorescence emission spectra between 300 and 800 nm were recorded with a JASCO FP 920 fluorescence detector using an excitation wavelength of 290 nm and quartz optical cells of 10 mm path length. Excitation and emission bandwidths were set at 18 and 40 nm, respectively. The protein concentrations were adjusted to 0.50 g/l. Corrections were made for the buffer blanks. The BSA/CGA solution mixtures (BSA-to-CGA mass ratio 0.50/0.01–0.10 g/l) were prepared freshly and the fluorescence was measured as described earlier.

Results and discussion

At a scan wavelength of 280 nm, both free BSA and BSA with bound CGA can be detected – if present. A comparison of the apparent sedimentation coefficients of the BSA/CGA(1:19) solution with the BSA solution investigated during the same run indicates for both scans a major component at 4.2 S (Table 1). In the case of the BSA/CGA(1:19) solution further components appear. The component with an essentially lower $s_{c,20°}$ value of 2.3–2.5 S ("shoulder"), which is not present in the BSA control sample, demonstrates a stronger change of the hydrodynamic behaviour (slower sedimentation).

That may be due to a more asymmetric particle, a "swollen" structure and/or partial unfolding of the BSA molecule. This observation is also confirmed by a larger proportion of this component with increasing concentration of CGA in the BSA/CGA(1:38) sample solution. The increase of the CGA concentration in the solution mixtures also results in more distinct components (maxima) owing to the binding of CGA to BSA. Proof of weak interactions between BSA and CGA is given by the fact that at 320 nm where only BSA with bound CGA can be registered (BSA does not absorb at 320 nm, CGA has an absorption maximum at 325 nm) sedimentation of a number of components from between 2 S to 11 S occurred [Table 1; BSA/CGA(1:19) solution sample]. The low $s_{c,20°}$ values of the major sedimenting component of 1.8–2.0 S (shoulder) and a second 3.1 S (maximum) component support the preceding discussion. Because of the large basic CGA absorbance of more than 1.5, BSA/CGA solution samples with molar ratios of 1:38 and 1:95 had to be scanned at a higher wavelength, but gave a consistent picture (Fig. 2). All the BSA/CGA solution samples exhibited maxima of the sedimentation coefficient distribution at about 2.3 S and a strong increase of the concentration of components below 2.3 S in the least-squares $g^*(s)$–$s_{c,20°}$ plot. Only 19 molecules of CGA per molecule of BSA produce this change in the hydrodynamic behaviour. Considering that BSA may interact with smaller molecules and ions owing to at least six binding regions on the protein molecule [18], there is an excess of CGA with about the 3 times number of CGA molecules per BSA molecule being present in the BSA/CGA(1:19) solution.

The BSA/CGA(1:19) and BSA/CGA(1:95) solutions show components in the region of the BSA absorption (4.2–4.3 S). This may be due to BSA molecules with a relatively small amount of bound CGA, so the sedimentation behaviour is not significantly different. Surprisingly the sample with a medium BSA-to-CGA

Table 1 Sedimentation coefficients of bovine serum albumin (*BSA*)/chlorogenic acid (*CGA*) solution mixtures of different BSA-to-CGA molar ratios. Analytical Ultracentrifugation conditions 45,000 rpm (≈160,000*g*), 20.0 °C

	Wavelength of scans (nm)	$s_{c,20°}$ values of the components (S)[a,b,c,d]						
		Components at decreasing concentration						
		1 (major component)	2	3	4	5	6	7
BSA/CGA(1:19)	280	4.2	5.2–5.4[b]	2.3–2.5[b]	7.3	8.9		
	320	1.8–2.0[b]	3.1	6.2	4.5	5.4–5.6[b]	8.3	10.8
BSA/CGA(1:38)	280	4.2	2.2	5.2–5.4	6.2–6.4	8.5	10.0	
BSA control	280	4.2	7–9.5[c]					

[a]Maxima from the least-squares $g^*(s)$–$s_{c,20°}$ diagrams
[b]Hidden maxima with "shoulder(s)"
[c]Hidden maxima without "shoulder(s)"
[d]All the results are from the same run; further runs show similar results
Evaluation program SEDFIT5 [12, 13]; method: least-squares $g^*(s)$; resolution 100; *F* ratio 0.68; $s_{c,20°}$ values of least-squares $g^*(s)$ values below 0.005 are not given. Molar masses: BSA 67,000, CGA 354.31; solvent KH_2PO_4/K_2HPO_4 buffer (pH 7.0; ionic strength 0.1 mol/l); BSA concentration in all solutions 0.50 g/l

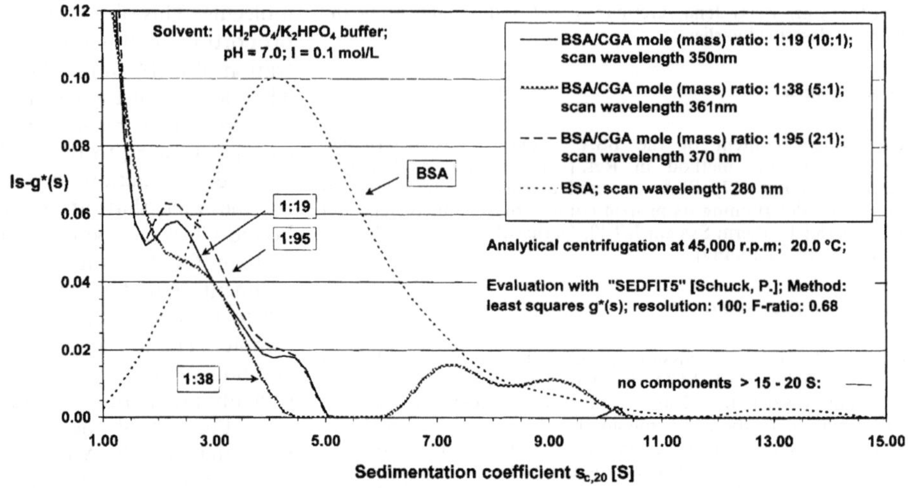

Fig. 2 Sedimentation coefficient distributions of bovine serum albumin (*BSA*)/CGA solution mixtures as a function of CGA concentration

molar ratio of 1:38 shows no component in the BSA region but two components at 7.1–7.3 and 9.1 S. This reveals that there is no simple relation between the formation of such components and the BSA-to-CGA molar ratio.

The results described could also be confirmed by further measurements of BSA/CGA solution mixtures of different molar ratios by including experiments at scan wavelengths in the BSA absorption region; the latter being made possible by means of CGA added to the reference sector.

Investigations on haze-active proteins and polyphenols [19, 20] propose a model for protein–polyphenol interactions, which includes the formation of different structures strongly depending on the concentrations of protein and polyphenol and their ratios. The authors proceed from a fixed number of binding ends of the polyphenol and binding sites of each protein. With excess polyphenol relative to protein, they assume all of the protein binding sites would be occupied, but the likelihood that bridging would occur would be low because each free polyphenol end would have a small chance of finding a free binding site on a protein molecule. They conclude, too, that this would result in small aggregates. Our results also indicate the formation of different structures depending on the CGA concentration. The BSA/CGA(1:19) and BSA/CGA(1:95) solutions show no components above 5 S (Fig. 2). The formation of greater aggregates consisting of two or more BSA molecules with bound CGA should be analogous to the aggregate formation of the native BSA (Fig. 2, small amount in the region of 13 S) and results in the formation of aggregates with $s_{c,20°}$ values greater than the $s_{c,20°}$ value of BSA (4.2 S). In the case of the BSA/CGA(1:38) solution the two maxima at 7.3 and 9.1 S indicate the formation of similar aggregates.

Higher molecular "impurities" could not be detected by additional runs performed at 3,000 rpm prior to the high-speed sedimentation velocity investigations.

The UV–VIS absorption spectra also support the occurrence of interactions between BSA and CGA. In order to check this with UV–vis absorption measurements on a spectrophotometer the product of the CGA concentration in the BSA/CGA solutions, *c*, and the cuvette layer thickness, *d*, was kept constant. In this case the number of absorbing molecules in the measured samples would not change and the absorbance values would remain constant. A change of the absorption properties of the BSA/CGA solution mixtures depending on the CGA concentration owing to interaction between BSA and CGA results in optical density (OD) values that are no longer constant.

The results in Fig. 3 of BSA/CGA solution mixtures with BSA-to-CGA molar ratios of 1:9.5, 1:19 and 1:95 illustrate a clear dependence on the CGA concentration. The highest OD values are shown by the BSA/CGA solution mixture with the lowest CGA concentration of 0.025 g/l. With increasing CGA concentration up to tenfold, the OD values decrease. This is due to the 5–8 times higher absorption level of CGA in comparison to BSA and the loss of "free" CGA by the binding to BSA through which the absorption properties of the solutions have changed. Pure CGA solutions show the opposite behaviour: increasing OD values with increasing CGA concentration. Similar UV–VIS absorption spectra were also obtained by wavelength scans recorded on the AUC at 3,000 rpm.

An intrinsic fluorescence study was performed to evaluate changes in tertiary structure caused by interactions of BSA with CGA. The maximum emission wavelength for unmodified BSA was 346 nm (Fig. 4, curve 1). The tryptophan residues are rather exposed on the protein surface, since its initial maximum of fluores-

Fig. 3 UV–vis absorption spectra of BSA/CGA solution mixtures measured at the same value of CGA concentration times cuvette layer thickness

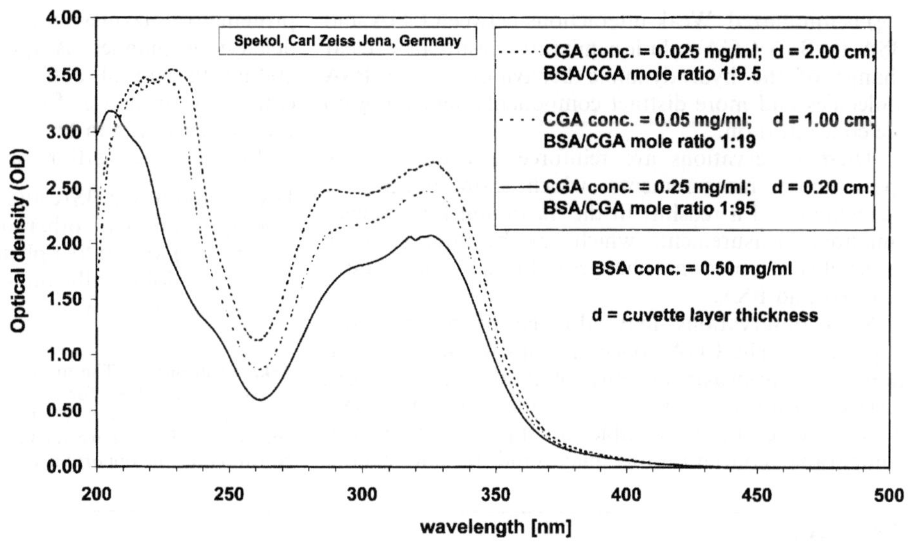

Fig. 4 Tryptophan fluorescence scans of BSA/CGA solution mixtures depending on the CGA concentration

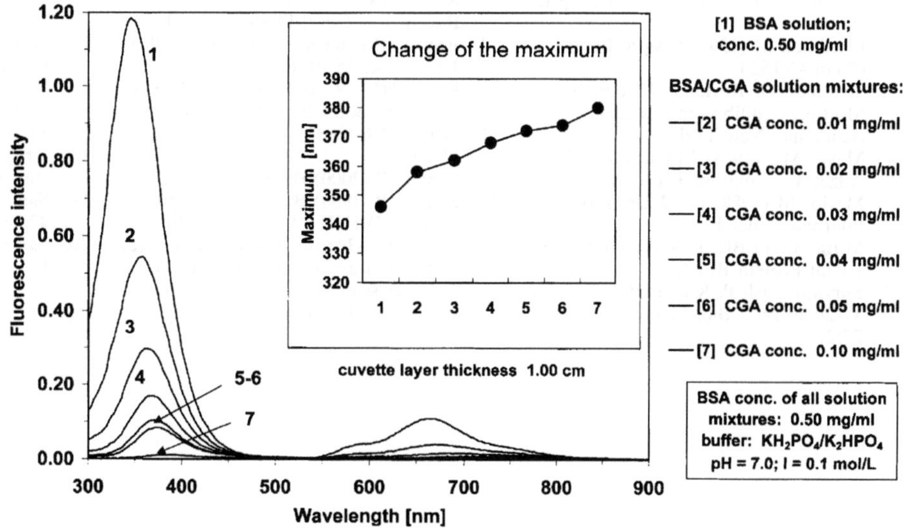

cence emission appears at a wavelength close to that of exposed tryptophan model systems. However, buried tryptophan groups in close proximity to polar residues may be capable of forming an excited-state complex (exciplex), resulting in emission spectra characteristic of exposed residues with peaks close to 350 nm. Both tryptophan residues (Trp3, Trp158) present in the sequence of BSA are in close proximity to lysine residues, and first one (Trp3) is located very near the N-terminal. This resulting high exposure can be confirmed by the tryptophan quenching, where a low BSA-to-CGA ratio (Fig. 4, curve 2) caused a corresponding strong decrease of tryptophan fluorescence. The maximum emission also decreased in intensity as a function of the CGA concentration and a redshift of the maximum emission was observed (Fig. 4). From the progressive quenching and the redshift observed in the maximum fluorescence emission of BSA, it can be inferred that conformational changes are induced by interactions of CGA and BSA.

Conclusions

Sedimentation velocity in the AUC has been shown to be a useful tool for the detection and assay of the "weak" interactions in BSA/CGA solutions, particularly when it is shown that under the chosen experimental conditions the formation of BSA/CGA derivatives (covalent bonding) is minimized. Weak interactions between CGA and BSA in BSA/CGA solution mixtures result in a clear change of the hydrodynamic behaviour of the BSA molecules and more distinct components depending on the concentration.

These observations are reinforced by fluorescence spectroscopy measurements, which show progressive quenching and a redshift in the maximum fluorescence emission, measurements which can be interpreted in terms of conformational changes induced by interactions of CGA and BSA.

Such observations may also be of physiological importance. The CGA concentrations in our investigations may approximate physiological concentrations resulting from coffee intake: if the interaction with BSA is nonspecific, it is reasonable to suppose that CGA at comparable concentrations may bind to endogenous proteins or enzymes and influence their functions in a deleterious manner. A quantity of 500 mg CGA per 200 ml (a cup of coffee) is equivalent to 2.5 g/l, a concentration that is, for example, 25 times higher than the concentration in the BSA/CGA(1:38) solution.

Further investigations should consider

Higher and lower BSA and CGA concentrations.
The application of other proteins (e.g., human serum albumin) and native phenolic compounds.
Other physiologically important pH values, including the type of buffer.

Acknowledgements The authors thank Arthur Rowe, University of Nottingham, UK, for critical reading and helpful discussion. The technical support of R. Kröck (Bergholz-Rehbrücke) is appreciated. Further we thank the German Institute for Human Nutrition for the opportunity to use the Optima XL-A AUC.

References

1. Clifford MN (1999) J Sci Food Agric 79:362
2. Friedman M (1997) J Agric Food Chem 45:1523
3. Steinhardt J, Reynolds JA (1969) Multiple equilibria in proteins. Academic, New York
4. Meyer MC (1968) J Pharm Sci 57:895
5. Adzet T, Camarasa J, Escubedo E, Merlos M (1988) Eur J Drug Metab Pharmacokinet 13:11
6. Muralidhara BK, Prakash V (1995) Int J Pept Protein Res 46:1
7. Suryaprakash P, Kumar RP, Prakash V (2000) Int J Biol Macromolecules 27:219
8. Rawel HM, Rohn S, Kroll J. (2000a) Recent Res Devel Phytochem 4:115
9. Rawel HM, Kroll J, Riese B (2000b), J Food Sci 65:1091
10. Rawel HM, Rohn S, Kruse H-P, Kroll J (2002) Food Chemistry 78:443
11. Green, AA (1933) J Am Chem Soc 55:2331
12. Schuck P (2000) Biophys J 78:1606
13. Schuck P (1998) Biophys J 75:1503
14. Stafford WF III (1992) In: Harding SE, Rowe AJ, Horton JC (eds) Analytical ultracentrifugation in biochemistry and polymer science. Royal Society of Chemistry, London, pp 359–393
15. Phillips DL (1962) J Assoc Comput Mach 9:84
16. Lowry OH, Rosebrough NJ, Farr AL, Randall J (1951) J Biol Chem 193:265
17. Markwell MAK, Haas SM, Bieber LL, Tolbert NE (1978) Anal Biochem 87:206
18. Kragh-Hansen U (1981) Pharmacol Rev 33:17
19. Siebert KJ (1999) J Agric Food Chem 47:353
20. Siebert KJ, Troukhanova NV, Lynn PY (1996) J Agric Food Chem 44: 80

Progr Colloid Polym Sci (2004) 127: 89–97
DOI 10.1007/b98018

R. Ludwig
T.K.D. Nguyen
H. Higuchi
J. Henniger

Modeling the transport of low molecular weight metal complexes across the liquid–liquid interface during extraction in a synthetic boundary cell by using the finite differences method

R. Ludwig (✉) · T.K.D. Nguyen
H. Higuchi
Institute of Chemistry/Inorganic and Analytical Chemistry, Freie Universität Berlin, Fabeckstrassee 34–36, 14195 Berlin, Germany
e-mail: rludwig@mail.chemie.fu-berlin.de
Tel.: +49-30-83852458
Fax: +49-30-83852424

J. Henniger
Institute of Nuclear and Particle Physics, Department of Radiation, Protection Physics, Technische Universität Dresden 01062 Dresden, Germany

Current address: H. Higuchi, Department of Applied Chemistry, Saga University, Saga 840-8502, Japan

Abstract We determine transport parameters such as the diffusion coefficient, D, the integral mass flux and the transfer coefficient across the liquid–liquid interface, α, in solvent extraction systems by using analytical ultracentrifugation, emphasizing extraction systems with chemical reaction at the liquid–liquid interface. If this reaction is fast, the transport process is diffusion-limited. The experimental data of dichromate extraction by tertiary amine represent such diffusion-limited transport. They can be described according to the second transport law with the parameters D, c, α and K_D by applying the model of Scott and Drickamer of diffusion-limited transport. However, by solving the second transport law by Laplace transformation as previously, we cannot take into account any chemical reaction kinetics, which plays an important role in many other systems, such as the extraction of Re(VII), which is also reported here. In order to be able to describe the transport across interfaces where chemical reactions take place, we developed a numeric solution based on finite differences.

Keywords Ultracentrifugation · Transport kinetics · Diffusion · Extraction · Finite differences method

Introduction

Mass transport in liquid–liquid extraction is governed by both thermodynamic and kinetic parameters. Both are equally relevant for understanding the extraction mechanism and for the layout of extraction devices. For determining kinetic parameters such as diffusion coefficients, D, in the aqueous and in the organic phase, the transport rate across the interface, α, the mass flux, J, and the integral mass flux, I, various techniques can be applied [1]. Examples are the stirred cell [2], single-drop methods [3, 4], a rotating disk [5] as well as analytical ultracentrifugation [6, 7]. The latter method is a direct one, yielding the parameters without extrapolation, and has some advantages over the other ones, namely, (1) a one-dimensional flux, which can be easily described by mathematic models, (2) an unstirred interface with definite area, owing to the centrifugal force, (3) the time of phase contact (overlay) is known exactly, (4) very small volumes, and (5) an optical system which avoids any interference with the extraction process.

Previously, we investigated diffusion and solvent extraction of Tc(VII) in the absence and presence of complexing cations such as U(VI) [8, 9]. We determined the concentration and its gradient at the interface from the refractive index gradients and compared these experimental values with calculated ones obtained by assuming a certain transfer coefficient, α. It turned out that the extraction process can be described well by the model of Scott, Tung, and Drickamer [10, 11, 12], if the transport across the interface is diffusion-limited. This model assumes diffusion in both phases and transfer across an interface of infinite thickness.

Table 1 Composition of the dichromate diffusion and extraction systems. Concentration c is given as the concentration where both phases meet (which is half of the initial one according to the error function). Aliquat 336 is a quaternary amine used in solvent extraction of metals

Dichromate system			
Diffusion, aqueous $Cr_2O_7^{2-}$		Diffusion, organic $[Cr_2O_7^{2-}(NRR_3'^{+})_2]$	
Metal ion	Matrix	Metal complex	Matrix
$c = 0.0096$ M ($c_{init} = 0.0192$ M); $c = 0.0024$ M ($c_{init} = 0.0048$ M)	0.2 M HCl	$c = 0.0096$ M	Toluene, 10 vol% Aliquat 336, 10 vol% *n*-decanol (equil. 0.2 M HCl)
Extraction			
Aqueous		Organic	
$c_{init} = 0.0192$ M $K_2Cr_2O_7$ in 0.2 M HCl		Toluene, 10 vol% Aliquat 336, 10 vol% *n*-decanol (equil. 0.2 M HCl)	

However, in some of the systems investigated the "best-fit" α values at the aqueous and organic sides of the interface were different from each other. We interpreted the latter systems as reaction-limited, because the model assumes only one definite α across the interface. Improving this model by considering interfacial reaction [13, 14] is one aim of the ongoing work.

In the original model, diffusion and transport equations were solved by Laplace transformation (LT). One of the advantages was the reasonably fast calculation speed. But it was not possible to take into account changing D and α values during an extraction run in cases where D is concentration-dependent or α changes owing to rate-limiting chemical reactions at the interface.

This motivated us to solve the transport equations of the Drickamer model numerically by the method of finite differences (FDM), which is closely related to the method of finite elements (FEM) [15, 16]. Although FEM is commonly applied to heat transport, a few examples regarding mass transport have been published [17, 18]. Therefore this paper also reports the successful application of the FDM to the one-dimensional mass transport during solvent extraction, together with kinetic and thermodynamic data on the extraction of Cr(VI) and Re(VII). These extraction systems were chosen in order to investigate in subsequent studies how cations which can form ion pairs or complexes with those ions in the aqueous phase influence their transport.

Experimental

Equipment

We used a model E analytical ultracentrifuge from Beckman Instruments, equipped with a UV–vis scanner and a Schlieren optical system. The latter was used for the experiments reported here and monitors changes in the refractive index gradient. Improvements from the original features are an analogue charge-coupled-device camera, a video recorder and printer, a digital caliper, an A4-size graphic tablet (Wacom), a program for calculating the peak area (Flexitrace) and a program for data treatment (Deltagraph).

The measuring cell contained a 1-cm thick carbon/epoxy double-sector synthetic boundary centerpiece with a manually deepened central capillary (V-shape) and the usual other components. A modified torch wrench was used to tighten the sapphire windows to the centerpiece without gaskets. The rotation speed was 8,000 rpm, which allows fast overlay through the central capillary but avoids sedimentation (maximum molecular weight of the complex is around 1,040 for Cr and 770 for Re). In aqueous and organic diffusion experiments, the metal-containing phase (120 μl) was overlaid with an equal volume of the Cr(VI)-free or the Re(VII)-free phase. Except for the diffusing metal ions or complexes, all other components were adjusted to equal concentrations with respect to both phases in order to avoid wrong refractive indices, including adjustment of the acid concentration by pre-equilibration. As we investigated multicomponent systems, the resulting diffusion coefficients are apparent ones. In extraction runs, the aqueous phase was overlaid with the organic phase having a composition as described in Tables 1, 2, and 3. Here the concentrations of water, acid, and in the Re system 2 also U(VI) were adjusted to equilibrium values between both phases prior to the experiment in order to see the refractive index gradient only from Cr or Re. Especially in extraction runs, the overlay had to be fast if extrapolation to the initial mass flux was desired, because the concentrations change rapidly at beginning. The temperature in the chamber was set to 298 ± 1 K.

Solutions

The compositions of the phases in the diffusion and extraction runs are given in Tables 1, 2, and 3. Aqueous solutions of metals were prepared by dissolving the corresponding salts in stock solutions and diluting them to the indicated concentrations with diluted acid. The aqueous metal concentrations were confirmed by inductively coupled plasma atomic emission spectroscopy at different emission wavelengths.

The organic phase before extraction contained the diluent, the extractant, a modifier, if indicated, to promote dissolution, and a second metal complex, if indicated, and an acid. The acid was introduced by equilibrating three times with aqueous acid having the initial concentration stated in the following. Thus, transport of any other component except the metal salt or complex was avoided, because they were already at equilibrium concentration. This is critical as we used the Schlieren optical system. The extractant concentration was chosen to be in sufficient excess over the metal ion concentration.

In separate experiments, the aqueous solutions were contacted batchwise with equal volumes of the corresponding organic phase

Table 2 Composition of the rhenium(VII) diffusion and extraction system 1

Rhenium(VII) system 1			
Diffusion, aqueous ReO_4^-		Diffusion, organic $[ReO_4^- NHR_3^+]$	
Metal	Matrix	Metal	Matrix
$c = 0.02$ M ($c_{init} = 0.04$ M)	3.0 M HNO_3	$c = 0.00786$ M ($c_{init} = 0.0157$ M)	Xylene, 10 vol% tridodecylamine, (equil. 3 M HNO_3)
Extraction			
Aqueous		Organic	
$c_{init} = 0.04$ M $KReO_4$ in 3 M HNO_3		Xylene, 10 vol% tridodecylamine, (equil. 3 M HNO_3)	

Table 3 Composition of the rhenium(VII) diffusion and extraction system 2

Rhenium(VII) system 2			
Diffusion, aqueous ReO_4^-		Diffusion, organic $[ReO_4^- NHR_3^+]$	
Metal	Matrix	Metal	Matrix
$c = 0.02$ M ($c_{init} = 0.02$ M)	3.0 M HNO_3, 0.08165 M U(VI)	$c = 0.01025$ M ($c_{init} = 0.0205$ M)	Xylene, 10 vol% tridodecylamine, 0.01732 M U(VI), (equil. 3 M HNO_3)
Extraction			
Aqueous		Organic	
$c_{init} = 0.05$ M $KReO_4$ in 3 M HNO_3 containing 0.08165 M U(VI)		Xylene, 10 vol% tridodecylamine, 0.01732 M complexed U(VI), (equil. 3 M HNO_3)	

Table 4 Measured diffusion coefficients and distribution ratios

System	$Cr_2O_7^{2-}$	ReO_4^- system 1	ReO_4^- system 2
$D_{aqueous}$ (cm^2s^{-1})	2.136×10^{-5} (0.0024 M) 1.79×10^{-5} (0.0096 M)	1.46×10^{-5} (0.02 M)	1.53×10^{-5} (0.025 M)
$D_{organic}$ (cm^2s^{-1})	1.76×10^{-6} (0.0096 M)	5.4×10^{-6} (0.00786 M)	6.05×10^{-6} (0.01025 M)
K_D	100 ± 15	0.667	0.603

by using Teflon-sealed 10-ml glass vials with shaking, followed by phase separation and metal content analysis of the aqueous phase. The metal content in the organic phase was calculated from the mass balance. The distribution ratios thus obtained are given in Table 4. These metal-loaded organic solutions were used for the diffusion runs as a source phase. The organic solutions containing Re were stored at 280 K. Those containing Cr were freshly prepared prior to use owing to slow reduction of Cr(VI). In the organic diffusion runs, the composition of the overlaying phase was corrected for the amount of extractant consumed owing to complexation in the source phase. Different from our previous study on tri-*n*-butyl phosphate complexes, the pure amine–metal complexes could not be isolated; therefore, the samples were prepared by the method described earlier. The experiments were repeated at least twice.

Small errors are introduced by the use of the metal salt. In the dichromate system, diffusion of potassium as counterions causes a slightly higher refractive index gradient, but its refractive increment is small compared with dichromate. In the case of perrhenate, its counterion is H^+ owing to partial protonation; its refractive increment is negligible.

Data treatment

As described previously [8, 9], the apparent diffusion coefficients, D, were determined from the slope of the plot according to Eq. (1), which was derived from the second law of transport:

$$\left(\frac{A_t}{H_{max}}\right)^2 = 4\pi(f_1 f_2)^2 Dt + b. \quad (1)$$

In this equation, H_{max} and A_t are Schlieren peak height and area respectively, f are optical amplification factors, t is the time after overlay, and b is an integration constant which contains the time difference between overlay and recorded start time. The reproducibility of the experiments was confirmed by comparing

A_t, which is the same for a given concentration of diffusing compound.

For the extraction runs, A_t and D, which were measured in diffusion runs before, are used to calculate the concentration gradient according to Eq. (2):

$$\left(\frac{dc}{dx}\right)_{x=0} = -\frac{c_{init}}{2\sqrt{\pi Dt}}. \tag{2}$$

Plotting the right term of Eq. (2) versus H_{max} of the diffusion run at the corresponding time gives a correlation which was applied to the extraction runs at comparable concentrations. A typical plot is shown in Fig. 1. In extraction runs, t was extrapolated from the plot of meniscus distance versus time.

In order to describe the mass transport, the concentration gradients determined during extraction runs at both sides of the interface were compared with theoretical values calculated according to a model for one-dimensional mass transport. Among the published models, such as by Chandrasekhar and Hoelscher [19] and by Scott, Tung, and Drickamer [10, 11, 12], we applied the latter one. It assumes the following mathematic boundary conditions (see also Fig. 2):

- Diffusion in the aqueous phase ($-a < x < 0$) and in the organic phase ($0 < x < a$) according to Eq. (3):

$$\left(\frac{\partial c}{\partial t}\right)_x = D\left(\frac{\partial^2 c}{\partial x^2}\right)_t. \tag{3}$$

- Diffusion across an interface of zero thickness; the hindrance to transport is expressed by the transfer coefficient α in Eq. (4), while the equilibrium is characterized by the distribution ratio K_D. The latter is defined as c_{org}/c_{aq}.

$$\vec{J}\left(c_{(x=-0,t)}, c_{(x=+0,t)}\right) = \alpha\left(c_{(x=-0,t)} - \frac{1}{K_D}c_{(x=+0,t)}\right) \tag{4}$$

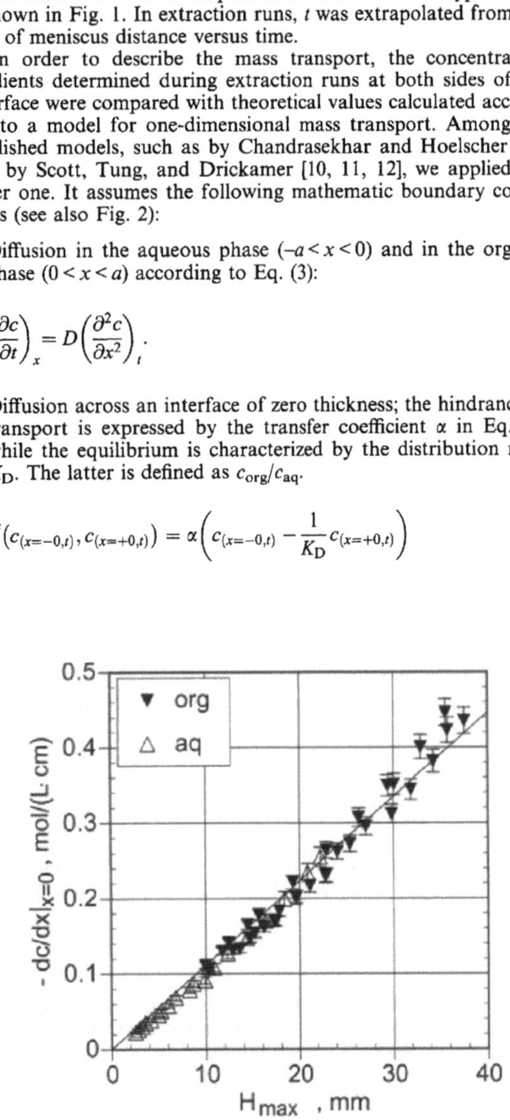

Fig. 1 Correlation between concentration gradient and Schlieren peak height at the synthetic boundary in diffusion runs. Aqueous system: 9.6 mM $Cr_2O_7^{2-}$ ($0.5c_{init}$) in 0.2 M HCl. Organic system: 9.6 mM Cr(VI) complex with organic ammonium ion in toluene containing 10% decanol and 10% extractant

The continuity at the interface is expressed by Eq. (5):

$$-D_1\left(\frac{\partial c}{\partial x}\right)_{x=-0,t} = -D_2\left(\frac{\partial c}{\partial x}\right)_{x=+0,t} = \vec{J}. \tag{5}$$

The concentration-dependency of D, especially in aqueous solutions, is not yet considered in the modeling ($D_{organic}$ is almost concentration-independent). Owing to the high K_D in the dichromate system, however, the concentration changes significantly. Owing to the fast extraction kinetics, we used $D_{aqueous}$ at the lowest concentration measurable by Schlieren optics (2.4 mM, Table 4) for subsequent modeling.

The solution of the differential equation system by means of LT was given in the literature [10]; a computer program described previously [20] applies this mathematic treatment. An improved solution is explained in the following section.

Results and discussion

The diffusion coefficients and distribution ratios measured in the systems investigated are given in Table 4; the concentration of the diffusing ion or complex ($0.5c_{init}$) is given in parentheses.

We interpret the different D values in terms of the different matrix compositions.

Dichromate extraction system

In the dichromate system, the extraction is described by the following equation, where an overbar denotes species in the organic phase and R and R' are the aliphatic substituents of the amine:

$$Cr_2O_7^{2-} + 2\left[\overline{NRR_3'^{+} \cdot Cl^-}\right] \rightleftharpoons \overline{Cr_2O_7^{2-} \cdot 2NRR_3'^{+}} + 2Cl^-$$

We observe Marangoni convection at the organic side of the interface for about 30 min after overlay, visible as a gray area due to turbulent flow. The thickness of this zone is 0.44 mm at beginning. As that phenomenon disappears at low metal concentration and prolonged time, we conclude it is caused by the

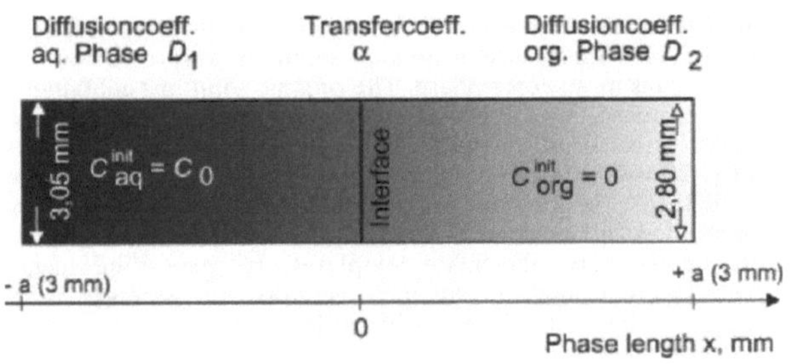

Fig. 2 Model of diffusion-limited transport according to Drickamer et al.

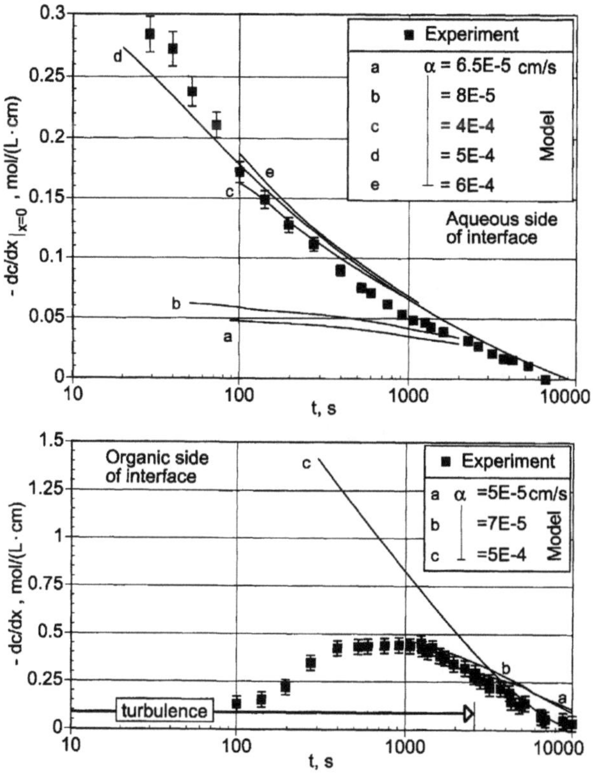

Fig. 3 Concentration gradient at both sides of the interface in the dichromate extraction system. For compositions see Table 1. Experimental data as well as the calculated data under variation of α. Note the turbulent zone in the organic phase which cannot be modeled

high flux into the organic side accompanied by an exothermic complexation by the extractant and low diffusion rates, rather than by interfacial tension. The concentration gradients at both sides of the interface in the dichromate system, determined from the Schlieren peak height as described in the Experimental section, are depicted in Fig. 3. During the turbulence at the organic side, the interface is invisible from there, but for reasons of comparison we included the observed concentration gradient at the sharp interface between the turbulent zone and the diffusion zone. The turbulent zone decreased in thickness and finally disappeared.

Rhenium(VII) extraction

In order to avoid turbulent flux we also investigated extraction systems with perrhenate, where K_D is significantly smaller. It extends our previous studies on pertechnetate. The following reactions of complexation, protonation and association describe the extraction process:

$$ReO_4^- + \overline{NHR_3^+ \cdot NO_3^-} \rightleftharpoons \overline{ReO_4^- \cdot NHR_3^+} + NO_3^-,$$

$$ReO_4^- + H^+ \rightleftharpoons HReO_4,$$

$$n ReO_4^- + UO_2^{2+} \rightleftharpoons [UO_2 \cdot (ReO_4)_n]^{2-n}.$$

The concentration gradient data at the interphase of both systems are combined in Fig. 4 with the compositions as stated in Tables 2 and 3. In system 1 the concentration gradients describe sharp maxima shortly after overlay. These maxima were not observed when U(VI) as a cation was added to the aqueous phase as in system 2; at $t < 13$ s the Schlieren curve was not resolved. The UO_2^{2+} in this system associates with Re(VII), thus competing with the extractant at the interface. However, we observe improved extraction of Re(VII), represented as a higher gradient in the presence of U(VI), which is only partly attributed to the higher initial Re(VII) concentration. Instead, the interfacial reaction dominates the transport rate as seen from the attempted modeling (vide infra).

The mass flux density, $\vec{J}$, and the integral mass flux, I, are accessible from the interfacial concentration gradient according to Eqs. (6) and (7):

$$\vec{J} = -D\left(\frac{dc}{dx}\right)_{x=0}, \tag{6}$$

$$\sum_{t=0}^{t_{max}} I = \sum_{t=0}^{t_{max}} \vec{J} dt. \tag{7}$$

$\vec{J}$ and I should be similar for the aqueous and the organic side of the interface in the absence of any surface-active compounds; however, slight differences are observed because very small concentrations (e.g. below 3 mM metal ion) cannot be detected reliably with the analytical system owing to optical distortions and coincidence with the baseline. Therefore it is desirable to describe the transport by a model and to compare modeled and experimental data at concentrations which can be measured reliably as described in the next subsection. The vector arrow will no longer be drawn as the transport is one-dimensional.

Modeling of the transport

The partial differential equations of the Drickamer model for mass transport were solved numerically for the following reasons:

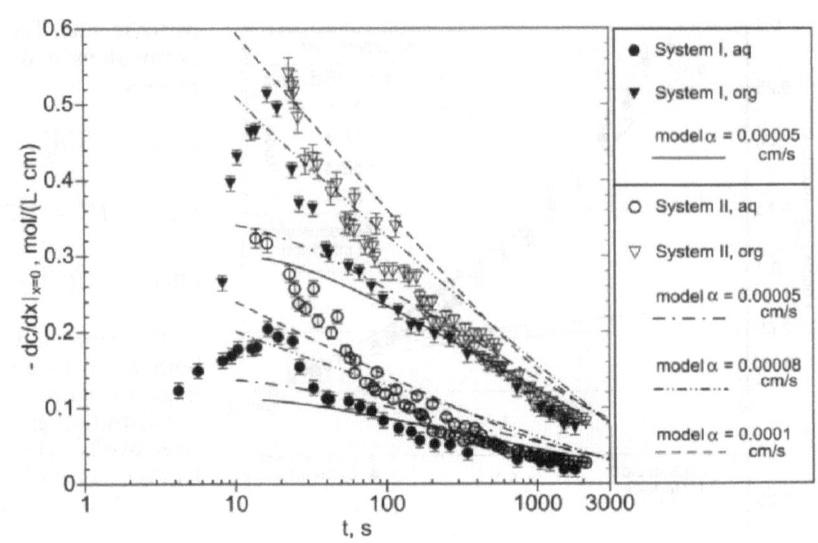

Fig. 4 Interfacial concentration gradient in the Re(VII) extraction systems 1 (*closed symbols*) and 2 (*open symbols*). Phase compositions are given in Tables 2 and 3. The *lines* represent the modeling with the α values indicated

1. The model with LT does not allow us to introduce a term for the interfacial reaction kinetics instead of α (which would mean α is not constant during the run).
2. The D values are concentration-dependent, but LT calculates with fixed values.
3. The cell is slightly conical, but LT assumes parallel side walls.
4. LT assumes infinite length in both directions starting at $x = 0$.
5. The solutions of the LT oscillate under certain conditions (high K_D, insufficient number of poles, short t).

Although in this paper we not address issues 1–3, the aim was to prove the FDM model can be applied to mass transport and that it gives correct results.

The numerical solution by the FDM in this one-dimensional transport considers each phase as consisting of many thin slices, so-called elements, as shown schematically in Fig. 6. Between neighboring elements i and $i+1$, the concentration changes almost linearly; the differential quotient for the gradient, G, can be approximated by that of the differences as expressed by Eq. (8), where x is the element distance (or thickness):

$$G_{(i \to i+1)} \approx \frac{[c(x_i) - c(x_{i+1})]}{\Delta x}. \tag{8}$$

We found 1,000 elements for each phase of 3-mm length to be a suitable number. In the source phase, the mass is transported by diffusion up to the interphase. J can be standardized to the (almost) constant area, A, between neighboring elements and is expressed by Eq. (9):

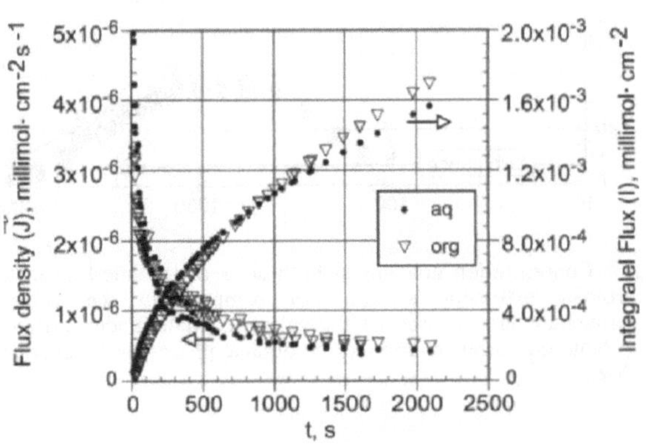

Fig. 5 Flux density and integral mass flux in the Re extraction system II

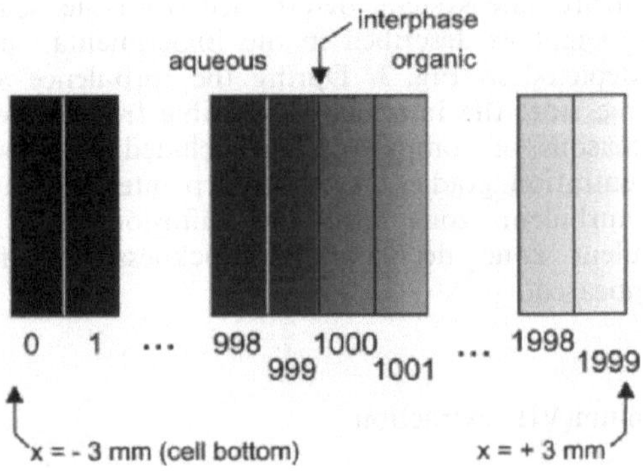

Fig. 6 A centrifuge sector during an aqueous–organic extraction run split into i finite elements for the purpose of finite differences calculations

$$J_A(i \to i+1) = \frac{J_{(i \to i+1)}}{dA} \approx D_1 G_{(i \to i+1)} \Delta t. \quad (9)$$

The time difference, Δt, is small, so the mass diffuses from element i to $n(i+1)$ but not yet to $i+2$ (stability criterion). The upper limit for Δt can be derived from Δx and D.

The mass balance in one phase is described by Eq. (10):

$$c_{i+1} = c_i - \frac{J_A}{\Delta x}. \quad (10)$$

The first element (cell bottom) is characterized by diffusion out of the element but no incoming mass. As an example for the programming (C+ +), Eq. (11) shows how the calculation is realized for this first element at increasing time ($t+ = \Delta t$):

$$c_{i=0}+ = \frac{D_1(c_{i=1} - c_{i=0})\Delta t}{\Delta x^2}. \quad (11)$$

The next aqueous 998 elements have input and output of mass by diffusion from their neighbors. The last aqueous element, which borders the interface, receives input by diffusion. If the purpose of the model were the description of only diffusion without an aqueous–organic interphase, then the calculation would continue as before. In liquid–liquid extraction, however, mass is leaving the last element by transport across the interface as expressed by Eq. (4). At the organic side, the first element receives mass according to Eq. (4) and releases it by diffusion, characterized by D_2. The subsequent organic elements are described by input and output due to diffusion up to the final one, which has no output.

Fig. 7 Concentration and concentration gradient profiles for the aqueous (*left*) and the organic phase (*right*) at distance x from the interphase, calculated by using $\alpha = 5 \times 10^{-4}$ cm s^{-1} for the dichromate extraction system (Table 1). The observed turbulent flow at the organic side is not considered in the calculation. *Arrows* point in the direction of increasing time after overlay

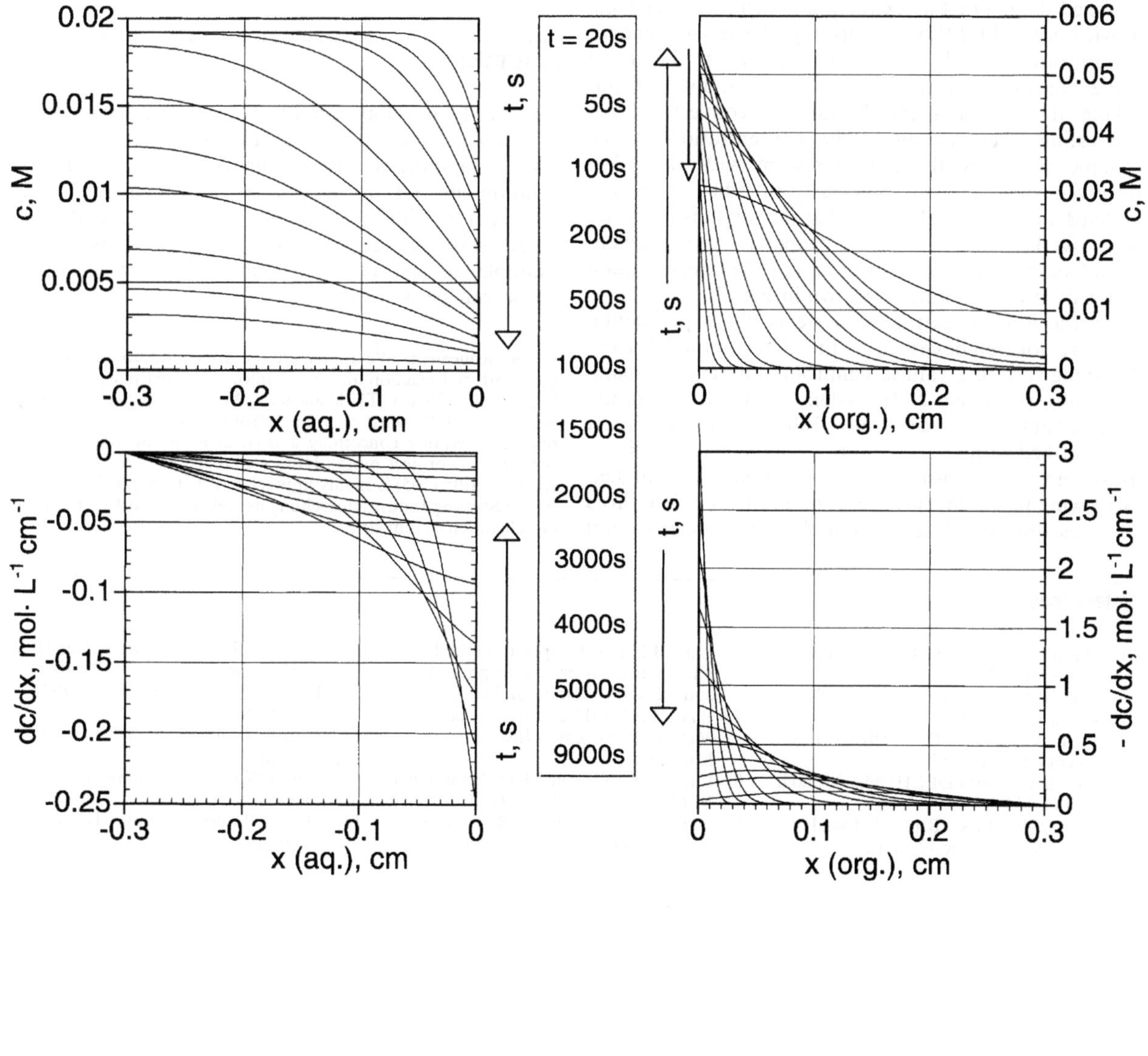

Selected results of the two calculation loops (time and distance) are stored in order to view the concentration and concentration gradient profiles at various times. The calculations are repeated with various α values until the experimental and modeled data agree with each other. Such variations are included in Fig. 3. The program requires about 5 min on a 2 GHz personal computer with a precision where the error as determined from the difference of aqueous phase output and organic input is well below 1%. We confirmed with identical parameters and boundary conditions that the same results are obtained from the LT and the new model calculation.

Comparison between experimental and modeled data

The best α value for the chromate extraction system is 5×10^{-4} cm s^{-1}. The small deviation at the aqueous side is interpreted as being caused by the turbulent flow at the organic side, which accelerates the output from the aqueous side. After the end of the turbulent flow, the organic side also shows good agreement with the modeled data. For this extraction system, a QuickTime movie is available [21] which displays the time-dependent concentration profile in the centrifuge cell as calculated using our program and the optimized α, but without taking into account the Marangoni convection. An interesting phenomenon is the concentration at the organic side, describing a maximum with time. We observed such maxima previously in other extraction systems and the model proves that the origin is not a too slow overlay but an inherent property. A two-dimensional view of the concentration and the concentration gradient at certain times is shown in Fig. 7.

In the Re(VII) extraction systems, the experimental data are not well described by the diffusion model. Even by trying various α values, significant differences remain, especially in system 2. In system 1, the concentration gradient at the organic side after 100 s can approximately be described with $\alpha = 5\times10^{-5}$ cm s^{-1} as depicted in Fig. 4. In system 2, the transfer coefficient appears to be higher, but owing to the chemical reaction at the interface this α changes during extraction and thus the model cannot determine it exactly. A similar discrepancy was observed for backextraction of Re(VII) from organic into aqueous solutions (data not shown). Therefore, incorporating the interfacial reaction into the model is one of our next goals. The range of α in the Re(VII) extraction is about 1 order of magnitude lower compared with the dichromate extraction, which underlines that an interfacial reaction limits the transport rate, at least at the beginning of the extraction process.

If the experimental data are plotted according to Crank, such as Ref. [19] (not shown), a linear relationship is found for the interfacial concentration gradient at the aqueous side when plotted versus $1/\sqrt{t}$. At the organic side the plot is nonlinear assuming either negligible or significant interfacial resistance, in the latter case by plotting the interfacial concentration gradient versus $\sqrt{t}$. We conclude therefrom that the complexation/decomplexation at the organic side or surface-active properties of the complex decrease the extraction rate. In the dichromate system, the plot according to Crank confirms the negligible interfacial resistance.

Conclusion

The newly developed mathematical treatment of the one-dimensional extraction process by the FDM represents significant improvements over the previous solution by LT in that it is an open model not restricted to diffusion-limited transport and allows a much wider range of boundary conditions. As an example, the extraction of dichromate into an organic phase can be well described.

Acknowledgements We appreciate the discussion about the role of the interfacial reaction and on coupled transport phenomena with Prof. Dr. W. Nitsch (TU München). We also highly appreciate the help of Dr. B. Plewinsky (on leave from BAM Berlin) in providing relevant literature concerning analytical ultracentrifugation. Support for this project by the DFG (reference no. Lu 573/5-1 + 2 in terms of SPP 1105) is highly appreciated. R.L. appreciates a grant by the Stiftung Industrieforschung as well as support by the Fonds der chemischen Industrie.

References

1. Hanna GJ, Noble RD (1985) Chem Rev 85:583–598
2. Nitsch W (1979) Ber Bunsenges Phys Chem 83:1171–1177
3. Licht W, Conway JB (1950) Ind Eng Chem 42:1151–1157
4. Nitsch W (1965) DECHEMA Monogr 55:143–153
5. Albery WJ, Fisk PR (1984) Faraday Discuss Chem Soc 77:53–65
6. Lloyd PH (1974) Optical methods in ultracentrifugation, electrophoresis, and diffusion. Clarendon Press, Oxford
7. Elias HG (1961) Ultrazentrifugen-Methoden. Beckman Instruments, Munich
8. Ludwig R, Nguyen TKD, Nabeshima M (2001) Radiochim Acta 89:169–177
9. Nabeshima M, Ludwig R, Nguyen TKD, Marx G (2000) Solvent Extr Res Dev Jpn 7:85–92
10. Scott EJ, Tung LH, Drickamer HGJ (1951) J Chem Phys 19:1075–1078
11. Tung LH, Drickamer HGJ (1952) J Chem Phys 20:6–9
12. Tung LH, Drickamer HGJ (1952) J Chem Phys 20:10–12
13. Nitsch W (1984) Faraday Discuss Chem Soc 77:85–96
14. Meares P (1984) Faraday Discuss Chem Soc 77:7–16

15. Marsal D (1989) Finite Differenzen und Elemente. Springer, Berlin Heidelberg New York
16. Faires JD, Burdon RL (2000) Numerische Methoden. Spektrum, Heidelberg
17. Chung J (1983) Finite Elemente in der Strömungsmechanik. Hanser, Munich
18. (a) Casulli V, Cheng RT Program TRIM-2D. University of Trient and US Geological Survey, Menlo Park; (b) Danish Hydraulic Institute Program MIKE11. The Federal Waterways Engineering and Research Institute Hamburg, www.hamburg.baw.de
19. Chandrasekhar S, Hoelscher HE (1975) AIChE J 21:103–109, and references therein
20. Gauglitz R (1995) Prog Colloid Polym Sci 99:199–208
21. http://link.springer.de/link/service/series/2882/

Progr Colloid Polym Sci (2004) 127: 98–112
DOI 10.1007/b98019

H. Durchschlag
P. Zipper

Modeling the hydration of proteins at different pH values

H. Durchschlag (✉)
Institute of Biophysics and Physical Biochemistry, University of Regensburg, Universitätsstrasse 31, 93040 Regensburg, Germany
e-mail: helmut.durchschlag@biologie.uni-regensburg.de
Tel.: +49-941-9433041
Fax: +49-941-9432813

P. Zipper
Physical Chemistry, Institute of Chemistry, University of Graz, Heinrichstrasse 28, 8010 Graz, Austria

Abstract The atomic coordinates of proteins, for example, of the enzyme lysozyme as a small model protein, can be used for calculating analytically the exact protein surface in terms of dot surface points and for visualizing both the atoms and the surface points. On the basis of known hydration numbers for the constituent amino acid (AA) residues as suggested by Kuntz and Hopfinger and applying site-specific hydration algorithms implemented in the program HYDCRYST, a modeling of individual water molecules placed at preferred sites on the protein surface can be performed. For this approach, both the bound water molecules and the atoms of the AA residues can be shaped by spheres of appropriate size ("bead modeling"). This also allows the prediction of structural and hydrodynamic molecular parameters by applying any kind of scattering and/or hydrodynamic modeling, in the latter case preferably by the program HYDRO. Use of distinct hydration numbers for different pH ranges allows us to simulate the hydration behavior for changing conditions. The hydration varies in different pH ranges, owing to the variable extent of ionization of the individual AA residues. The transition from neutral to acidic pH values leads to diminished values for the hydration, accompanied by a decrease in the radius of gyration and enhanced values for sedimentation and diffusion coefficients. The comparison of the predicted values for hydration and molecular parameters with observed data is satisfactory within the accuracy of the available experimental data. The prediction of the behavior at alkaline pH values is impeded by insufficient knowledge of the hydration numbers of the AA residues in the basic pH range and the unavailability of experimental reference data.

Keywords Lysozyme · Protein hydration · pH dependence · Advanced modeling · Hydrodynamics

Introduction

Proteins are naturally occurring polymers of high molar mass, generally consisting of 20 different types of amino acids (AAs) which are bonded in a linear chain. For the majority of proteins the arrangement of the AAs is sufficient to determine under physiological conditions the 3D structure of the folded molecule in its native state, thus representing the lowest energy configuration of the molecule [1, 2].

The interaction of a protein with the solvent environment is important for many properties, such as protein hydration and stability, protein–protein association, ligand binding, biological activity (e.g., catalytic activity

Table 1 Hydration numbers for amino acid (AA) residues in different pH ranges, as proposed by Kuntz [72] and Hopfinger [73]. For comparison also the intrinsic pK_a values of the ionizable groups found in proteins [2] are added. The AA residues are arranged alphabetically according to their three-letter code

AA residue		Ionizable group[a]		Hydration number (N_w per AA)					
Three-letter code	One-letter code	Group	pK_a [2]	Kuntz [72]			Hopfinger [73]		
				pH 4[b]	pH 6–8	pH 12[b]	pH 4[b]	pH 6–8	pH 12[b]
Ala	A				1.5			1.0	
Arg	R	δ-Guanido	12.0		3	3		2.3	2.3
Asn	N				2			2.2	
Asp	D	β-Carboxyl	3.9–4.0	2	6		3.8	6.5	
Cys	C	Thiol	9.0–9.5		1				
Gln	Q				2			2.1	
Glu	E	γ-Carboxyl	4.3–4.5	2	7.5		3.6	6.2	
Gly	G				1			1.1	
His	H	Imidazole	6.0–7.0		4			2.8	
Ile	I				1			0.8	
Leu	L				1			0.8	
Lys	K	ε-Amino	10.4–11.1		4.5	4.5		5.3	1.2
Met	M				1			0.7	
Phe	F				0			1.4	
Pro	P				3			0.9	
Ser	S				2			1.7	
Thr	T				2			1.5	
Trp	W				2			1.0	
Tyr	Y	Phenolic hydroxyl	10.0–10.3		3	7.5		2.1	
Val	V				1			0.9	

[a]The peptide backbone is chemically relatively inert: The pK_a values for the terminal amino and carboxyl groups vary over a wide range (6.8–8.0 for the α-amino group and 3.5–4.3 for the α-carboxyl group), depending on the nature of these two AA residues [2]
[b]Only values which differ from the values at neutral pH are given, unless explicitly cited by the author

of enzymes), and chelate effects. Hydration is one of the major driving forces for folding a polypeptide chain into a native protein conformation and for the occurrence of manifold reactions in the usual solvent water. Water stabilizes protein structures and allows motions of their components.

The protein backbone, however, is chemically rather inert, with the exception of the behavior of the terminal α-amino and α-carboxyl groups, which exhibit pK_a values of about 7.4 and 3.9, respectively [2]. Internal peptide bonds normally do not exchange hydrogen.

The chemical properties of the different AA side chains differ over a wide range. The existence of several ionizable groups in proteins (Table 1) is fundamental for our understanding of the protein behavior at different pH values of the surrounding medium. The carboxylic groups of the side chains of the acidic AA residues aspartic acid and glutamic acid have pK_a values of about 4; these groups are ionized under physiological conditions (in contrast to the corresponding amide residues asparagine and glutamine). Similarly, the amino group of the basic AA residue lysine with a pK_a value of about 11 is ionized to a large extent in a physiological environment, whereas the guanido group of the basic residue arginine with a pK_a value of about 12 causes nearly complete ionization under the conditions where proteins can usually maintain their native structure. The imidazole side chain of histidine residues possesses a pK_a value of about 7; it is a strong base at neutral pH, which, however, loses its nucleophilicity upon protonation. The hydroxyl group of the phenolic ring of tyrosine residues has a pK_a value of about 10 and is, in general, ionized at alkaline pH values. Finally, the thiol group of cysteine residues with a pK_a value of about 9 is ionized at slightly alkaline pH values.

Overall, the occurrence of ionizable groups observed in proteins allows the affiliated AA residues to participate in a multitude of reactions, thereby modulating these reactions depending on the pH value of the surrounding medium. Frequently, the protonated forms of ionizable groups are less reactive or even unreactive or other reactions prevail with increasing numbers of nonionized groups.

Many proteins tend to unfold at pH values less than 5 or greater than 10, several not until much more extreme pH values are applied [2, 3, 4]. Nonionized groups buried in the folded protein conformation may become ionized only after exposure to the solvent. In particular, this holds for histidine and tyrosine residues, which may cause unfolding at acidic and alkaline pH values, respectively.

In native, water-soluble proteins the vast majority of ionizable groups (aspartic acid, glutamic acid, lysine, arginine) are located on the protein surface and are, therefore, exposed to the solvent. Polar AA residues are able to form hydrogen bonds to water molecules surrounding the protein envelope. Of course, maximum hydrophilicity of AA side chains is achieved with ionized residues [5]. By contrast, nonpolar residues cannot participate in hydrogen bonding; they rather interact via hydrophobic effects.

Water-soluble proteins are surrounded by a layer of tightly bound water molecules, with properties that differ significantly from those of the bulk water [2, 6, 7, 8, 9, 10, 11, 12, 13, 14, 15, 16]: higher density, greater heat capacity, pronounced order, and lower mobility of the preferentially bound water molecules are the most extraordinary characteristics. Irrespective of the tight binding of water molecules to the protein surface, they exchange permanently with the bulk solvent. The residence times of bound water molecules, however, differ from those of bulk water molecules [15].

About 0.35 g water is bound per gram of protein, corresponding to approximately two water molecules per AA residue [13]. The exact amount of water binding is highly influenced by the number and the nature of the AA residues on the protein surface, by its topography, and by the experimental conditions, such as pH value or presence of electrolytes (salts).

Owing to the fact that besides polar AAs appreciable numbers of hydrophobic patches are also located on the protein surface, preferentially bound water molecules do not cover entirely the protein surface [13, 14, 15, 17, 18, 19, 20, 21, 22]. This finding, of course, is a serious obstacle for any attempt to quantitatively describe hydrated protein structures and excludes the exact validity of shell models, in addition to uncertainties in our knowledge of the peculiarities of bound waters.

Modeling hydrated protein structures is a necessary requirement for our understanding of the protein behavior in aqueous solution, including interpretation, comparison, and prediction of scattering and hydrodynamic parameters in solution. Anyhow, serious modeling approaches have to take into account somehow the hydration contributions. This goal may be achieved by different approaches (summarized, for example, in Refs. [15, 17, 23, 24, 25, 26, 27, 28]), all of them having their merits. Simple whole-body approaches make use of molecular data that inherently contain the contribution of bound water (e.g., hydrated volume), apply qualified assumptions regarding hydration, or use some kind of rescaling. Multibody approaches ("bead modeling") also employ a rescaling procedure or model a water shell. Modern modeling approaches start from the atomic coordinates and exploit the information of comprehensive databases where the AA sequence and the atomic coordinates of several thousands of proteins are stored. Again, hydration is taken into account by scaling factors or by construction of an appropriate water shell surrounding the anhydrous protein. Recently, special hydration algorithms have been developed which place individual water molecules on the protein surface [13, 14, 15, 17, 18, 19]. For the special case of 3D reconstructions retrieved from electron microscopy, bead modeling may be used too, preferably in conjunction with density thresholds derived from an analysis of voxel-density distributions [19, 29, 30]. Ab initio modeling based on small-angle scattering data allows the prediction of low-resolution 3D structures; the hydration is taken into account by fitting to scattering curves which inherently contain hydration contributions in the case of X-ray scattering; for a comparison of the results obtained by these approaches see Ref. [31].

For our modeling studies we used lysozyme, an enzyme well characterized from the physicochemical point of view; its crystallographic structure is known to high resolution. This protein has already been the subject of several modeling attempts (e.g., Ref. [27]), including our hydration strategies by placing water molecules at definite molecular sites [15, 32]. While the previous modeling results are valid for neutral pH only, this study is the first attempt to consider also the different hydration contributions at varying pH conditions. The study also includes a comparison of structural and hydrodynamic parameters, predicted for different pH ranges.

Lysozyme

Lysozyme is a widely occurring hydrolase. It acts as a bacteriolytic enzyme, i.e., it affords protection against bacterial invasion by hydrolyzing proteoglycans of the bacterial cell wall. In vertebrates it is found particularly in egg white, saliva, tears, and mucosae.

The lysozyme from hen egg white, like other animal lysozymes, consists of a single chain of 129 AA residues. Its atomic coordinates [33] were obtained from the Protein Data Bank (PDB) (accession code 2LYZ) [34] and the molar mass ($M = 14.313$ kg/mol) was retrieved from the AA sequence data (entry P00698) stored in the SWISS-PROT data bank [35]. Lysozyme is a basic protein with an isoelectric point of about 9.3, mainly caused by 27 charged AA residues: 18 positively charged residues (11 arginine, six lysine, one histidine) and nine negatively charged residues (seven aspartic acid, two glutamic acid).

The scientific literature was thoroughly examined for available small-angle X-ray and neutron scattering (SAXS, SANS) data and for hydration and hydrodynamic reference data of lysozyme, in particular for the following molecular properties: partial specific volume, $\bar{v}$, hydrated volume, V, hydration, δ_1, radius of gyration,

R_G, sedimentation coefficient, s, and translational diffusion coefficient, D.

Table 2 lists a variety of experimental results obtained in different laboratories in the last 60 years and under different experimental conditions with respect to protein concentration, c, solvent components, pH, and temperature; obviously erroneous or incompatible results have been discarded. The table impressively demonstrates the considerable scatter of data, even if we restrict our inspection to data extrapolated to zero concentration. When considering X-ray and neutron data, however, we also have to be aware of the fact that usually X-ray data include hydration contributions, while neutron data are devoid of these contributions [6]. Remaining differences in the data may be addressed to differing changes in solvent components (salts) and pH, in addition to different homogeneity of the samples under analysis. One experiment was performed at pH 2, most of the experiments between pH 3 and 5, and several between pH 5 and 7.

There is a lot of evidence that, in the case of lysozyme solutions, only the pH interval between 3 and 5 guarantees the monomeric enzyme state, while experiments near or above neutrality rather correspond to aggregated enzyme forms, particularly to dimers [36, 37, 38, 39, 40, 41, 42]. Investigations at highly acidic pH, where the enzyme unfolds, are not the subject of this study. Similarly, no realistic reference data are available for alkaline pH values because of enzyme dimerization. For obvious reasons, the crystal data for the monomeric, folded enzyme conformation cannot be transferred to unfolded monomeric or dimerized enzyme structures. The modeling approaches to be performed will therefore mainly focus on the slightly acidic and neutral pH regimes.

Unfortunately, particularly in the older literature, some experimental details are missing in the cited references, for example, concerning extrapolation to infinite dilution, solvent, pH, and temperature. This, of course, hampers a clear-cut comparison of observed and predicted data. For the modeling results to be analyzed, the following considerations were taken into account.

For $\bar{v}$, we used a value of 0.702 cm^3/g [43], determined at pH 3, where the enzyme is known to be monomeric and stable; a nearly identical value (determined at pH 5.4) has also been used in sedimentation experiments [36]. However, we additionally checked the use of a consensus value (Table 3), derived from averaging over more than 20 values described in the literature, in its influence to modify s. In the pH range of the stability of the enzyme (pH 3–5), no significant changes of $\bar{v}$ values have to be expected [42]; similarly no significant changes between isomolal or isopotential specific volume quantities have been observed [43]. By contrast, at the extremes of the pH scale, where acid or alkaline denaturation occurs, nonlinear changes of specific volumes of proteins (frequently decreases) and differences between isomolal and isopotential volume quantities would be of importance [44, 45, 46].

The experimental values for V, δ_1, R_G, s, and D (Table 2) were used to derive consensus values (Table 3), in order to allow the most scrupulous comparison with the predicted data sets. In deriving experimental consensus values, however, averaging over values at extraordinary conditions (such as high protein or salt concentrations) was avoided. That also reminds us of the common practice in sedimentation analysis that nonideality and charge effects may be reduced by adding neutral salts [47, 48]. However, the presence of considerable concentrations of electrolytes in the solvent can influence the ionization of certain AA residues significantly, thus leading to prominent changes in the preferential binding/exclusion of salt and water to/from proteins [49, 50, 51, 52, 53, 54].

In context of changes in pK_a values, it also has be borne in mind that these may be achieved by some other factors influencing the protonization of AA residues, among them the neighborhood of charged protein groups being the most important [2, 55]. In the case of lysozyme, an anomalous ionization of an acidic carboxylic group, owing to the proximity of another charged AA residue, has been reported [42], leading to a reversible pH-dependent dimerization of lysozyme. As a consequence of acid denaturation, the pK_a values of several groups may alter [56]. Changes in pK_a values may also be reflected by determination of the net charge of the protein: for lysozyme at pH 4.5 a net charge of 14.1 ± 1 has been estimated [57], in agreement with the previously mentioned number of charged AA residues and consideration of the fact that not all of them are fully protonated at this pH value.

Modeling

Modeling proteins from the atomic coordinates

Modeling proteins from the atomic coordinates means applying a multibody approach and starting from a huge number of coordinates listed in a database [12, 25, 26, 27, 29, 30, 31, 58, 59, 60, 61, 62, 63, 64, 65, 66, 67, 68]. The molecular volume of the protein can be calculated by summing up the volumes of the AA residues according to the methods in Refs. [45, 46, 69]. Hydration contributions were taken into account by means of special hydration algorithms, which allow the placement of individual water molecules on the protein surface [13, 14, 15, 17, 18, 19]. Of course, the exact protein surface must be calculated first before any kind of hydration algorithm may be applied.

Table 2 Survey of experimental structural and hydrodynamic data for lysozyme from hen egg white under different experimental conditions, as retrieved from a critical inspection of the relevant literature

Parameter	Value	*c* (g/l)	Solvent	pH	Temperature (°C)	Reference
$\bar{v}$ (cm^3/g)	0.702	$c \to 0$	1 mM HCl, 0.1 M KCl	3.0	20	Lee and Timasheff [43]
	0.703	13–27	0.02 M acetate buffer, 0.15 M KCl	5.4	20	Sophianopoulos et al. [36]
	0.688–0.734				20–25	Colvin [87], Durchschlag [45, 46]
V_{SAXS} (nm^3)	19.8		0.15 M NaCl	3.8		Krigbaum and Kügler [37]
	20.4	$c \to 0$	0.15 M NaCl	5.7	25	Luzzati et al. [88]
	24.2 ± 0.4	$c \to 0$	0.15 M NaCl	3.8		Pessen et al. [89]
V_{SANS} (nm^3)	17.8	$c \to 0$	H_2O/D_2O variation	3.65	23	Stuhrmann and Fuess [90]
$\delta_{1,SAXS}$ (g/g)	0.317	$c \to 0$	0.15 M NaCl	3.8		Pessen and Kumosinski [91]
	0.32	50–100				Ritland et al. [92]
	0.33 ± 0.02	$c \to 0$	0.15 M NaCl	3.8		Pessen et al. [89]
δ_1 (g/g)	0.29					Ling [93]
	0.3		0.02 M acetate buffer, 0.15 M KCl	5.4	20	Sophianopoulos et al. [36]
	0.3	$c \to 0$	$I = 0.6$–1.0 M ($MgCl_2$)	4.0	20	Grigsby et al. [40]
	0.31–0.34					Rupley and Careri [22]
	0.34					Kuntz [72]
	0.34–0.52					Kuntz and Kauzmann [77]
	0.57				20	Squire and Himmel [94]
$\delta_{1,calc}$ (g/g)	0.32[a]			6–8		Hopfinger [73]
	0.335[b]			6–8		Kuntz [72], Kuntz and Kauzmann [77]
	0.35[c]			6–8		Durchschlag and Zipper [13]
	0.36[c]			6–8		Kuntz [72], Kuntz and Kauzmann [77], Hopfinger [73]
$R_{G,SAXS}$ (nm)	1.43 ± 0.03	$c \to 0$	0.15 M NaCl	3.8		Krigbaum and Kügler [37]
	1.43 ± 0.02	$c \to 0$	0.15 M NaCl	3.8		Pessen et al. [89]
	1.52 ± 0.02	$c \to 0$	0.15 M NaCl	5.7	25	Luzzati et al. [88]
	1.54 ± 0.02	$c \to 0$	0.040 M sodium acetate, 0 or 0.15 M NaCl, H_2O or D_2O	3.8		Svergun et al. [10]
	1.55 ± 0.05	10–20	0.05 M sodium acetate buffer	4.5	18	Ducruix et al. [95]
	1.60	50–100				Ritland et al. [92]
	1.63 ± 0.06	10–20	Varying ionic strength (1–880 mM) of different anions/cations	4.5	18	Ducruix et al. [95]
$R_{G,SANS}$ (nm)	1.24 ± 0.02	$c \to 0$	0.040 M sodium acetate buffer, D_2O	4.25		Svergun et al. [10]
	1.25	20–60	D_2O	4.7	18	Niimura et al. [96]
	1.38 ± 0.02	$c \to 0$	0.040 M sodium acetate buffer, 0.15 M NaCl	4.25		Svergun et al. [10]
	1.38	$c \to 0$	H_2O/D_2O variation	3.65	23	Stuhrmann and Fuess [90]
s (10^{-13} s)	1.87 ± 0.02	3–8	0.05 M Phosphate	6.8	20	Colvin [87]
	1.9	5–15	0.15 M NaCl		20	Alderton et al. [97]
	1.90		0.15 M NaCl	5.7	25	Luzzati et al. [88]
	1.91	$c \to 0$	0.02 M acetate buffer, 0.15 M KCl	5.4	20	Sophianopoulos et al. [36]
D (10^{-7} cm^2/s)	10.2 ± 0.4	$c \to 0$	0.1 M sodium acetate buffer	4.2	20	Eberstein et al. [98]
	10.3	$c \to 0$	0.04 M sodium acetate buffer, 2% NaCl	4.2	25	Mikol et al. [99]
	11.4	$c \to 0$	0.04 M sodium acetate buffer, 0.8 M $(NH_4)_2SO_4$	4.2	25	Mikol et al. [99]
	11.5	$c \to 0$	0.04 M sodium acetate buffer, 1.0 M $(NH_4)_2SO_4$	4.2	25	Mikol et al. [99]
	10.4 ± 0.1	3–8	0.05 M Phosphate	6.8	20	Colvin [87]
	10.5	$c \to 0$	$I = 0.6$ M ($MgCl_2$)	4.0	20	Grigsby et al. [40]
	11.4	$c \to 0$	$I = 0.8$ M ($MgCl_2$)	4.0	20	Grigsby et al. [40]
	10.6	$c \to 0$	$I = 1.0$ M ($MgCl_2$)	4.0	20	Grigsby et al. [40]
	10.6 ± 0.2	3–20	0.01 M phosphate buffer, 0.1 M NaCl	6.0	20	Foord et al. [100]
	10.6 ± 0.1	10	0.1 M sodium acetate buffer	4.2	20	Dubin and coworkers [38, 101]
	10.7		0.05 M KCl	4.0	20	Barel et al. [102]
	11.2	10	0.15 M NaCl		20	Alderton et al. [97]
	11.2	$c \to 0$	0.02 M acetate buffer, 0.15 M KCl	5.4	20	Sophianopoulos et al. [36]
	11.5 ± 0.3	60		5.6	25	Dubin et al. [103]
	11.5	10	0.20 M KCl	2.0	25	Nicoli and Benedek [104]

Table 2 (continued)

Parameter	Value	c (g/l)	Solvent	pH	Temperature (°C)	Reference
	11.8		0.15 M NaCl	5.7	25	Luzzati et al. [88]
	10.9 ± 0.5 (consensus)					Durchschlag and Zipper [15]
	11.1 (consensus)					Kuntz and Kauzmann [77], Allison and Tran [39]

[a]Hopfinger model; all AA residues assumed fully exposed
[b]Kuntz model; corrected for buried AA residues
[c]Kuntz model; all AA residues assumed fully exposed

Calculation of protein surfaces from the atomic coordinates

The exact surface topography of proteins can be calculated analytically by the programs MSROLL [70] or SIMS [71], producing a smooth "dot surface" by using a rolling-ball mechanism. In our previous work [13, 14, 15, 17, 18, 19, 28, 32] we tested the available programs and found both programs very successful for creating a huge number of dot surface points, N_{dot}, and normal vectors. In this study we use the program SIMS [71], which applies two probe spheres of different size, thereby producing a very smooth surface. Our calculations exploit the atomic coordinates included in the original PDB file, however, ignoring the coordinates of the 101 crystallographically found waters [15, 33]. In the present study, we use the input parameters which in a series of tuning studies turned out to be useful for our purposes: dot density $d_{dot} = 500\ 1/nm^2$, probe radius $r_{probe} = 0.145$, 0.1475, 0.150, or 0.1525 nm (for the water volumes $V_w = 0.0245$, 0.0257, 0.0269, or 0.0284 nm^3, respectively, used in the subsequent program HYDCRYST), and a smoothing probe sphere of $r_{sm} = 0.04$ nm.

Modeling individual water molecules on the protein surface

The surface points and normal vectors, created by the surface calculation programs MSROLL or SIMS on the basis of the atomic coordinates, may be used as a pool of hypothetical positions where water molecules, in principle, might be located. Special hydration algorithms, however, are required to select appropriate positions out of this pool of points. This is done by our program HYDCRYST [14, 15], which uses the information provided by MSROLL or SIMS.

The number of water molecules assigned to each accessible AA residue was based on the hydration numbers of Kuntz [72] or Hopfinger [73], described in more detail in the following section, and a minimum distance between adjacent waters ($d_w = 0.290$–0.305 nm). The selected water molecules were then approximated as spheres of volume V_w (0.0245–0.0284 nm^3). Thus, the molecular volume of bound water was taken as being lower by about 5–20% than the bulk water (0.0299 nm^3). Computations for different V_w values are required, since, at present, no clear-cut decision in favor of a definite density of bound water can be made, presumably owing to the fact that this value is only an average/default value which is influenced by the composition and the nature of the AA residues on the protein surface.

Besides the variation of V_w, we have already tested a large number of further tuning and fine-tuning parameters for our hydration approaches [13, 14, 15, 17, 18, 19]. Our current calculations were performed by means of an improved version of HYDCRYST which evaluates the PDB file in a two-pass algorithm, starting at the carboxyl terminus of the protein. The proper amount of water binding was adjusted by scaling factors (f_K, f_H), acting on the original hydration numbers (see next paragraph) suggested by Kuntz [72] or Hopfinger [73] which do not discriminate between the AA residues on the protein surface or in the interior. By variation of f_K or f_H between about 0.8 and 5.0, different stages of water binding to the protein [minimum, intermediate (realistic), and maximum hydration] can be simulated appropriately.

Correct choice of hydration numbers for individual AA residues

Hydration numbers have been derived by several authors from experiments or theoretical considerations. Owing to different ionization states of the AA residues on the protein surface under varying pH conditions, the applied numbers are only valid in certain pH ranges, preferably near neutrality. Out of a series of existing compilations, only the numbers given by Kuntz [72] and Hopfinger [73] present a relatively comprehensive list of data for all the AA residues (Table 1). Both compilations, however, do not distinguish between AA residues on the protein surface or in the interior.

The hydration numbers given by Kuntz [72] contain all 20 AA residues; they are based on NMR

Table 3 Comparison of calculated parameters for hydrated lysozyme models as obtained by the program HYDCRYST adapted for different pH ranges and different input parameters (for further input parameters see Modeling section), together with a survey of the experimental values (for detailed references see Table 2). The calculation of the *s* values was based on a value for $\bar{v}$ of 0.702 cm^3/g (Table 2); use of a consensus value of 0.715 ± 0.015 cm^3/g would yield *s* values lower by about 4%; all other predicted values are unperturbed by the assumed value for $\bar{v}$)

Experiment/model	Input		Output						
	V_w (nm^3)	f_K	V (nm^3)	N_w	δ_1 (g/g)	$R_{G,SAXS}$ (nm)	$R_{G,Hydro}$ (nm)	$s\times10^{13}$ (s)	$D\times10^7$ (cm^2/s)
Experiment			19.8–24.2		0.29–0.34	1.43–1.55		1.87–1.91	10.2–11.8
Consensus			21.5 ± 2.4		0.32 ± 0.02	1.50 ± 0.06		1.90 ± 0.02	10.9 ± 0.5
Unhydrated model			16.95	0	0.000	1.451	1.386	2.153	12.30
Hydrated KA models									
Kuntz approach: all AAs[a]				237	0.298				
Kuntz approach: accessible AAs[b]				217	0.273				
HYDCRYST	0.0245	1.00	22.00	206	0.259	1.503	1.477	1.921	10.97
	0.0245	1.25	23.13	252	0.317	1.517	1.499	1.893	10.81
	0.0245	1.50	24.01	288	0.363	1.530	1.517	1.878	10.72
	0.0245	1.75	24.79	320	0.403	1.537	1.528	1.862	10.63
	0.0257	1.00	22.07	199	0.250	1.493	1.479	1.914	10.93
	0.0257	1.25	23.33	248	0.312	1.506	1.505	1.886	10.77
	0.0257	1.50	24.20	282	0.355	1.518	1.525	1.868	10.66
	0.0257	1.75	24.97	312	0.393	1.524	1.536	1.858	10.61
	0.0269	1.00	22.22	196	0.247	1.482	1.483	1.905	10.88
	0.0269	1.25	23.52	244	0.307	1.492	1.508	1.883	10.75
	0.0269	1.50	24.43	278	0.350	1.500	1.526	1.867	10.66
	0.0269	1.75	25.26	309	0.389	1.505	1.537	1.855	10.59
	0.0284	1.00	22.60	199	0.250	1.468	1.488	1.902	10.86
	0.0284	1.25	24.05	250	0.315	1.474	1.514	1.870	10.68
	0.0284	1.50	24.90	280	0.352	1.479	1.534	1.854	10.59
	0.0284	1.75	25.59	304	0.383	1.481	1.543	1.847	10.55
HYDCRYST: mean ± S.D.									
f_K = 1.25–1.50			23.9 ± 0.6	265 ± 18	0.334 ± 0.023	1.50 ± 0.02	1.52 ± 0.01	1.87 ± 0.01	10.7 ± 0.1
f_K = 1.00–1.75			23.8 ± 1.2	260 ± 43	0.328 ± 0.054	1.50 ± 0.02	1.51 ± 0.02	1.88 ± 0.02	10.7 ± 0.1
Hydrated KN models									
Kuntz approach: all AAs[a]				276	0.347				
Kuntz approach: accessible AAs[b]				256	0.322				
HYDCRYST	0.0245	1.00	22.61	231	0.291	1.510	1.488	1.915	10.94
	0.0245	1.25	23.69	275	0.346	1.524	1.509	1.884	10.76
	0.0245	1.50	24.55	310	0.390	1.536	1.526	1.871	10.68
	0.0245	1.75	25.21	337	0.424	1.542	1.535	1.857	10.60
	0.0257	1.00	22.74	225	0.283	1.500	1.493	1.905	10.88
	0.0257	1.25	23.94	272	0.342	1.513	1.517	1.874	10.70
	0.0257	1.50	24.74	303	0.381	1.523	1.535	1.862	10.63
	0.0257	1.75	25.41	329	0.414	1.528	1.543	1.854	10.58
	0.0269	1.00	22.87	220	0.277	1.487	1.496	1.898	10.84
	0.0269	1.25	24.08	265	0.334	1.497	1.519	1.876	10.71
	0.0269	1.50	24.97	298	0.375	1.504	1.536	1.860	10.62
	0.0269	1.75	25.59	321	0.404	1.508	1.543	1.850	10.57
	0.0284	1.00	23.37	226	0.284	1.471	1.501	1.890	10.79
	0.0284	1.25	24.56	268	0.337	1.476	1.524	1.863	10.64
	0.0284	1.50	25.42	298	0.375	1.481	1.543	1.850	10.56
	0.0284	1.75	25.84	313	0.394	1.482	1.548	1.846	10.54
HYDCRYST: mean ± S.D.									
f_K = 1.25–1.50			24.5 ± 0.6	286 ± 18	0.360 ± 0.022	1.51 ± 0.02	1.53 ± 0.01	1.87 ± 0.01	10.7 ± 0.1
f_K = 1.00–1.75			24.3 ± 1.1	281 ± 39	0.353 ± 0.049	1.51 ± 0.02	1.52 ± 0.02	1.87 ± 0.02	10.7 ± 0.1
Hydrated KB models									
Kuntz approach: all AAs[a]				290	0.364				

Table 3 (continued)

Experiment/model	Input		Output						
	V_w (nm³)	f_K	V (nm³)	N_w	δ_1 (g/g)	$R_{G,SAXS}$ (nm)	$R_{G,Hydro}$ (nm)	$s \times 10^{13}$ (s)	$D \times 10^7$ (cm²/s)
Kuntz approach: accessible AAs[b]				270	0.339				
HYDCRYST	0.0245	1.00	22.61	231	0.291	1.510	1.488	1.915	10.94
	0.0245	1.25	23.69	275	0.346	1.524	1.509	1.884	10.76
	0.0245	1.50	24.55	310	0.390	1.536	1.526	1.871	10.68
	0.0245	1.75	25.21	337	0.424	1.542	1.535	1.857	10.60
	0.0257	1.00	22.76	226	0.284	1.500	1.493	1.904	10.87
	0.0257	1.25	23.94	272	0.342	1.513	1.517	1.874	10.70
	0.0257	1.50	24.74	303	0.381	1.523	1.534	1.862	10.63
	0.0257	1.75	25.41	329	0.414	1.528	1.543	1.854	10.58
	0.0269	1.00	22.90	221	0.278	1.488	1.496	1.898	10.84
	0.0269	1.25	24.08	265	0.334	1.497	1.519	1.876	10.71
	0.0269	1.50	24.97	298	0.375	1.504	1.536	1.860	10.62
	0.0269	1.75	25.59	321	0.404	1.508	1.543	1.850	10.57
	0.0284	1.00	23.37	226	0.284	1.471	1.501	1.890	10.79
	0.0284	1.25	24.56	268	0.337	1.476	1.524	1.863	10.64
	0.0284	1.50	25.42	298	0.375	1.481	1.543	1.850	10.56
	0.0284	1.75	25.84	313	0.394	1.482	1.548	1.846	10.54
HYDCRYST: mean ± S.D.									
f_K = 1.25–1.50			24.5 ± 0.6	286 ± 18	0.360 ± 0.022	1.51 ± 0.02	1.53 ± 0.01	1.87 ± 0.01	10.7 ± 0.1
f_K = 1.00–1.75			24.4 ± 1.1	281 ± 39	0.353 ± 0.049	1.51 ± 0.02	1.52 ± 0.02	1.87 ± 0.02	10.7 ± 0.1
Hydrated HA models									
Hopfinger approach: all AAs[a]				233	0.294				
Hopfinger approach: accessible AAs[b]				217	0.273				
HYDCRYST	0.0245	1.00	21.85	200	0.252	1.498	1.470	1.937	11.06
	0.0245	1.25	22.81	239	0.301	1.509	1.487	1.911	10.91
	0.0245	1.50	23.42	264	0.332	1.517	1.498	1.903	10.87
	0.0245	1.75	24.40	304	0.383	1.530	1.517	1.872	10.69
	0.0257	1.00	22.02	197	0.248	1.489	1.473	1.936	11.05
	0.0257	1.25	23.07	238	0.300	1.500	1.493	1.897	10.83
	0.0257	1.50	23.76	265	0.334	1.508	1.508	1.887	10.77
	0.0257	1.75	24.71	302	0.380	1.519	1.528	1.865	10.65
	0.0269	1.00	22.12	192	0.242	1.480	1.476	1.921	10.97
	0.0269	1.25	23.27	235	0.296	1.488	1.496	1.896	10.83
	0.0269	1.50	23.97	261	0.329	1.493	1.510	1.883	10.75
	0.0269	1.75	24.97	298	0.375	1.501	1.529	1.863	10.64
	0.0284	1.00	22.55	197	0.248	1.466	1.479	1.917	10.95
	0.0284	1.25	23.74	239	0.301	1.471	1.502	1.887	10.77
	0.0284	1.50	24.45	264	0.332	1.474	1.516	1.876	10.71
	0.0284	1.75	25.33	295	0.371	1.478	1.533	1.857	10.60
HYDCRYST: mean ± S.D.									
f_K = 1.25–1.50			23.6 ± 0.5	251 ± 14	0.315 ± 0.017	1.50 ± 0.02	1.50 ± 0.01	1.89 ± 0.01	10.8 ± 0.1
f_K = 1.00–1.75			23.5 ± 1.1	249 ± 39	0.314 ± 0.049	1.50 ± 0.02	1.50 ± 0.02	1.89 ± 0.03	10.8 ± 0.1
Hydrated HN models									
Hopfinger approach: all AAs[a]				257	0.324				
Hopfinger approach: accessible AAs[b]				241	0.303				
HYDCRYST	0.0245	1.00	22.22	215	0.271	1.503	1.477	1.930	11.02
	0.0245	1.25	23.08	250	0.315	1.513	1.493	1.906	10.88
	0.0245	1.50	23.67	274	0.345	1.520	1.503	1.897	10.83
	0.0245	1.75	24.57	311	0.391	1.531	1.520	1.870	10.68
	0.0257	1.00	22.40	212	0.267	1.494	1.482	1.924	10.99
	0.0257	1.25	23.38	250	0.315	1.505	1.501	1.891	10.80
	0.0257	1.50	23.99	274	0.345	1.511	1.513	1.883	10.75
	0.0257	1.75	24.82	306	0.385	1.520	1.529	1.865	10.65

Table 3 (continued)

Experiment/model	Input		Output						
	V_w (nm^3)	f_K	V (nm^3)	N_w	δ_1 (g/g)	$R_{G,SAXS}$ (nm)	$R_{G,Hydro}$ (nm)	s×10^{13} (s)	D×10^7 (cm^2/s)
	0.0269	1.00	22.47	205	0.258	1.482	1.483	1.915	10.93
	0.0269	1.25	23.54	245	0.308	1.490	1.501	1.892	10.80
	0.0269	1.50	24.22	270	0.340	1.496	1.515	1.880	10.73
	0.0269	1.75	25.10	303	0.381	1.502	1.531	1.861	10.63
	0.0284	1.00	22.92	210	0.264	1.468	1.488	1.909	10.90
	0.0284	1.25	23.97	247	0.311	1.472	1.507	1.883	10.75
	0.0284	1.50	24.62	270	0.340	1.475	1.518	1.874	10.70
	0.0284	1.75	25.39	297	0.374	1.479	1.534	1.857	10.60
HYDCRYST: mean ± S.D.									
f_K=1.25–1.50			23.8 ± 0.5	260 ± 13	0.327 ± 0.016	1.50 ± 0.02	1.51 ± 0.01	1.89 ± 0.01	10.8 ± 0.1
f_K=1.00–1.75			23.8 ± 1.0	259 ± 36	0.326 ± 0.045	1.50 ± 0.02	1.51 ± 0.02	1.89 ± 0.02	10.8 ± 0.1
Hydrated HB models									
Hopfinger approach: all AAs[a]				233	0.293				
Hopfinger approach: accessible AAs[b]				216	0.272				
HYDCRYST	0.0245	1.00	21.66	192	0.242	1.498	1.469	1.937	11.06
	0.0245	1.25	22.54	228	0.287	1.509	1.487	1.911	10.91
	0.0245	1.50	23.15	253	0.318	1.516	1.498	1.903	10.87
	0.0245	1.75	24.03	289	0.364	1.528	1.517	1.872	10.69
	0.0257	1.00	21.86	191	0.240	1.490	1.473	1.936	11.05
	0.0257	1.25	22.84	229	0.288	1.501	1.493	1.897	10.83
	0.0257	1.50	23.43	252	0.317	1.507	1.508	1.887	10.77
	0.0257	1.75	24.28	285	0.359	1.516	1.528	1.865	10.65
	0.0269	1.00	21.90	184	0.232	1.479	1.476	1.921	10.97
	0.0269	1.25	22.95	223	0.281	1.486	1.496	1.896	10.83
	0.0269	1.50	23.62	248	0.312	1.492	1.510	1.883	10.75
	0.0269	1.75	24.46	279	0.351	1.499	1.529	1.863	10.64
	0.0284	1.00	22.26	187	0.235	1.466	1.479	1.917	10.95
	0.0284	1.25	23.29	223	0.281	1.470	1.502	1.887	10.77
	0.0284	1.50	23.97	247	0.311	1.473	1.516	1.876	10.71
	0.0284	1.75	24.76	275	0.346	1.477	1.533	1.857	10.60
HYDCRYST: mean ± S.D.									
f_K=1.25–1.50			23.2 ± 0.5	238 ± 13	0.299 ± 0.017	1.49 ± 0.02	1.50 ± 0.01	1.89 ± 0.01	10.8 ± 0.1
f_K=1.00–1.75			23.2 ± 1.0	237 ± 35	0.298 ± 0.045	1.49 ± 0.02	1.50 ± 0.02	1.89 ± 0.03	10.8 ± 0.1

[a]Overall hydration, assuming all AA residues fully exposed
[b]Considering only accessible AA residues

experiments on a variety of polypeptides [72, 74] and are valid for the pH range between 6 and 8. A few values are additionally given for the acidic pH range at about pH 4: the hydration for aspartic acid and glutamic acid is definitely lower than at neutrality. For the basic range between 10 and 12 some values are also mentioned: arginine (pH 10), lysine (pH 10–11), and tyrosine (pH 12). While the values for arginine and lysine at alkaline pH do not differ from the values near neutrality, the value for ionized tyrosine is substantially higher than for the uncharged form at neutral pH. The meaning of these values may be understood on the basis of the pK_a values of the ionizable groups observed in proteins [2]. Any changes of the values for cysteine and histidine at alkaline pH are not mentioned.

The hydration numbers suggested by Hopfinger [73] come from theoretical calculations describing the thermodynamic behavior of water–polypeptide group interactions [75, 76]. At neutral pH a full list of hydration values, except the value for cysteine, is also given. For our calculations employing the Hopfinger model we used a value of 1 for cysteine, adopted from Kuntz [72]. Again, for aspartic acid and glutamic acid, lower values for the acidic pH range are given, while for the basic range the value for lysine is substantially lower than that at neutral pH. For the basic pH range no values for the charged forms of tyrosine and cysteine or for histidine are listed.

A comparison of the hydration numbers given by Kuntz and Hopfinger for the neutral pH range reveals large similarities for the majority of values, but signifi-

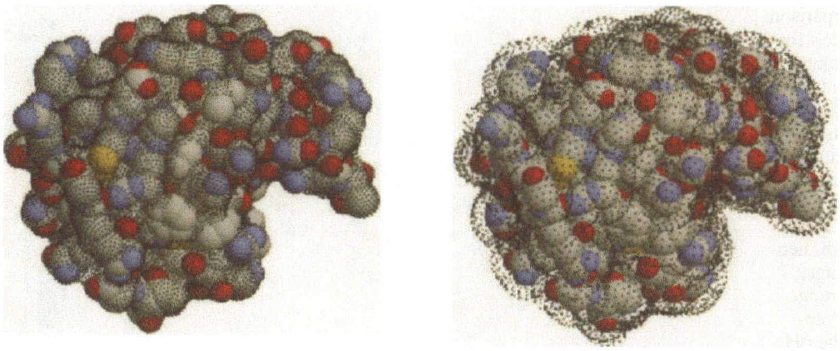

Fig. 1a, b Space-filling model for anhydrous lysozyme and molecular dot surface points created by the program SIMS **a** for the anhydrous contour and, in combination with the program HYDCRYST, **b** for the contour of potential water points. All dot surface points were calculated from the atomic coordinates of the enzyme. In **a** only the dot surface points on or outside the spheres representing the amino acid (AA) residues are visible (in a few cases they are located slightly underneath the surface of the spheres). The basic protein atoms derived from the Protein Data Bank file (2LYZ) are shown in space-filling format and in the usual Corey–Pauling–Koltun colors (C in *light gray*, O in *red*, N in *light blue*, S in *yellow*) and the dot surface points and potential water points are colored *black*

cant differences for proline and phenylalanine. The values for aspartic acid and glutamic acid differ significantly at acidic pH. The most striking difference, however, can be observed for the basic pH range: While Kuntz [72] reports a significant (plausible) increase in the value for tyrosine, Hopfinger [73] presents an extraordinary diminution of the value for lysine.

Both sets of hydration numbers have been successfully used by many authors [13, 72, 73, 77, 78, 79] to predict water binding to several proteins (overall hydration), though the original approaches assume that each AA residue is fully exposed to the solvent. This assumption is obviously incorrect; it leads to an overestimate of the predicted hydration (in the case of lysozyme about 10% [72]; cf. the $\delta_{1,\mathrm{calc}}$ values in Table 2). More realistic predictions have to use the surface AA residues only, instead of the total number of AA residues. Consequently, the hydration numbers to be applied to the surface residues have to be modified (normalized) appropriately, to be in agreement with the experiments [13].

Visualization of models

The anhydrous model created, the dot surface points, and the water molecules created in the following were visualized by the program RASMOL [80].

Prediction of structural and hydrodynamic parameters of hydrated proteins

The structural parameters (V, R_G) were derived from the atomic coordinates as in previous papers [13, 14 15, 17, 18, 19]. The resultant values for the radius of gyration ($R_{G,\mathrm{SAXS}}$, $R_{G,\mathrm{Hydro}}$) are significantly different for different procedures and input parameters applied for their determination. Hydrodynamic calculations use bead radii, r_b, whereas SAXS data are usually evaluated from both r_b and the electron density of the beads [17].

The prediction of hydrodynamic parameters (s, D) made use of the approach pioneered by García de la Torre and coworkers [27, 81], applying version 13 of the program HYDRO. For our calculations the input routine of the program was adapted to facilitate the handling of a large number of data sets. A "filling model" strategy [27, 82] was applied to retrieve both scattering and hydrodynamic quantities without changing the computation strategy [15] and to allow also the discovery of internal water molecules, finally leading to spheres ("beads") of equal density but unequal size. The presentation of atoms by spheres corresponding to their volumes causes an appreciable number of overlapping nonequal spheres [83]. The problem of overlapping was handled by our ad hoc expression for the interaction tensor [66, 67, 83] which is already implemented in version 13 of HYDRO [84].

Computational requirements

For our calculations personal computers with Intel Pentium III or AMD Athlon processors with 1 GHz and 512 MB random access memory were used. The computation time for calculations with HYDRO amounted to about 10 min for one hydrated lysozyme model; the time for calculations with all other programs was much lower.

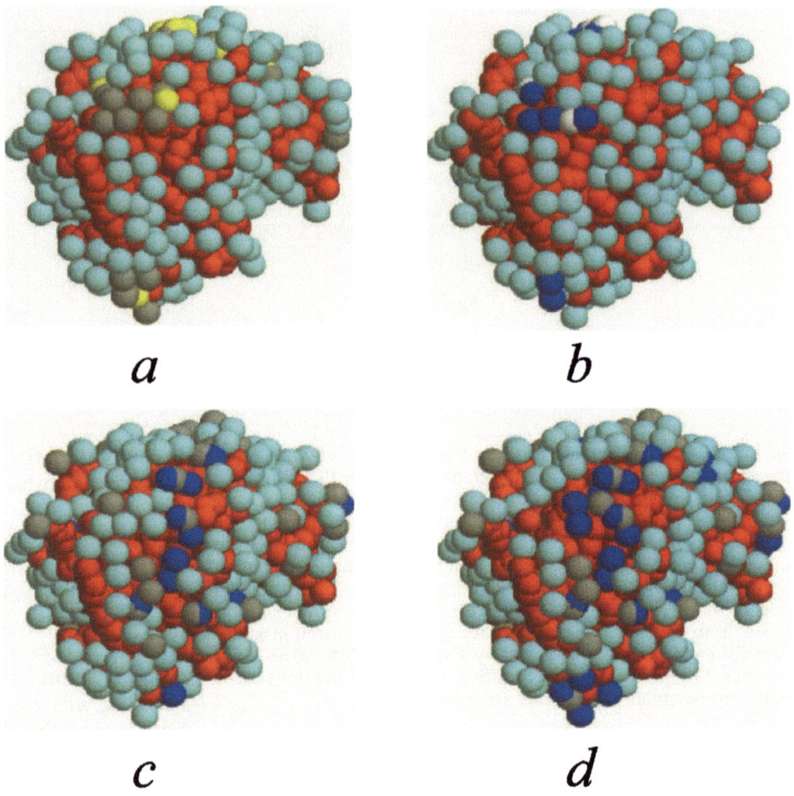

Fig. 2a–d Pairwise comparison of space-filling AA models for hydrated lysozyme, obtained by usage of the hydration numbers suggested by Kuntz (K) or Hopfinger (H) for the individual AA residues and the neutral (N) and acidic (A) pH ranges. **a** KN-KA: Kuntz approach, neutral versus acidic pH range; **b** HN-HA: Hopfinger approach, neutral versus acidic pH range; **c** KN-HN: neutral pH range, Kuntz versus Hopfinger approach; **d** KA-HA: acidic pH range, Kuntz versus Hopfinger approach. Each image represents the superposition of two models. All protein atoms are displayed in *red*. Bound water molecules common to both models compared (position tolerance 0.01 nm) are highlighted in *cyan*, whereas waters present in only one model are given in *gray* (K), *blue* (H), *yellow* (KA), or *light gray* (HA). The models are based on water volumes $V_w = 0.0257\ nm^3$ and scaling factors (f_K, f_H) of 1.25

Results and discussion

Space-filling models and dot surfaces of the anhydrous protein

Space-filling models of anhydrous lysozyme were created on the basis of the PDB structure. The atoms of the enzyme (except the hydrogens) are presented by spheres and are given in the usual Corey–Pauling–Koltun colors in Fig. 1. Application of the surface calculation program SIMS allowed us to envelope the entire protein by a dense network of surface points. Figure 1a displays the anhydrous protein together with the dot surface points. The figure demonstrates that a smooth surface consisting of a vast number of points has indeed been produced. The points are located exactly on the atoms of the various AA residues. Exploiting the normal vectors to these surface points gives the same number of potential sites of water points, but about 0.15-nm distant from the surface points of the anhydrous protein. This is impressively visualized in Fig. 1b, together with the space-filling image of the anhydrous protein.

Models for the hydrated protein

Application of the program HYDCRYST allowed the assignment of the potential water points to the AA residues on the protein surface (including any kind of inner surface). The site-specific hydration algorithm utilizes various reasonable criteria for the selection of water molecules out of the pool of potential water points, together with several input parameters, including those for the water volume (V_w) and the scaling factors (f_K, f_H) for the hydration numbers. The program thus reduces the initial number of surface dots (about 29,000 in the case of lysozyme, depending on d_{dot}) to a reasonable value (about 280), in full agreement with the experimentally observed hydration δ_1 (Table 2).

Figure 2 illustrates selected space-filling models for hydrated lysozyme, obtained after usage of the hydration numbers for different pH ranges. The figure shows a pairwise comparison of the bound water molecules, obtained after application of the hydration numbers for the neutral and acidic pH range suggested by Kuntz [72] and Hopfinger [73]. The comparison of the hydrated models at different pH values reveals significant differences, after application of both the Kuntz and the Hopfinger numbers. The models valid for the acidic pH range are covered with a lower number of water molecules. The models based on the Kuntz numbers resemble those derived from the Hopfinger values; remaining slight differences at specific surface sites are caused by different hydration numbers for specific AA residues. Because of conflicting and lacking numbers for hydration in the basic pH range, we abstain from a graphical representation of these results. Furthermore, as already mentioned, the enzyme is not stable at alkaline pH and, consequently, no scattering or hydrodynamic reference data are available for this pH range.

A comparison of the hydration and structural and hydrodynamic parameters predicted for the hydrated protein

A quantitative comparison of some representative data for the acidic, neutral, and basic pH range is summarized in Table 3. The underlying data were obtained after application of HYDCRYST and subsequent prediction of structural and hydrodynamic parameters by HYDRO. Previous studies [15] have revealed that scaling factors between 1.25 and 1.5 should be preferred for lysozyme, to simulate a reasonable hydration of the protein. The examples given in the table, therefore, are restricted to the most relevant range of scaling factors (f_K,f_H) between 1.0 and 1.75, though our calculations covered a much wider range (11 scaling factors in the range 0.8–5.0).

A critical inspection of the data in the table reveals that all input parameters tested (V_w, f_K, f_H, hydration numbers by Kuntz and Hopfinger for the AA residues in different pH ranges) lead to slightly different results. The influence of the water volume, however, is marginal. It is also moderate for the scaling factors, if varied in the given narrow range between 1 and 1.75.

The averaged results (V_w=0.0245–0.0284 nm^3, f_K or f_H=1.00–1.75) obtained on the basis of the Kuntz hydration numbers suggest a hydration δ_1 of about 0.33, 0.35, and 0.35 g/g for the acidic, neutral, and basic pH ranges (for short, KA, KN, KB approaches), respectively. This corresponds to about 260, 280, and 280 water molecules, N_w. The corresponding values after applying the Hopfinger hydration numbers for the same pH ranges (for short, HA, HN, HB approaches) are 0.31, 0.33, and 0.30 g/g and about 250, 260, and 240 water molecules. Nearly identical results are obtained if for averaging only the scaling factors 1.25–1.50 are used. Thus, the average values for δ_1 (or N_w) at acidic pH (KA, HA) are slightly lower than the values at neutrality (KN, HN), whereas the values at basic pH exhibit different behavior for the two sets of hydration numbers used in this study. Within the accuracy of the experimental data, the experimentally observed consensus value for δ_1 of about 0.32 g/g for the pH range between 3 and 7 agrees well with the values calculated for the acidic and neutral pH ranges. If, with all caveats in mind, we predict the values for the basic pH range, the results are in disagreement. The KB values are similar to the KN values, whereas the HB values are distinctly lower than the values obtained by the HN approach.

Quantifying the structural (V, R_G) and hydrodynamic (s, D) parameters yields results in agreement with the changes for δ_1 or N_w (Table 3). The values for the radii of gyration ($R_{G,SAXS}$, $R_{G,Hydro}$) decrease in the acidic pH range (KA, HA) compared with the values at neutrality (KN, HN), whereas the values for s and D increase at acidic pH in comparison with the values at neutral pH. Again the behavior at basic pH disagrees for the two approaches (KB, HB). In addition, a comparison with experimental consensus values shows agreement between predicted and observed values for the pH range between 3 and neutrality, but, caused by the scatter of experimental results, does not allow us to discern between the results found at specific pH values.

Use of scaling factors f_K and f_H of about 5 (not shown) results in a higher coverage of the protein with water molecules. Because of the larger number of water molecules bound to the protein surface, differences between the approaches and the results achieved in different pH ranges disappear.

A comparison of the hydration predicted for individual AA residues

The results for one selected example (V_w=0.0257 nm^3, f_K or f_H=1.25) are compared with respect to the number of hydrated AA residues and the total number of bound water molecules, both for the Kuntz and for the Hopfinger hydration numbers and acidic, neutral, and basic pH ranges (Fig. 3). Certain AA residues, such as arginine (R), asparagine (N), aspartic acid (D), glutamic acid (E), lysine (K), and serine (S), are hydrated to a larger extent than the others, after application of both types of hydration numbers used. Among this group preferably the charged AAs are found which favor an appreciable amount of bound water (Table 1).

A critical assessment of the amount of water bound by the AA residues involved primarily in hydration discloses that the four representatives of charged AA residues (arginine, aspartic acid, glutamic acid, and lysine) are responsible for about 45% of the total amount of water binding (Table 4). The six AA residues mentioned earlier together account for about 68% of the total hydration effects.

Both the histograms in Fig. 3 and the summary in Table 4 reflect the significant differences observed for the

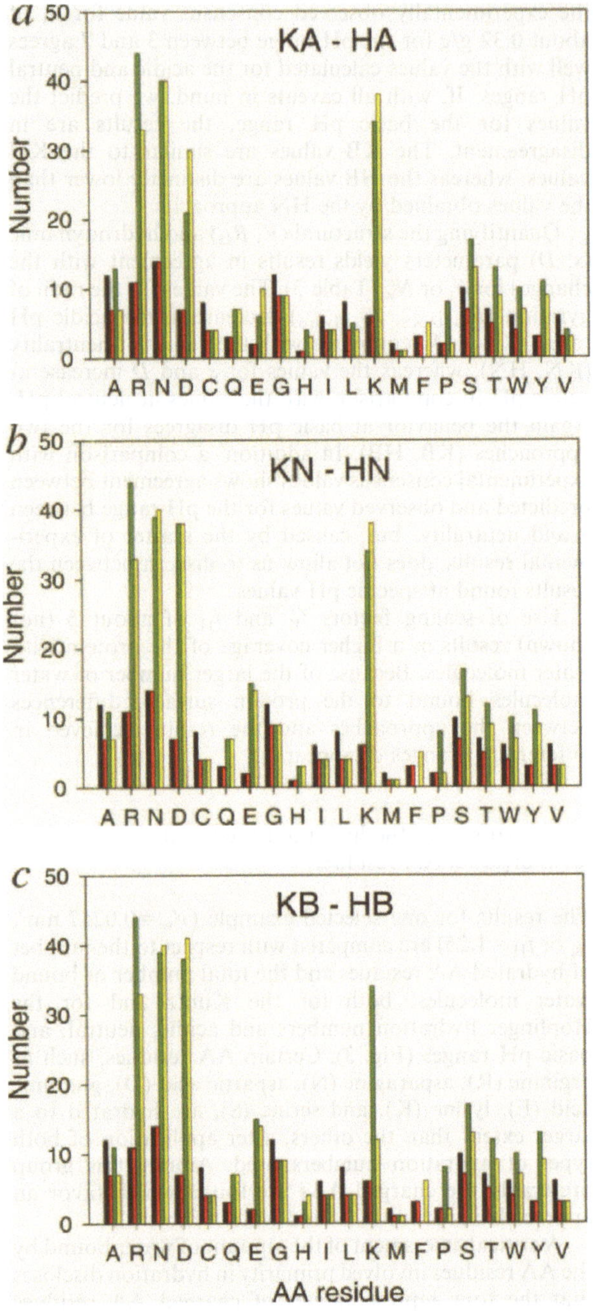

◄ **Fig. 3a–c** Histograms showing the total number of AA residues of lysozyme (*black bars*), the number of hydrated AA residues (*red bars*), together with the total number of bound water molecules after application of the Kuntz (*green bars*) and Hopfinger (*yellow bars*) hydration numbers. The amount of hydration was obtained by HYDCRYST (V_w=0.0257 nm^3, f_K or f_H=1.25). The AA residues are given in alphabetical order of their three-letter code; for short, however, they are symbolized by the one-letter code

Kuntz and Hopfinger approaches at acidic (KA, HA) and neutral pH (KN, HN), and the contradictory behavior at alkaline pH (KB, HB) for lysine, already mentioned in the context of the data presented in Table 3.

Table 4 Number and percentage of water molecules bound to the AA residues contributing most to the hydration of lysozyme (V_w=0.0257 nm^3, f_K or f_H=1.25), compared with the total number of bound water molecules

Model	Number and percentage of bound waters				Total number of bound waters
	Arg, Asp, Glu, Lys[a]		Arg, Asp, Glu, Lys, Asn, Ser[b]		
	N_w	Percentage	N_w	Percentage	N_w
KA	100	40.3	161	64.9	248
KN	127	46.7	188	69.1	272
KB	127	46.7	188	69.1	272
HA	106	44.5	161	67.6	238
HN	120	48.0	175	70.0	250
HB	97	42.4	152	66.4	229

[a] This group comprises 26 AA residues
[b] This group comprises 50 AA residues, 48 residues are accessible

Conclusions

Water binding to proteins and other biomolecules is of central importance for our understanding of their behavior in solution. The precise prediction of structural and hydrodynamic parameters of proteins requires us to take the atomic coordinates of the protein into account and to consider also appropriately the contributions of preferentially bound water molecules. The progressive development of site-specific hydration algorithms [13, 14, 15, 17, 18, 19] currently allows us to describe biophysically relevant hydration sites on the protein envelope, though various other approaches using rescaling or water shells are sufficient to predict molecular parameters quantitatively. Anyhow, neglect of water contributions (i.e., the use of anhydrous protein models) for predictive purposes is discouraged.

Application of modern surface calculation programs such as SIMS [71] together with our hydration modeling approach HYDCRYST [14, 15] to existing atomic coordinates yields realistic numbers of bound water molecules and presents possible residence sites on the protein surface. Though water molecules form a constantly fluctuating water network covering the protein surface and though there is little preference for water molecules to occupy particular surface sites, the concept of hydration sites seems to be useful for describing the water structure around proteins [15, 85, 86].

Among the input parameters required for HYDCRYST, the values for the water volume, V_w, scaling factors f_K and f_H, and hydration numbers for AA residues are the most important ones. However, averaging over V_w and most plausible values for the respective scaling factors give realistic hints for actual hydration phenomena. This was done in the present study to describe the water binding to lysozyme in different pH ranges, but outside the extreme pH ranges where denaturation and unfolding occur.

Usage of the hydration numbers suggested by Kuntz [72] and Hopfinger [73], which are based on quite different approaches, yields comparable results for the neutral and acidic pH ranges. Compared with neutrality, the hydration is less expressed at acidic pH values. Consequently, R_G values predicted for the acidic range should decrease, while s and D values should increase. In contrast to these results, the predictions for the basic pH range are rather conflicting when using the Kuntz or Hopfinger hydration numbers. Unless more appropriate hydration numbers for this pH range are available, no realistic predictions for the basic range can be given. This deficiency cannot be circumvented by a comparison of predicted and observed molecular parameters because of the instability of the enzyme at alkaline pH values.

The large scatter of experimental values, observed in the whole pH range, is a general obstacle for a quantitative comparison of observed and predicted structural and hydrodynamic parameters. Since, on the other hand, the differences between the parameters predicted for different pH values are rather small, the precise anticipation of these quantities is of subordinate importance if aiming at the absolute values. By contrast, any predictive statements on water binding to special surface topographies is of high biophysical relevance, for example, for describing the behavior of proteins and constituent AA sites in context with changing pH conditions, site-directed mutagenesis experiments, and the construction of tailored enzymes.

In the present study, modeling of protein hydration was achieved by transferring the hydration properties of individual AA residues to the residues on the surface of a chosen protein, applying a variety of strategies for modeling the exact protein surface and preferentially bound water molecules, in addition to qualified assumptions for the properties of the water molecules bound to the protein surface. Of course, all peculiarities of proteins (properties of certain AA residues or bound water molecules), deviating severely from the average behavior, affect the results to be obtained. For example, the aforementioned changes of pK_a values of individual AA residues, occurring, for example, as a consequence of adjacent charged groups or environmental changes owing to the presence/absence of electrolytes in the solvent, may influence the hydration of individual protein sites. Whether the hydration strategy applied, which obviously works satisfactorily in the cases studied so far, can be applied quite generally and to what extent has to be established in the future. Further studies in this direction have to be performed with additional proteins of different size and hydration binding behavior, including a large variety of additional experimental results from different fields of research.

Acknowledgements The authors are much obliged to Y.N. Vorobjev, J. García de la Torre, and R.A Sayle for use of the programs SIMS, HYDRO, and RASMOL, respectively.

References

1. Anfinsen CB (1973) Science 181:223–230
2. Creighton TE (1993) Proteins. Structures and molecular properties, 2nd edn. Freeman, New York
3. Lapanje S (1978) Physicochemical aspects of protein denaturation. Wiley, New York
4. Jaenicke R (1981) Annu Rev Biophys Bioeng 10:1–67
5. Thanki N, Thornton JM, Goodfellow JM (1988) J Mol Biol 202:637–657
6. Perkins SJ (1986) Eur J Biochem 157:169–180
7. Gerstein M, Tsai J, Levitt M (1995) J Mol Biol 249:955–966
8. Gerstein M, Chothia C (1996) Proc Natl Acad Sci USA 93:10167–10172
9. Murphy LR, Matubayasi N, Payne VA, Levy RM (1998) Folding Design 3:105–118
10. Svergun DI, Richard S, Koch MHJ, Sayers Z, Kuprin S, Zaccai G (1998) Proc Natl Acad Sci USA 95:2267–2272
11. Ebel C, Eisenberg H, Ghirlando R (2000) Biophys J 78:385–393
12. Perkins SJ (2001) Biophys Chem 93:129–139
13. Durchschlag H, Zipper P (2001) Biophys Chem 93:141–157
14. Durchschlag H, Zipper P (2002) J Phys Condens Matter 14:2439–2452
15. Durchschlag H, Zipper P (2003) Eur Biophys J 32:487–502
16. Chaplin M (2003) Water structure and behavior. http://www.sbu.ac.uk/water/
17. Durchschlag H, Zipper P (2002) Prog Colloid Polym Sci 119:131–140
18. Zipper P, Durchschlag H (2002) Physica A 304:283–293
19. Zipper P, Durchschlag H (2002) Physica A 314:613–622
20. Richards FM (1977) Annu Rev Biophys Bioeng 6:151–176
21. Blake CCF, Pulford WCA, Artymiuk PJ (1983) J Mol Biol 167:693–723
22. Rupley JA, Careri G (1991) Adv Protein Chem 41:37–172
23. Kumosinski TF, Pessen H (1985) Methods Enzymol 117:154–182
24. Harding SE (1995) Biophys Chem 55:69–93
25. Perkins SJ, Ashton AW, Boehm MK, Chamberlain D (1998) Int J Biol Macromol 22:1–16
26. Byron O (2000) Methods Enzymol 321:278–304

27. García de la Torre J, Huertas ML, Carrasco B (2000) Biophys J 78:719–730
28. Durchschlag H, Zipper P (2002) Prog Colloid Polym Sci 119:121–130
29. Zipper P, Durchschlag H (2000) J Appl Crystallogr 33:788–792
30. Zipper P, Krebs A, Durchschlag H (2002) Prog Colloid Polym Sci 119:141–148
31. Zipper P, Durchschlag H (2003) J Appl Crystallogr 36:509–514
32. Durchschlag H, Hefferle T, Zipper P (2003) Radiat Phys Chem 67:479–486
33. Diamond R (1974) J Mol Biol 82:371–391
34. Berman HM, Westbrook Z, Feng Z, Gilliland G, Bhat TN, Weissig H, Shindyalov IN, Bourne PE (2000) Nucleic Acids Res 28:235–242
35. Bairoch A, Apweiler R (2000) Nucleic Acids Res 28:45–48
36. Sophianopoulos AJ, Rhodes CK, Holcomb DN, Van Holde KE (1962) J Biol Chem 237:1107–1112
37. Krigbaum WR, Kügler FR (1970) Biochemistry 9:1216–1223
38. Dubin SB, Clark NA, Benedek GB (1971) J Chem Phys 54:5158–5164
39. Allison SA, Tran VT (1995) Biophys J 68:2261–2270
40. Grigsby JJ, Blanch HW, Prausnitz JM (2000) J Phys Chem B 104:3645–3650
41. Nesmelova IV, Skirda VD, Fedotov VD (2002) Biopolymers 63:132–140
42. Sophianopoulos AJ, Van Holde KE (1964) J Biol Chem 239:2516–2524
43. Lee JC, Timasheff SN (1974) Biochemistry 13:257–265
44. Durchschlag H, Jaenicke R (1982) Biochem Biophys Res Commun 108:1074–1079
45. Durchschlag H (1986) In: Hinz H-J (ed) Thermodynamic data for biochemistry and biotechnology. Springer, Berlin Heidelberg New York, pp 45–128
46. Durchschlag H (2003) In: Hinz H-J (ed) Landolt–Börnstein new series biophysics – proteins, vol VII/2A. Springer, Berlin Heidelberg New York, pp 4/1–157
47. Tanford C (1961) Physical chemistry of macromolecules. Wiley, New York
48. Schachman HK (1959) Ultracentrifugation in biochemistry. Academic, New York
49. Casassa EF, Eisenberg H (1964) Adv Protein Chem 19:287–395
50. Eisenberg H (1976) Biological macromolecules and polyelectrolytes in solution. Clarendon Press, Oxford
51. Eisenberg H (1981) Q Rev Biophys 14:141–172
52. Eisenberg H (1990) In: Saenger W (ed) Landolt–Börnstein new series biophysics – nucleic acids, vol VII/1C. Springer, Berlin Heidelberg New York, pp 257–276
53. Timasheff SN (1998) Adv Protein Chem 51:355–432
54. Timasheff SN (2003) In: Hinz H-J (ed) Landolt–Börnstein new series biophysics – proteins, vol VII/2A. Springer, Berlin Heidelberg New York, pp 5/1–23
55. Van Holde KE, Johnson, WC, Ho PS (1998) Principles of physical biochemistry. Prentice Hall, Upper Saddle River, NJ
56. Sophianopoulos AJ, Weiss BJ (1964) Biochemistry 3:1920–1928
57. Behlke J, Ristau O (1999) Biophys Chem 76:13–23
58. Müller JJ (1983) J Appl Crystallogr 16:74–82
59. Müller JJ (1991) Biopolymers 31:149–160
60. Durchschlag H, Zipper P, Wilfing R, Purr G (1991) J Appl Crystallogr 24:822–831
61. Müller JJ, Schrauber H (1992) J Appl Crystallogr 25:181–191
62. Beavil AJ, Young RJ, Sutton BJ, Perkins SJ (1995) Biochemistry 34:14449–14461
63. Durchschlag H, Zipper P, Purr G, Jaenicke R (1996) Colloid Polym Sci 274:117–137
64. Byron O (1997) Biophys J 72:408–415
65. (a) Spotorno B, Piccinini L, Tassara G, Ruggiero C, Nardini M, Molina F, Rocco M (1997) Eur Biophys J 25:373–384; (b) Erratum Eur Biophys J 26:417
66. Zipper P, Durchschlag H (1997) Prog Colloid Polym Sci 107:58–71
67. Zipper P, Durchschlag H (1998) Biochem Soc Trans 26:726–731
68. García de la Torre J (2001) Biophys Chem 93:159–170
69. Durchschlag H, Zipper P (1997) J Appl Crystallogr 30:803–807
70. Connolly ML (1993) J Mol Graph 11:139–141
71. Vorobjev YN, Hermans J (1997) Biophys J 73:722–732
72. Kuntz ID (1971) J Am Chem Soc 93:514–516
73. Hopfinger AJ (1977) Intermolecular interactions and biomolecular organization. Wiley, New York
74. Kuntz ID (1971) J Am Chem Soc 93:516–518
75. Hopfinger AJ (1971) Macromolecules 4:731–737
76. Hopfinger AJ, Battershell RD (1976) J Med Chem 19:569–573
77. Kuntz ID Jr, Kauzmann W (1974) Adv Protein Chem 28:239–345
78. Lee JC, Timasheff SN (1979) Methods Enzymol 61:49–57
79. Prakash V, Timasheff SN (1985) Methods Enzymol 117:53–60
80. Sayle RA, Milner-White EJ (1995) Trends Biochem Sci 20:374–376
81. García de la Torre J, Navarro S, López Martínez MC, Díaz FG, López Cascales JJ (1994) Biophys J 67:530–531
82. Carrasco B, García de la Torre J (1999) Biophys J 75:3044–3057
83. Zipper P, Durchschlag H (1999) Prog Colloid Polym Sci 113:106–113
84. Carrasco B, García de la Torre J, Zipper P (1999) Eur Biophys J 28:510–515
85. Bon C, Lehmann MS, Wilkinson C (1999) Acta Crystallogr Sect D 55:978–987
86. Henchman RH, McCammon JA (2002) J Comput Chem 23:861–869
87. Colvin JR (1952) Can J Chem 30:831–834
88. Luzzati V, Witz J, Nicolaieff A (1961) J Mol Biol 3:367–378
89. Pessen H, Kumosinski TF, Timasheff SN (1971) J Agric Food Chem 19:698–702
90. Stuhrmann HB, Fuess H (1976) Acta Crystallogr Sect A 32:67–74
91. Pessen H, Kumosinski TF (1985) Methods Enzymol 117:219–255
92. Ritland HN, Kaesberg P, Beeman WW (1950) J Chem Phys 18:1237–1242
93. Ling GN (1972) In: Horne RA (ed) Water and aqueous solutions. Wiley, New York, pp 663–700
94. Squire PG, Himmel ME (1979) Arch Biochem Biophys 196:165–177
95. Ducruix A, Guilloteau JP, Riès-Kautt M, Tardieu A (1996) J Cryst Growth 168:28–39
96. Niimura N, Minezaki Y, Ataka M, Katsura T (1994) J Cryst Growth 137:671–675
97. Alderton G, Ward WH, Fevold HL (1945) J Biol Chem 157:43–58
98. Eberstein W, Georgalis Y, Saenger W (1994) J Cryst Growth 143:71–78
99. Mikol V, Hirsch E, Giegé R (1990) J Mol Biol 213:187–195
100. Foord R, Jakeman E, Oliver CJ, Pike ER, Blagrove RJ, Wood E, Peacocke AR (1970) Nature 227:242–245
101. Dubin SB, Feher G, Benedek GB (1973) Biochemistry 12:714–720
102. Barel AO, Prieels JP, Maes E, Looze Y, Léonis J (1972) Biochim Biophys Acta 257:288–296
103. Dubin SB, Lunacek JH, Benedek GB (1967) Proc Natl Acad Sci USA 57:1164–1171
104. Nicoli DF, Benedek GB (1976) Biopolymers 15:2421–2437

Progr Colloid Polym Sci (2004) 127: 113–118
DOI 10.1007/b98020

S.E. Harding
E. Longman
A. Ortega
K. Kreusel
S.B. Tendler
K. King
J. García de la Torre

Use of the sedimentation coefficient for modelling antibodies. Refinements to the crystallohydrodynamics approach

S.E. Harding (✉) · E. Longman
National Centre for Macromolecular Hydrodynamics, School of Biosciences, University of Nottingham, Sutton Bonington LE12 5RD, UK
e-mail: steve.harding@nottingham.ac.uk
Tel.: +44-115-9516148

A. Ortega
J. García de la Torre
Departamento de Quimica Fisica, Facultad de Quimica, Universidad de Murcia, 30071 Murcia, Spain

K. Kreusel
S.B. Tendler
Pharmaceutical Sciences, University of Nottingham, Nottingham NG7 2GG, UK

K. King
Pfizer Limited, Sittingbourne, Kent ME9 8AG, UK

Abstract In a previous paper (Carrasco B, García de la Torre J, Davis KG, Jones S, Athwal D, Walters C, Burton DR, Harding SE (2001) *Biophysical Chemistry* 93:181–196) we introduced the concept of "Crystallohydrodynamics" for modelling the conformation of antibodies in solution and allowing for the so-called hydration problem. We now make some further improvements to allow for the hydrated dimensions of the domains before assembling models for the intact antibody using the algorithms HYDROSUB and SOLPRO. The procedure is illustrated by application to a human immunoglobulin construct of the subclass IgG4 from the mouse IgG1 (γ4P).

Keywords Bead-shell · Uniqueness · Hydration

Introduction

Although high-resolution crystal structures are readily available for the "fragment crystallisable" (Fc) and "fragment antigen-binding" (Fab) (also crystallisable) domains of the IgG and now IgE classes of immunoglobulins, the structure determination for intact, immunologically active antibodies has proved elusive, except for a few cases: hinge deleted mutants or antibodies with a short hinge region [1, 2, 3]. A further drawback is that the upper limit of molar mass in high-resolution NMR studies is around 50,000 Da, thus ruling this approach out for the study of the high-resolution structure of the intact molecule for molar masses of more than 146,000 Da.

Hydrodynamic methods can give low-resolution structural information in terms of orientation of domains. An early successful attempt was the first demonstration, using hydrodynamic bead modelling and the algorithm TRV [4], that IgE was cusp shaped [5]. The classical hydration problem (namely hydrodynamic parameters like the frictional ratio also depend on the time-averaged hydration as well as the conformation) was dealt with by comparison of the hydrodynamic properties of the IgE molecule with those for the hingeless mutant IgGMcg molecule, whose crystal structure was known.

An improved method of dealing with the hydration problem was given for modelling the IgG subclasses in the "crystallohydrodynamics" approach introduced earlier [6], which employed

1. The latest bead–shell approach (HYDRO/SOLPRO) for surface modelling the domains as surface ellipsoids.
2. The known crystal structures of the IgG Fab and Fc domains, which, when combined with measured

hydrodynamic properties of these domains, allowed a better estimate of the apparent time-averaged hydration of the domains (and, hence, from a weighted average, the intact antibody) to be made.

δ_{app} is referred to as "time-averaged" in the sense that hydration is a dynamic rather than a static process. It is referred to as an "apparent" hydration because besides volume effects it also includes the contribution of all other nonconformational and nonmass factors to the frictional coefficient. The value calculated for δ_{app} depends not only on the true hydration but is also affected by the fact that the domains are not true ellipsoid structures, the domains have considerable surface rugosity, and small imperfections in the bead model approximation exist (the hydrodynamic parameters for a bead and bead–shell model cannot be calculated exactly, as they can for ellipsoids). The δ_{app} for intact IgG antibody molecules was thus evaluated and this value, when combined with the experimentally measured translational frictional ratio of the intact antibody molecule, provided a route to obtaining the solution conformation of the intact molecule.

We now present a further refined approach which takes into account the hydrated dimensions of the domains.

Low-resolution information obtained

As before we would like to stress at the outset that although high-resolution information can be entered into hydrodynamic analysis, high-resolution information is not returned, a limitation which unfortunately is not always appreciated. Furthermore, an important drawback that limits the application of hydrodynamic methods is the influence of nonconformational parameters on the hydrodynamic properties – the most important of these is the time-averaged effect of water association with the protein or "hydration" – which is very difficult to estimate with any precision. Another drawback is that the most sophisticated shape for which hydrodynamic parameters can be calculated exactly is still the ellipsoid [7, 8]: although the overall conformation of antibody domains may be represented by equivalent ellipsoids, the intact immunologically active structure cannot be represented by either axially symmetric (prolate/oblate "ellipsoids of revolution") or centrally symmetric ("general triaxial ellipsoid") shapes. Good approximations are available in terms of multiple-sphere array or "bead models", and have been thoroughly checked against ellipsoids, yielding errors of no worse than around 2% for parameters based on the translational frictional property. Bead modelling strategies can therefore be applied with considerable confidence for representing domain orientations of antibodies.

For a typical globular protein, fine structural details (crevices, pockets, protrusions, etc.) can make a relatively large contribution to the hydrodynamics. However, for multisubunit structures (antibodies are a paradigmatic example) it seems evident that the main aspect is the arrangement of the subunits: whether or not there is a hinge (which may be flexible), and whether the conformation is more open or closed. As mentioned before, there is the complicating effect of hydration; therefore, it is really justified to reduce the complexity of the problem, by making structural approximations for the subunits, thus allowing us to concentrate on their spatial arrangement. This approach also greatly facilitates the modelling of the flexibility between domains [9].

Background to modelling methods

For the last 2 decades, bead modelling approaches have been popular for modelling the hydrodynamic properties of macromolecules, with the overall shape of the macromolecule being represented as an array of spheres of either uniform or nonuniform size [4, 10, 11, 12, 13, 14]. An alternative approach is to represent only the surface of the particle with identical "minibeads", in what is known as a bead–shell model. Such bead–shell models are not only valuable for calculating hydrodynamic parameters, but a decrease in the size of the beads enables a closer representation of the surface [15]. The hydrodynamic properties of multiple sphere arrays – beads or shells, although approximate, can be calculated to a high but finite degree of accuracy (normally better than around 2%).

In the studies of Carrasco and coworkers [6, 16] the main parameter used for representing the solution conformation for the human IgG antibodies was the Perrin function, P, or "translational frictional ratio due to shape". The Perrin function is a universal shape parameter, meaning it is independent of the size of the particle, and therefore the shape of the particle alone determines the value. In order to obtain the Perrin function of a particle, its sedimentation coefficient must first of all be combined with the partial specific volume and molar mass (molecular weight), in order to gain the translational frictional ratio, f/f_o. The translational frictional ratio is defined as "the ratio of the frictional coefficient of the macromolecule to that of a sphere of the same mass and anhydrous volume". In terms of experimental parameters

$$f/f_o = [M(1-\bar{v}\rho_o)/N_A s^o_{20,w}]/[6\pi\eta_0(3M\bar{v}/4\pi N_A)^{1/3}], \quad (1)$$

in which M is the molar mass (grams per mol), $\bar{v}$ is the partial specific volume (millilitres per gram), N_A is Avogadro's constant (6.0221×10^{23} mol^{-1}) and $s^o_{20,w}$ is the sedimentation coefficient (seconds or Svedbergs) corrected to the standard conditions of density (ρ_0, grams per millilitre) and viscosity (η_0, poise) of water at 20.0 °C, and extrapolated to infinite dilution.

In order to find the Perrin function value from the translational frictional ratio, the time-averaged apparent hydration value (δ_{app}, in grams of water per gram of protein) for the particle must be known:

$$P = (f/f_o)[1 + (\delta_{app}/\bar{v}\rho_o)]^{-1/3}. \quad (2)$$

With the hydration value already being available for the Fab domain [16], Carrasco et al. [6] used the same approach to estimate the hydration value for the Fc domain. This approach was to fit an inertial triaxial ellipsoid to the surface of the crystal structure of the Fc domain. The two axial ratios a/b and b/c [6, 16] from this ellipsoid are then entered into the routine ELLIPS2 [8], which specifies the value of P. When piecing the molecule together, ellipsoids cannot be used directly, but have to be converted to surface bead–shell models for the hydrodynamically equivalent ellipsoids of revolution. Carrasco and coworkers [6, 16] showed that the calculated P values for the bead–shell models were in excellent agreement with those for ellipsoid models. Combining the frictional ratio of the particle with its Perrin function gives the apparent hydration value, δ_{app} (Eq. 2). The apparent hydration value for the intact antibody can therefore be found once the apparent hydration values for the Fc and Fab domains are known. P, and the experimental f/f_o for each domain then yield δ_{app}. δ_{app} for the whole antibody is approximately $[2\delta_{app}(\text{Fab}) + \delta_{app}(\text{Fc})]/3$. This value, combined with the experimentally measured f/f_o for the intact antibody then yields an experimental P, which can then be compared with the P values calculated for the various bead models.

Using this procedure Carrasco et al. [6] found, for example, IgG4 to be effectively T-shaped and hingeless, resembling the hingeless mutant antibody IgGMcg, shown to be in agreement with X-ray crystallography [1].

The refined crystallohydrodynamics approach

The same approach as introduced before [6] is utilised to estimate the apparent time-averaged hydration of an IgG antibody, i.e., from representing the crystallographic structure of the individual domains as a smooth ellipsoidal surface shell–bead model, calculating the Perrin translational frictional ratio due to shape, P, and comparing this with the experimentally obtained translational frictional ratio, f/f_o, to give the apparent time-averaged hydration, δ_{app}, for each domain. Hence δ_{app} of the intact IgG can be estimated on the basis of IgG having two Fabs and one Fc. This value is necessary to calculate the experimental P from f/f_o for the intact antibodies, whose crystal structures are generally not known. The refinement we make here is that when modelling the assembled domains, even though P is size-independent, we take into account the effect of different hydrations on the relative dimensions of the domains, i.e., we find the dimensions of the ellipsoids of revolution, whose hydrodynamic properties mirror those of the antibody fragments they are representing, and the algorithm HYDROSUB is used to specify the surface bead coordinates of the ellipsoidal surfaces used for the Fc and Fab domains.

Finding the relative dimensions of the hydrated domains

For an ellipsoid of revolution of axial ratio $p = a/b$ ($a > b$), with a prolate ellipsoid having semiaxial dimensions (a,b,b), and an oblate ellipsoid having the dimensions (a,a,b), a and b can be obtained provided the axial ratio and the hydrated volume of the particle, V_{hyd}, are known:

$$V_{anh} = M\bar{v}/N_A, \quad (3)$$

$$V_{hyd} = (M\bar{v}/N_A)(1 + \delta/\bar{v}\rho). \quad (4)$$

For a prolate ellipsoid,

$$V_{hyd} = (4/3)\pi b^3 p. \quad (5)$$

For an oblate ellipsoid,

$$V_{hyd} = (4/3)\pi a^3/p. \quad (6)$$

The axial ratio p is not to be confused with the Perrin function P. We have to assume the effect of hydration does not change the axial ratio significantly of course, but if we are using the value for p obtained from the Perrin function and δ (approximated as δ_{app}) both according to the crystallohydrodynamic procedure we can obtain an estimate for a and b for both the Fab domain (represented as a prolate model) and the Fc domain (represented as an oblate model).

Method for construction of bead models: HYDROSUB

Another improvement to our earlier [6] approach is that we can now take advantage of using the computer program HYDROSUB [15], for the construction of the bead models. HYDROSUB has the capacity to build models of multisubunit macromolecules, by modelling them as structures composed of ellipsoids of revolution and cylinders: the coordinate data can then be interfaced directly into the algorithm SOLPRO for the calculation of the hydrodynamic properties and shape functions – including the Perrin P parameter, corresponding to the bead model and domain arrangement.

To build the bead–shell model, the program requires the input of four pieces of information. For each ellipsoid, HYDROSUB requires (1) the dimensions a and b, (2) the three Cartesian coordinates that define the position of the centre, and (3) the two polar angles θ and ϕ that define the orientation of the main axis of the ellipsoid. There is another improvement on the earlier approach where we used three Euler angles [6]; we have realized (and this is implemented in HYDROSUB) that for an axisymmetric particle the orientation is defined by just two spherico-polar angles, θ and ϕ (Fig. 1).

We now provide an example of how the IgG model for HYDROSUB can be constructed (Fig. 1). Fc is represented as an oblate ellipsoid, and then we place its equatorial circle on the (y,z)-plane, with its centre on the negative part of the z-axis, at a distance, d_3, from the origin equal to (or slightly greater than, if we wish to consider some spacing, hinge, etc.) the longest semiaxis. Thus, its coordinates are $x_{c3}=0$, $y_{c3}=0$, $z_{c3}=-d_3$. With this placement, the main axis of the oblate Fc is along the x-axis, and therefore its polar angles are $\theta_3=90$, $\phi_3=0$. Now we have to situate the two Fabs (represented as prolate ellipsoids), which will be situated above the (x,y) horizontal plane. The essential choice is that of the angles. For one of the Fab ellipsoids we adopt certain values of θ_1 and ϕ_1. The distance from the centre of the ellipsoid, d_1, to the origin will be equal to or greater than the longest semiaxis of the prolate ellipsoid, and the coordinates of the centre are $x_{c1}=d_1\sin\theta_1\cos\phi_1$, $y_{c1}=d_1\sin\theta_1\sin\phi_1$ and $z_{c1}=d_1\cos\theta_1$. For the other Fab, d_2 is the same, and in principle there will be a variability in the choices of θ_2 and ϕ_2. However, we restrict ourselves to symmetric configurations, having a Y or a T shape, where the essential parameter is the angle, β, between the two arms. Then with the values of θ_1 and ϕ_1, the value of β, and the condition that Fc is on the plane of the angle or bisecting the angle, then the polar angles θ_2 and ϕ_2 can be deduced from geometric arguments. Of course, any other choice of axes, and any other initial placement of the first subunit, can be made. The details of this description would change, but the procedure would be equivalent.

The program then builds the bead–shell model by stacking rings of minibeads, in the direction of the symmetry axis, varying the radius of the rings to form an ellipsoidal shape (or not varying this radius in the case of constructing a cylinder). Thus, the result is a structure in which the ellipsoidal surface is formed by the central points of the minibeads.

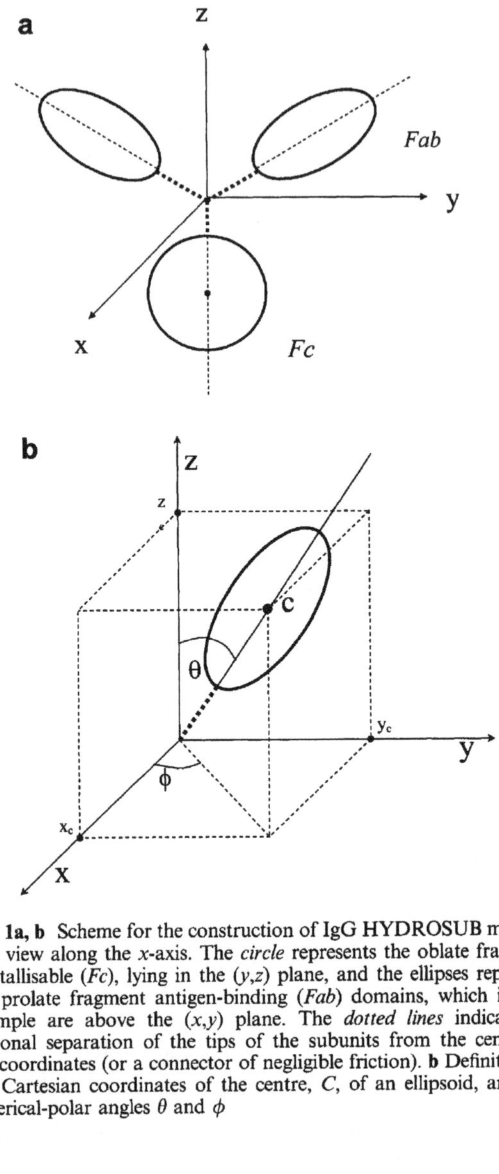

Fig. 1a, b Scheme for the construction of IgG HYDROSUB models. **a** A view along the x-axis. The *circle* represents the oblate fragment crystallisable (*Fc*), lying in the (y,z) plane, and the ellipses represent the prolate fragment antigen-binding (*Fab*) domains, which in this example are above the (x,y) plane. The *dotted lines* indicate an optional separation of the tips of the subunits from the centre of the coordinates (or a connector of negligible friction). **b** Definition of the Cartesian coordinates of the centre, C, of an ellipsoid, and the spherical-polar angles θ and ϕ

The number of beads, N, is clearly dependent upon the number of minibeads, and this affects the shape of the ellipsoidal surface produced and therefore the resulting properties produced by the program. In order to overcome this problem, the program calculates the value of each property, for a range of bead sizes, and then extrapolates forward to an infinite number of beads. One restriction is that the time taken to run a program to construct a shell model is proportional to N^3, and as a result the program works with a maximum of 2,000 minibeads. Once the bead–shell model of the antibody is constructed, an output file produced by HYDROSUB can be run with SOLPRO

Table 1 Theoretical SOLPRO Perrin function values, *P*, obtained for antibody bead–shell models with varying angles θ and ϕ. Unless otherwise stated, the position of the fragment crystallisable (*Fc*) domain is represented by the orientation angles θ = 90 and ϕ = 90. β represents the angle between the main axis of the ellipsoids representing the fragment antigen-binding (*Fab*) domains. The spaces in the models represent the distances between the Fc ellipsoid and each Fab ellipsoid. Entries shown in *bold* correspond to those models which fit the experimental range of values of *P* between 1.195 and 1.230 for γ4P

Space	Angles between main axis of ellipsoids	P
None	β = 90, θ = 90, ϕ = 45/315	1.177
	β = 80, θ = 90, ϕ = 40/320	1.170
	β = 120, θ = 90, ϕ = 60/300	1.180
	β = 120, θ = 90, ϕ = 60/300 (oblate, θ = 0, ϕ = 90)	1.173
	β = 180, θ = 90, ϕ = 90/270	1.15
	θ = 70, ϕ = 45/315	1.168
	θ = 70, ϕ = 40/320	1.165
	β = 90, θ = 45, ϕ = 90/270	1.107
5 Å	**β = 90, θ = 90, ϕ = 45/315**	**1.208**
	β = 80, θ = 90, ϕ = 40/320	**1.200**
	β = 45, θ = 45, ϕ = 57.24/122.77	1.131
	θ = 45, ϕ = 90/270	1.135
	β = 90, θ = 50, ϕ = 22.62/157.38	1.166
	β = 180, θ = 90, ϕ = 0	1.160
	β = 90, θ = 90, ϕ = 45/315 (oblate, θ = 0, ϕ = 90)	**1.199**
	β = 120, θ = 90, ϕ = 60/300	**1.204**
8 Å	**β = 90, θ = 90, ϕ = 45/315**	**1.225**
10 Å	β = 90, θ = 90, ϕ = 45/315	1.237
	β = 120, θ = 90, ϕ = 60/300	1.246
	β = 180, θ = 90, ϕ = 90/270	**1.208**
	β = 90, θ = 90, ϕ = 45/315 (oblate, θ = 0, ϕ = 90)	**1.229**
20 Å	β = 90, θ = 90, ϕ = 45/315	1.294
25 Å	β = 90, θ = 90, ϕ = 45/315	1.320
	β = 120, θ = 90, ϕ = 60/300	1.330
	β = 180, θ = 90, ϕ = 90/270	1.289

Fig. 2 One of the bead–shell models with a Perrin function value matching that of the experimentally measured value for the IgG4 chimeric antibody γ4P. Domain angles θ=90, ϕ=45, 10 Å between domains (shown as frictionless spacers). Theoretical P=1.229. Experimental range of allowed values P_{exp}= 1.195-1.230

[17] in order to obtain the full spectrum of hydrodynamic parameters.

If the calculated Perrin function matches the experimental Perrin function of the antibody, the model represents a potential conformation for the antibody. This procedure may then be repeated many times, to investigate many different domain positions, and identify models fitting the experimental data.

Case study: an IgG4 construct

We illustrate the refined procedure by application to an IgG4 construct, known as γ4P [18] with the following hydrodynamic properties as reported by Longman et al. [19]: $s^{o}_{20,w}$=(6.80 ± 0.10) S, M=147 kDa and $\bar{v}$ = 0.730 (millilitres per gram). According to Eq. (1) this gives a range of values (allowing for experimental error) for f/f_0 of between 1.456 and 1.499, and using the estimate for the time-averaged apparent hydration for IgGs found previously of 0.59, we estimate the experimental Perrin function, P_{exp}, as between 1.195 and 1.230. We can then check off from a table of values for *P* for the various models to see which ones fit the data.

The various combinations of the θ and ϕ angles used for the construction of candidate bead–shell models using HYDROSUB and their resultant Perrin function values obtained from SOLPRO are shown in Table 1. Models representing antibodies with no spacing between the domains all provide *P* values significantly below the P_{exp} values determined. However, by including frictionless linkers between the domains, the *P* value is seen to increase, as expected for more extended structures. A bead–shell model with a similar *P* value to that of the γ4P chimera, and therefore potentially representing the structure of this chimera in solution, is shown in Fig. 2. Other models matching the (γ4P) chimera data contained either 5-Å spaces between the domains and 90–120° between Fab domains or 10 Å between the domains and 90–180° between Fab domains and are specified in bold type in Table 1. The spacing of domains in the final, more open bead–shell model structure shows similarity to those previously produced for murine IgG1 and IgG2a molecules through X-ray crystallography [2, 3].

It will be patently clear from Table 1 that more data are required to be specific about the configuration since a wide range of models will have *P* values agreeing with the experimental *P* values: this is the classical "uniqueness problem" [4, 7]. Data from viscosity and low-angle X-ray scattering are required, but this is a subject for a further study.

Acknowledgements The authors acknowledge support from grant BQU2000-0229 from the Direccion General de Investigacion (Ministerio de Ciencia y Tecnología, Spain) and Pfizer Limited.

References

1. Rajan SS, Ely KR, Abola EE, Wood MK, Colman PM, Athay RJ, Edmundson AB (1983) 3- Mol Immunol 20:787–799
2. Harris LJ, Larson SB, Hasel KW, McPherson A (1997) Biochemistry 36:1581–1597
3. Harris LJ, Skaletsky E, McPherson A (1998) J Mol Biol 275:861–872
4. García de la Torre J (1989) In: Harding SE, Rowe AJ (eds) Dynamic properties of biomolecular assemblies. Royal Society of Chemistry, Cambridge, UK, pp 3–31
5. Davis KG, Glennie M, Harding SE, Burton DR (1990) Biochem Soc Trans 18:935–936
6. Carrasco B, García de la Torre J, Davis KG, Jones S, Athwal D, Walters C, Burton DR, Harding SE (2001) Biophys Chem 93:181–196
7. Harding SE (1989) In: Harding SE, Rowe AJ (eds) Dynamic properties of biomolecular assemblies. Royal Society of Chemistry, Cambridge, UK, pp 32–56
8. Harding SE, Horton JC, Cölfen H (1997) Eur Biophys J Biophys Lett 25:347–359
9. García de la Torre J, Perez Sanchez HE, Ortega A, Hernandez JG, Fernandes MX, Diaz FG, Lopez Martinez MC (2003) Eur Biophys J 32:477–486
10. Byron O (1992) Solution studies on the conformation and assembly of the monoclonal antibody B72.3. University of Nottingham, UK
11. Byron O (2001) Methods Enzymol 321:278–304
12. Carrasco B, Harding SE, García de la Torre J (1998) Biophys Chem 74:127–133
13. Carrasco B (1998) Propiedades de macromoleculas rigidas en disolucion: Modelos, metodos computacionales y analysis de los datos experimentales. Universidad de Murcia, Spain
14. Carrasco B, García de la Torre J (1999) Biophys J 76:3044–3057
15. García de la Torre J, Carrasco B (2002) Biopolymers 63:163–167
16. Carrasco B, García de la Torre J, Byron O, King D, Walters C, Jones S, Harding SE (1999) Biophys J 77:2902–2910
17. García de la Torre J, Carrasco B, Harding SE (1997) Eur Biophys J 25:361–372
18. Kreusel KM, Adair JR, Eeley NRA, Davies MC, Jackson DE, Roberts CJ, Tendler SJB, Williams PM (1994) J Vac Sci Technol B 12:1517–1520
19. Longman EJ, Kreusel KM, Tendler SJB, Fiebrig I, King K, Adair JR, O'Shea P, Ortega A, García de la Torre J, Harding SE (2003) Eur Biophys J 32:503–510

Progr Colloid Polym Sci (2004) 127: 119–125
DOI 10.1007/b98021

A.D. Molina-García

Bacteriophage fd: exactly how flexible?

A.D. Molina-García (✉)
Instituto del Frío, C.S.I.C.,
José Antonio Novais 10,
Ciudad Universitaria,
28040 Madrid, Spain
e-mail: ifrm111@if.csic.es
Tel.: +34-91-5445607
Fax: +34-91-5493627

Abstract The particles from filamentous bacteriophages are generally considered to be flexible. This assumption originally arose from the observation of electron micrographs where the virions appeared as long and randomly bent filaments. How accurately does this aspect of filamentous viruses represent the behaviour in solution? While the conditions of microscopic experimentation are very different from those of solution or the natural environment, other observations may correspond to more rigid particles. Hydrodynamical measurements are a help to ascertain the actual degree of flexibility of these particles in the more real "three dimensional" solution conditions. Employing sedimentation equilibrium and sedimentation velocity data for bacteriophage fd, and mathematical modelling, we explore the behaviour of these particles here. The relative insensitivity of the data pair (sedimentation coefficient–translational diffusion coefficient) so obtained is confirmed and the superiority of the conformational information contained in other parameters, such as the rotational diffusion coefficient and the intrinsic viscosity. Despite this lack of sensitivity, the straighter model conformations satisfy better the experimental data, which can be an indication of a relatively rigid viral particle.

Keywords Filamentous bacteriophage · Flexibility · Conformation

Introduction

Filamentous bacteriophage fd, a nearly micron-long viral particle, is typical of innovirus bacteriophages. The capside of this virus is formed by a single molecule of circular ssDNA and several thousands of identical capsomers, apart from some minor proteins [1, 2, 3, 4]. A difference from with the well-known plant virus tobacco mosaic virus is the lack of structural independence of its capside and capsomers from the nucleic acid component. So, while both empty capsides (completely assembled protein helicoidal structures) and soluble protein monomers are found in the tobacco mosaic virus, the fd protein capsomer is not stable in solution by itself, and capside assembly is only achieved after induction by the DNA molecule. This is related to a number of electrostatic interactions taking place among charged groups in each protein subunit and the DNA molecule [5, 6].

Filamentous viruses are parasites of a number of bacteria. The infection process involves penetration through the pili (employed by male bacteria to transfer genetic material) and they do not induce cell lysis or otherwise death, as they can abandon the cell without membrane disruption. Phages have a significant role in the physiology of natural bacteria and they are employed to transform them, as a simple way to introduce genetic material. Other uses of these viruses are the immunologic studies known as "phage display" [1, 7].

Different studies on bacteriophage fd and other filamentous bacterial viruses, carried out by a range of techniques, primarily analytical ultracentrifugation and electron microscopy, yielded their morphological properties [8, 9, 10, 11, 12]. The agreement for the fd molecular weight and the overall dimensions is generally good. The same can be said for some hydrodynamic parameters, such as sedimentation or diffusion coefficients [13, 14, 15]. Even more refined and detailed structural information is available, allowing a full description of the protein subunits and their interactions with the DNA molecule [16, 17, 18].

The viral particles of filamentous viruses are traditionally considered as flexible. This assumption derives primarily from their aspect of randomly bent filaments in electron micrographs. Nevertheless, electron microscopy subjects samples to harsh treatment and drastic alterations from the conditions of their solution state or the "natural" ones, where their interaction with bacterial cells takes place. The conditions of phage-display assay, where an epitope of interest is expressed in the outer surface of the modified capsomeric protein for the study of its interaction with antibodies, are also of interest [7, 19]. Data on its flexibility in these conditions are not in complete agreement, but some evidence points to particles relatively rigid in solution, or at least more of which could be expected form their observed micrographic aspect [11, 20, 21]. Other data obtained by indirect methods are indicative of a higher degree of flexibility [22].

The information that can be obtained from ultracentrifugation studies alone is somehow limited and can be insufficient to characterise some conformational properties, such as flexibility, degree of elongation or even relatively small length variations in a particle [23]. These types of properties are better described by other parameters, such as the rotational diffusion coefficient or the intrinsic viscosity. In this article, and as a test to the sensitivity of these techniques to small conformational peculiarities, solely from ultracentrifugation data (equilibrium sedimentation molecular weight combined with sedimentation coefficient) it is endeavoured to find reasonable limits for the possible conformations of the fd particle and to gain some information on its flexibility.

Methods

A number of conformational models were built in order to compare the fit of the predicted hydrodynamic properties to the available experimental bacteriophage fd data. This will also allow us to evaluate how useful ultracentrifugation data are for investigating conformational differences. Three criteria were followed in model selection:

1. Widely different shapes covering a range of radii of gyration.
2. Hinge shapes for a range of angles.
3. A weakly bent shape following some of the conformations observed in electron microscopy.

Several planar (two-dimensional) conformations of distinct gyration radius, R_G (as a measurement of their "bending degree") are shown in Fig. 1 (conformations a–e). Starting form a straight rod conformation (a), several forms were generated: a 90 ° bent conformation (b), a closed loop (actually rectangular, c), a "saw" shape (d), and a planar spiral (e). Following the same scheme as in conformation b, a rod with a hinge at its midpoint, other models were built for hinge angles 130 ° (f), 140 ° (g), 150 ° (h), 160 ° (i) and 170 ° (j) (not shown). Two three-dimensional conformations were considered: a double loop (two perpendicular planar loops, k), two views of which are depicted in Fig. 2, and a weakly bent conformation which was constructed after the observation of electron micrographs of filamentous viruses (Fig. 3, l). The virion in the electron micrograph used as a model for the conformation constructed here was actually not fd, but negatively stained fl bacteriophage (only different in a few nucleotides) [3]. The degree of bending observed was extended to a third dimension, as all particles in the micrographs are planar, which is, presumably, an artefact. All model figures presented here are for illustrative purposes and to allow comparison with the electron microscopy images. A simple computing routine was employed for this purpose and it was faithful to the backbone of the model, while the relations and contacts among beads were not accurately represented.

The models were built as strings of equal size beads, for simplicity. Taking into consideration the comparatively large errors in the parameters to be fitted and the uncertainties in the particle dimensions, more accurate modelling was deemed unnecessary. The use of equal size beads introduces an error in the particle length, as it is constrained to an exact multiple of its diameter. This

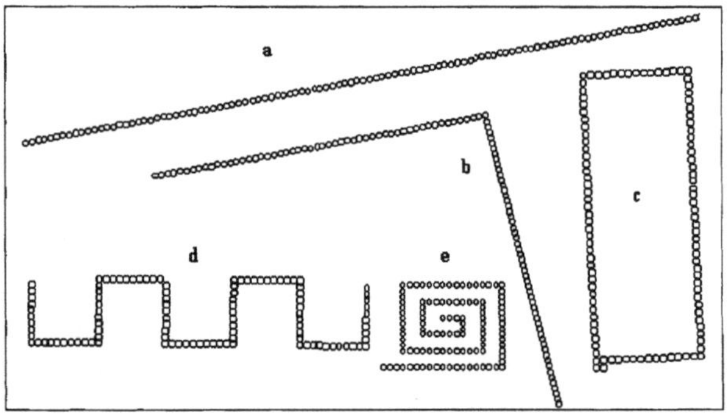

Fig. 1 Sketch of some of the model conformations studied (only the backbone is accurately represented, while bead contacts are just approximate). The labelling *a–e* is used to designate each conformation. Shapes *f-j* follow the same scheme as that for conformation *b*, but the angle between both hemiparticles is larger than 90 ° (see text)

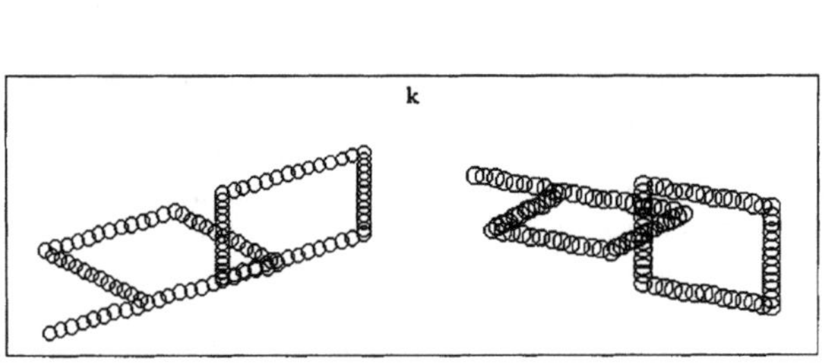

Fig. 2 Two schematic views of a three-dimensional model conformation (*k*)

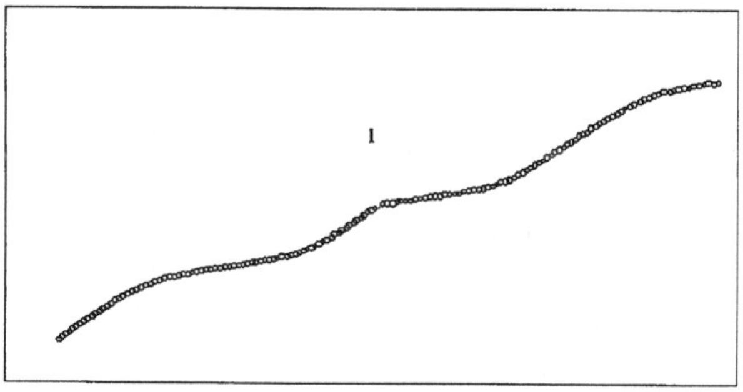

Fig. 3 Sketch of a model conformation (l): three-dimensional shape constructed after a filamentous virus electron micrograph [3], adding an extra degree of bending as the third dimension

could, again, have been avoided by the addition of an exact length "cap" constructed of smaller beads, but instead the error so introduced was evaluated and was found to be less than 1% for all parameters derived (but for intrinsic viscosity), and was considered acceptable.

Whenever possible, "unrealistic" conformations priming straight lines were used. This did not introduce special errors, as the aim was to compare the general degree of bending rather than structural details. In addition, it avoids the problem of the possible alteration of particle dimensions by bending. The bending of macroscopic rods is often the cause of length and diameter alterations. In building a macromolecular scale model whether the beads are to be kept tangential or if the centre-to-centre distance has to be adjusted is unclear and will depend on the inner particle structure. As the straighter shapes are the more likely conformations for real fd particles, the invariance of the length was considered essential for comparison. In this case consecutive beads were in tangential contact with each of two neighbours, so the contour length, or rather the model particle length following the centre of each bead, is invariant with each particular conformation.

This simple modelling obviously corresponds to a uniform mass distribution. fd particles are known to have a protein assembly on one end, which can contribute to inhomogeneities, but it has not been accounted for. The models were altered (by changing the bead diameter and/or the number) for testing alternative dimensions of the particles.

Despite its design for nonflexible particles, considering the degree of rigidity expected for fd bacteriophage, the program HYDRO [24] was employed to calculate the hydrodynamic parameters. The coordinates of the beads corresponding to the different model conformations and dimensions were fed into the program, together with the solution properties corresponding to the experimental hydrodynamic data. A range of molecular weights, M, and partial specific volumes, v_p, were also employed.

Results and discussion

Among the different data available for geometric and hydrodynamic properties of bacteriophage fd, this study was centred primarily on those of Table 1, as they keep a higher degree of homogeneity among them, both with respect to the viral isolates employed and with the experimental techniques and measurement conditions employed [5, 11, 14]. Other, slightly different, reported

Table 1 Geometric and hydrodynamic parameters of fd bacteriophage virions employed for fitting the different conformations

Parameter	Selected value[a]
L (nm)	866 ± 37 [5]
d (nm)	8.0 ± 0.4 [11]
M (10^6)	15.1 ± 1.5[b] [14]
v ($cm^3\ g^{-1}$)	0.720 ± 0.002 [11]
$D_{t(20,w)}$ ($10^{-8}\ cm^2\ s^{-1}$)	2.37 ± 0.29 [14]
$s^o_{20,w}$ (10^{13} s)	40.6 ± 0.8 [14]

[a]Selected because all values correspond to the same viral isolate and/or similar experimental methodology and conditions
[b]Weight-average molecular weight (corrected for nonideality)

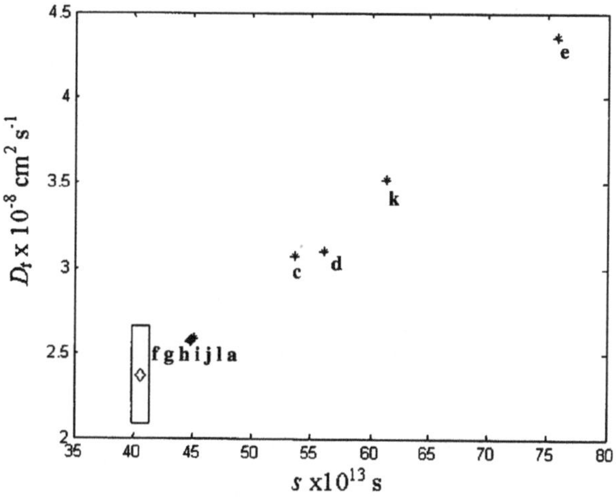

Fig. 4 s and D_t data pair (*asterisks*) obtained for each model conformation. The M and v_p employed for the calculations were 15.1 and 0.720 $cm^3\ g^{-1}$, respectively. Each conformation is denoted by its label. The experimental s and D_t values are shown by a *rhombus* and the *box* around this point is the experimental error interval

data were, nevertheless, considered. The homogeneity of the viral particle preparations was conveniently checked by different methods (i.e., light scattering, sharpness of ultracentrifugation boundary).

When the spatial coordinates of the different model conformations adjusted to the "optimal" length, L, (866 nm) and diameter, d, (8 nm) were fed to HYDRO, a set of hydrodynamic properties was obtained. Figure 4 shows a comparison of the values for the sedimentation coefficient, s, and the translational diffusion coefficient, D_t, so obtained with those derived experimentally from ultracentrifugation measurements (D_t derived from application of the Svedberg equation to the combination of s and the molecular weight, M, obtained from equilibrium sedimentation) [14].For simplicity, throughout the article, the symbol s will be employed for the sedimentation coefficient extrapolated to infinite dilution and corrected to water and 20 °C, instead of $s^o_{(20,w)}$, and

D_t will be employed for the translational diffusion coefficient extrapolated to infinite dilution and corrected to water and 20 °C, instead of $D^{o}_{t(20,w)}$ All strongly bent conformations yielded values quite different from the experimental data and their error area. On the other hand, several weakly bent model forms were quite similar in their s and D_t, despite their conformational differences.

A value of $(2.30 \pm 0.04) \times 10^{-8}$ cm^2 s^{-1} (after buffer and temperature corrections) reported for D_t [11] was within the error margin of Table 1, and, as can be seen in Fig. 4, does not improve the fit. The alternative s value, $(41.7 \pm 0.3)\times10^{13}$ s after correction [11] would not alter the general data agreement. Similar agreement could be found with a set of data obtained for the very similar bacteriophage M13 [$s=(43.0 \pm 2)\times10^{13}$ s, $D_t=(2.33 \pm 0.1)\times10^{-8}$ cm^2 s^{-1}, $M=(16.5 \pm 1)\times10^{6}$] [15]. The shift in s, providing a better fit, was compensated by the displacement of the calculated sedimentation coefficient when the higher molecular weight was employed.

When the s and D_t values so obtained were represented versus R_G, as a measurement of the degree of bending of a given conformation (Fig. 5), a similar conclusion was drawn.

For further study, four conformations were chosen among those better fitting the experimental data: a (straight rod), l (weakly bent), j (170 ° hinge) and, for a comparison of how sensitive ultracentrifugation data could be to this angle, b, the 90 ° hinge conformation. Although the agreement for the physical properties of fd viral particles is generally good, a certain degree of uncertainty exists for all parameters. The molecular weight employed here is not discordant from those in the literature. These go from the multitechnical 14.3-14.7×10^6 [12] to the hydrodynamically obtained value of $(16.4 \pm 0.6)\times10^6$ [11]. Even when accepting the 15.1×10^6 value [14], the wide error margins (1.5×10^6) must not be forgotten. Figure 6 shows how small variations in molecular weight were reflected in changes in the calculated sedimentation coefficient. This variation (s decreased by approximately 1% for a 1% increase in M) was much larger than the differences originating from the distinct conformations considered, which is indicative of the difficulty to distinguish among them solely with s and D_t as experimental parameters.

The particle length is also open to discussion. Other reported values range from 870 to 940 nm [8, 9, 10, 11, 12, 25]. They were obtained by a variety of techniques, including electron microscopy and combinations of hydrodynamic measurements. For small length variations (and using the a – straight rod – conformation for calculations, though other similarly extended conformations yielded essentially equivalent results – data not shown) a 1% increment in the length caused 0.81 and 0.78% decreases in s and D_t, respectively. Again, a better fit could be obtained for the model when the length was

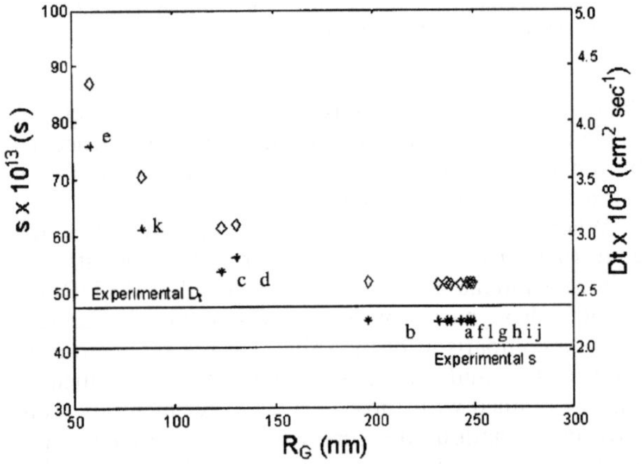

Fig. 5 The dependence of sedimentation (*asterisks*) and translational diffusion coefficient (*rhombus*) on the radius of gyration for each model conformation. The M and v_p employed for the calculations were 15.1 and 0.720 cm^3 g^{-1}, respectively. The labels identify each conformation, while *horizontal lines* show the experimental s and D_t values.

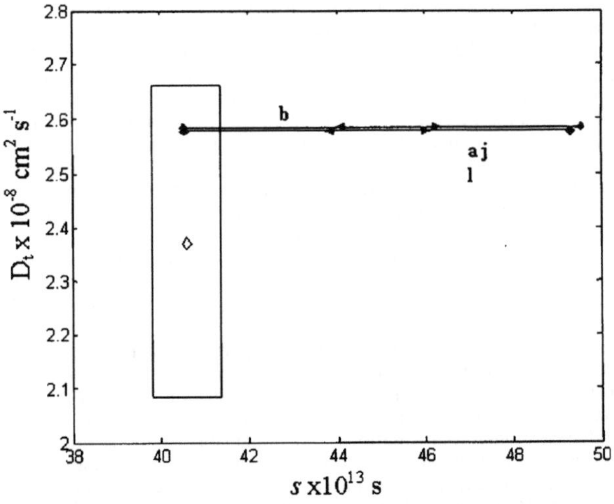

Fig. 6 The dependence of the s and D_t data pairs obtained for model conformations a, b, j and l, as a function of M and v_p. The *lines* between *black squares* show the intervals for M ranging from 13.6×10^6 to 16.6×10^6 (corresponding to the experimental error 1.5×10^6), keeping v_p as 0.720 cm^3 g^{-1}. The *lines* between *arrowheads* show the intervals for v_p ranging form 0.7128 to 0.7272 cm^3 g^{-1} (corresponding to a 1% variation). Conformations are denoted by their labels. The experimental s and D_t values are shown by a *rhombus* and the box around this point is the experimental error interval. It can be appreciated how the calculated s and D_t for the four different conformations are very close (for conformations a, j and l, the differences in s and D_t are less than 0.05%). Small changes in molecular weight and partial specific volume have a much more noticeable effect than significant conformational alterations

so adjusted, but little difference was observed among the diverse conformations. In this respect, the variation in the calculated hydrodynamic parameters caused by the addition or removal of one bead (for conformation a) corresponded to 0.9% L alteration and was reflected in approximately 0.7% changes in s and in D_t.

The diameter, on its side, though more disparate in percentage (from 6 to 9 nm [8, 21, 25]) and in spite of its quadratic influence on the whole particle volume, has a lesser effect, owing to the highly asymmetric particle shape. A 1% increase in d produced a decrease in s of only 0.09% and in D_t of 0.19% (again, for the a conformation).

Other hydrodynamic parameters more related to conformational alterations appeared to be more sensitive to length variations. A 1% length increase caused a 2.7% increase in the intrinsic viscosity and a 1.1% increase in the rotational diffusion coefficient. Also, the three "closer" conformations in terms of s and D_t (**a**, **j** and **l**), differing by less than 0.5% in both s and D_t, showed a difference in the intrinsic viscosity of 10% and in the rotational diffusion coefficient as high as 100%.

The value of the partial specific volume employed, 0.720 ± 0.008 $cm^3\ g^{-1}$ [11] can also be criticised. An uncertainty of 1% in v_p is usually considered. The value employed was obtained at similar conditions as those used for the ultracentrifugation experimental studies (15 mM KCl, 15 mM sodium phosphate buffer, pH 7) [14], so further hydration corrections were not considered necessary. Anyhow, the data were very sensitive to this parameter. Figure 7 shows how the fit to s and D_t is dramatically improved as v_p is increased to around

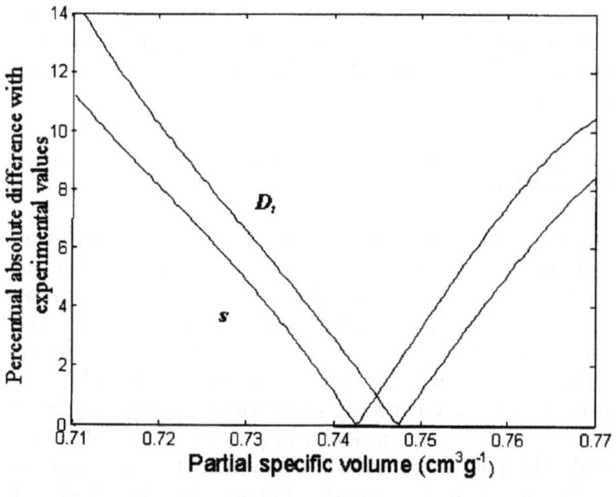

Fig. 7 Dependence of the difference between s and D_t experimental and calculated values on the partial specific volume employed for the calculations. M was kept constant at 15.1×10^6. A straight rod, a, conformation with $L = 866$ nm and $d = 8$ nm was assumed

0.765 $cm^3\ g^{-1}$. The experimental values of D_t were also altered, according to v_p, as it is employed in the Svedberg equation for obtaining D_t [14]. Meanwhile, corrections to water and 20 °C, which also made use of the partial specific volume, are less sensitive and a 1% change in v_p caused an alteration in s smaller than 0.6%. A value of 0.76 $cm^3\ g^{-1}$ may be unreasonably high, and again, the wide error margins for D_t and s must be remembered.

fd viral particles, whatever their intrinsic degree of flexibility, are constrained to a certain degree of "straightness", by the existing external charge. It has been evaluated as 10 e nm^{-1} at pHs close to 7 [26]. A strong dependence of solution factors such as buffer, counterions or macromolecular crowding can thus have a dramatic effect in this repulsion-driven conformation, by shielding the contribution of the electric repulsion. Another restraint would be that derived from the need to keep all the interprotein and protein–DNA interactions within the particle. fd particles are considered as very stable and the forces between neighbouring protein subunits in the virions as especially strong [27]. The viral particle has been described as rigid, stable and ordered after IR linear dichroism studies [21]. It must be noted that electron micrographs show no sharp bends in fd particles, unless particle breakage is found [1, 3, 28, 29, 30, 31].

Physical situations where the conformation of fd particles could have a noticeable effect include electron microscopy, phage display, interactions with the host and studies of highly concentrated solutions. In all these cases some of the conditions so carefully avoided in hydrodynamic studies are present. Electron microscopy is likely to alter particle hydration and the solvation layer. Counterion concentration can wildly change in the water film remaining around the particle before total drying, which can alter completely the electrostatic stability of its conformation. The metallic conductive base is, no doubt, having a role in this behaviour, and so the dyes, exerting a surface tension component on the particle, and even the effects of the electron beam could erase any memory of previous shapes. The two-dimensional character of the conformations seen in electron microscope studies is an exponent of how much alteration has actually taken place.

Both phage-display and physiological interactions, such as those with hosts, are of biological interest. The bacterial host pili are contacted endwise, while phage-display can use the whole accessible viral surface for its interaction with antibodies, enzymatic systems, etc. Alternatively, a minor capside protein can be employed for expression of the desired epitope or protein domain, and in this case also a terminal approach to bind to the target will take place [7, 32]. A highly bent particle would be inconvenient for both cases, as it will reduce the accessibility to the active particle end or to part of the surface area. Anyhow intermolecular interactions are, per se, a cause of alteration of factors such as solvation layers and counterion shielding, and the actual behaviour of the virion would not be fully predictable. In hydrodynamic studies it the assumption of ideality has been made, i.e., noninteracting bodies. Highly concentrated solutions of fd have also been studied often. Their properties are of as much interest as that of a monodisperse homogeneous particle and similar observations to those reported in more dilute conditions, in terms of conformation and flexibility, have been made [e.g. 33, 34, 35, 36, 37]. But the macromolecular crowding situation has little to do with that existing in dilute solutions.

A persistence length of 2,000 ± 200 nm has been reported for the fd viral particle, obtained by (static) electron microscopic data and (dynamic) birefringence high-concentration measurements [36, 38]. For the very similar bacteriophage M13 a close value to 2,200 ± 200 nm has been obtained [15]. While hydrodynamic studies have accurately modelled fd by a rigid rod with a small length reduction [11], other studies found a much shorter persistence length (of about 760 nm on average) [22]. When considering a simple model such as the hinge conformation described here (b-j), the calculation of the persistence length can be easily made using Eq. (1) [39]:

$$\langle R^2 \rangle = 4PL\{1 - (2P/L)[1 - \exp(-L/2P)]\}, \qquad (1)$$

where R is the end-to-end (straight line) distance, P the persistence length and L the contour length (taken here as the constant particle length).

This allows the visualisation of particles of $P = 2{,}000$ nm as hinge conformations with an average angle of 149 °. Actually, the persistence length (Fig. 8) is very sensitive to small changes is R, in the region close to a P-to-L ratio above 1. The data in Ref. [22], $P = 760$ nm, would be fulfilled by a hinge conformation of 132 °. The data given in this article are illustrative of the difficulty to ascertain among such closely related possibilities from a hydrodynamic point of view. Not only could the three conformations more similar in behaviour (a straight, j 170 ° hinge, and l weakly bent) not be distinguished even with the better accuracy ultracentrifugation data, but even a conformation so different as b, a 90 ° hinge conformation, was also closer to the other solutions than the error uncertainty. In terms of persistence length, conformations with an average state such as the j 170 ° hinge would have $P = 10{,}000$ nm, while that of 90 °, b, will have $P = 170$ nm. These conformations have been found to be indistinguishable by the use of their s and D_t.

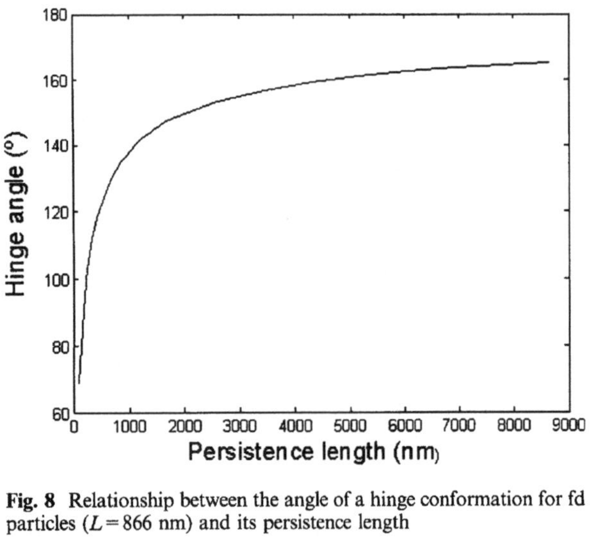

Fig. 8 Relationship between the angle of a hinge conformation for fd particles ($L = 866$ nm) and its persistence length

The previously noted observation of the strong dependence of s with the particle length (for nearly straight particles) readdresses the fact of the conservation of this length in comformers. The tightly bound capsomers of this virion have the role, among others but may be as the principal one, to protect the DNA molecule from the attack of nucleases and undesired agents that could bind to the nucleic acid. In a bend, a resistance to separation of the protein subunits on the "outer" face of the particle could be imagined. But it may be easier to imagine a stronger resistance to compression in the inner face. A particle bending with a conservation of the length in its inner face (while in the cases considered here that conserved was the length of the central axis) would have an increased effective length that could also alter its hydrodynamic properties. Such small bending as a 170 ° angle would yield, in a particle of the dimensions of fd ($L = 866$ nm and $d = 8$ nm), an increase in L of 0.7 nm, which would alter s and D_t by about a 0.64%, which is a small value, but larger than the differences observed for the conformational bending per se.

Conclusions

A major conclusion of this article is that parameters such as the intrinsic viscosity or the rotational diffusion coefficient are required for discrimination among relatively similar particle conformations, while the parameters that can be obtained solely form ultracentrifugation, s and D_t, are not sensitive enough. The intrinsic viscosity is highly sensitive to both the angle of bending in the hinge model and particle length changes, while the rotational diffusion coefficient is sensitive to angle but not to length. Owing to this lack of discriminating power of s and D_t, only those conformations with a high degree of bending can be discarded, while small uncertainties or alterations in molecular weight, partial specific volume or particle length can be responsible for much larger s and D_t changes than those associated with different weakly bent conformations.

Acknowledgements. The collaboration in previous related work of S.E. Harding, R.N. Perham, J. García de la Torre, F.G Díaz and D. Rowitch is acknowledged, as well as the comments of A. Rowe and C. Ebel.

References

1. Day LA, Marzec CJ, Reisberg S, Casadevall A (1988) Annu Rev Biophys Chem 17:509
2. Makowski L (1985) In: Jurnak FA, McPherson A (eds) Biological macromolecules and assemblies: virus Structures, vol I. Wiley, New York, pp 203–253
3. Webster RE, Lopez J (1985) In: Casjens S (ed) Virus structure and assembly. James & Barlett, Boston, pp 235–267
4. Marvin DA (1988) Curr Opin Struct Biol 8:150
5. Hunter GJD, Rowitch DH, Perham RN (1987) Nature 327:252
6. Rowithch DGH, Hunter GJ, Perham, RN (1988) J Mol Biol 204:663
7. Hogenboom HR, de Brüine AP, Hufton SE, Hoet RM, Arends JW, Roovers RC (1998) Immunotechnology 4:1
8. Banner DVC, Nave C, Marvin DA (1981) Nature 289:814
9. Frank H, Day LA (1970) Virology 42:144
10. Day LA (1966) J Mol Biol 15:395
11. Newman JL, Swinney HL, Day LA (1977) J Mol Biol 116:539
12. Berkowitz SA, Day LA (1976) J Mol Biol102 531
13. Marvin DA, Hohn B (1969) Bacteriol Rev 33:172
14. Molina-García AD, Harding SE, Díaz FG, Garcia de la Torre J, Rowitch D, Perham RN (1989) Biophys J 63:1293
15. Song L, Kim U-G, Wilcoxon J, Schurr M (1991) Biopolymers 31:547
16. Marvin DA, Hale RD, Nave C, Helmer-Citterich M (1994) J Mol Biol 235:260
17. Caspar DL, Makowski L (1981) J Mol Biol 145:611
18. Blanch EW, Hecht L, Syme CD, Volpetti V, Lomonossoff GP, Nielsen K, Barron LD (2002) J Gen Virol 83:2593
19. Perham RN, Terry TD, Willis AE, Greenwood J, di Marzo Veronese F, Appella E (1995) FEMS Microbiol Rev 17:25
20. Chen FC, Koopmans G, Wiseman RL, Day LA, Swinney HL (1980) Biochemistry 19:1373
21. Fritzsche H, Cross TA, Opella SJ, Kallenbach NR (1981) Biophys Chem 4:283
22. Lau AWC, Lin K-H, Yodh AG (2002) Phys Rev E 66:020401–1
23. Harding SE (1995) Biophys Chem 55:69
24. García de la Torre J, Navarro S, López Martínez MC, Díaz FG, López Cascales J (1994) Biophys J 67:530
25. Day LA, Wiseman RL (1978) In: Denhardt DT, Dressler D, Ray DS (eds) The single stranded DNA phages. Cold Spring Harbor Laboratory, New York, pp 605–625
26. Zimmermann K, Hagedorn J, Heuck CC, Hinrichsen M, Ludwig J (1986) J Biol Chem 261:1653
27. Stopar D, Spruijt RB, Wolfs CJAM, Hemminga MA (1997) Biochemistry 36:12268
28. www.ncbi.nlm.nih.gov/ICTVdb/ICTVdB/35000000.htm
29. Ray DS (1968) In: Fraenkel-Conrat H (ed) Molecular basis of virology, ACS monographs. Reinhold, New York, pp 222–254
30. Goodheart CR (1969) An introduction to virology. Saunders, Philadelphia
31. Fraenkel-Conrat H (1974) In: Fraenkel-Conrat H, Wagner RR (eds) Comprehensive virology, vol I. Plenum, New York, pp
32. Verhaert RMD, Beekwilder J, Olsthoorn R, van Duin J, Quax WJ (2002) J Biotechnol 96:103
33. Barrientos LG, Louis JM, Gronenborn AM (2001) J Magn Reson 149:154
34. Dogic Z, Fraden S (2000) Langmuir 16:7820
35. Tang J, Fraden S (1995) Liq Cryst 19:459
36. Tang J, Fraden S (1996) Biopolymers 39:13
37. Purdy KR, Dogic Z, Fraden S, Ruhm A, Lurio L, Mochrie SG (2003) Phys Rev E Stat Nonlin Soft Matter Phys 67:031708
38. Maeda T, Fujime S (1985) Macromolecules 18:2430
39. Rivetti C, Guthold M, Bustamante C (1996) J Mol Biol 264:919

Progr Colloid Polym Sci (2004) 127: 126–135
DOI 10.1007/b98022

P. Zipper
A. Krebs
H. Durchschlag

Comparative modeling of giant annelid hemoglobins

P. Zipper (✉)
Physical Chemistry, Institute of Chemistry,
University of Graz, Heinrichstrasse 28,
8010 Graz, Austria
e-mail: peter.zipper@uni-graz.at,
Tel.: +43-316-3805410,
Fax: +43-316-3809850

A. Krebs
Structural and Computational Biology
Programme, European Molecular Biology
Laboratory, Meyerhofstrasse 1,
69117 Heidelberg, Germany

H. Durchschlag
Institute of Biophysics and Physical
Biochemistry, University of Regensburg,
Universitätsstrasse 31,
93040 Regensburg, Germany

Abstract The information provided by small-angle X-ray scattering, electron microscopy and hydrodynamics for the giant hemoglobins from different annelid species (*Lumbricus terrestris*, *Macrobdella decora*, *Eudistylia vancouverii*) has been compared. The results reveal that these hemoglobins exhibit slight differences in their fine structure, particularly with respect to the presence of a cavity in the central part of the proteins. Results have been obtained by application of several advanced modeling strategies. The low-resolution shape and internal structure of the hemoglobins under analysis were restored ab initio from experimental small-angle scattering data by means of the program DAMMIN using simulated annealing. To improve the results, however, the strict ab initio modeling approach had to be replaced with a procedure which exploited additionally the knowledge of symmetry and anisometry of the proteins. Moreover, 3D reconstructions from electron microscopy were advantageously used as templates for the analysis of scattering data by DAMMIN. For all the models obtained, hydrodynamic data (*s*, *D*) were predicted by use of the program HYDRO and compared to literature data. In all cases satisfactory agreement between observed and predicted values could be achieved. The results led to consensus models for the various hemoglobins.

Keywords Extracellular hemoglobins · Advanced modeling techniques · Small-angle X-ray scattering · Electron microscopy · Hydrodynamics

Introduction

Hemoglobin occurs in all kingdoms of living organisms. Among the invertebrates, the hexagonal bilayer (HBL) structures of the annelid hemoglobins represent a summit of structural complexity. These hemoglobins are giant multisubunit molecules of molecular masses of about 3.5 MDa. Electron microscopy (EM) and small-angle X-ray scattering (SAXS) studies of the hemoglobins from different annelid species have provided information on their 3D structure [1, 2, 3, 4, 5, 6]. The crystal structure at 0.55-nm resolution of the hemoglobin from *Lumbricus terrestris* has been reported recently [7]. Numerous studies revealed that these hemoglobins look very similar; however, they presumably exhibit slight differences in their fine structure. This is also suggested by consensus models derived from the SAXS studies and by models based on 3D reconstructions from cryogenic EM (cryo-EM) [4, 5]).

Previous studies on hemoglobin from *L. terrestris* have already established this protein as a model system for scattering and hydrodynamic modeling approaches [8, 9, 10, 11]. Reconstructions of low-resolution 3D solution structures of this protein can be achieved from 1D SAXS data by classical (trial-and-error) or more advanced (ab initio) approaches. The ab initio recon-

Table 1 Comparison of experimental and calculated structural and hydrodynamic parameters of annelid hemoglobins

	Lumbricus terrestris	*Macrobdella decora*	*Eudistylia vancouverii*
Experimental[a]			
$R_{G,SAXS}$ (nm)	10.71 ± 0.02 [4, 20]	10.75 ± 0.01 [4]	10.73 ± 0.09 [4]
V (nm^3)	6,200 ± 200 [4, 20]	5,800 ± 300 [4]	5,100 ± 100 [4]
s (10^{-13} s)	58.9 ± 0.2 [22]; 61.1 [23, 24]	≈60 [6]	≈60 [6]
D (10^{-7} cm^2/s)	1.61 ± 0.05 [22]		
Mean values of ab initio models			
Total number of models	49	21	34
Number of models selected	27	11	11
r_b (nm)	0.8	0.8	0.8
N_b	1,826 ± 44	1,662 ± 44	1,505 ± 17
$R_{G,hydro}$ (nm)	10.70 ± 0.01	10.78 ± 0.01	10.78 ± 0.02
s (10^{-13} s)	61.5 ± 0.8	60.4 ± 0.8[b]	61.4 ± 1.0[b]
D (10^{-7} cm^2/s)	1.60 ± 0.02	1.57 ± 0.02	1.60 ± 0.03
Mean values of electron-microscopy-biased models			
Beads on lattice points:			
Number of models	16	14	14
r_b (nm)	0.8 and 0.6	0.8 and 0.6	0.8 and 0.6
N_b	1,785 ± 43 and 3,959 ± 80	1,620 ± 31 and 3,493 ± 33	1,449 ± 19 and 3,045 ± 56
$R_{G,hydro}$ (nm)	10.69 ± 0.01	10.77 ± 0.01	10.76 ± 0.01
s (10^{-13} s)	62.8 ± 0.4	62.7 ± 0.6[b]	62.7 ± 0.6[b]
D (10^{-7} cm^2/s)	1.64 ± 0.01	1.64 ± 0.02	1.63 ± 0.02
Beads on local centers of gravity:			
Number of models	4	4	4
r_b (nm)	0.8	0.8	0.8
N_b	2,132 ± 59	1,701 ± 46	1,786 ± 40
$R_{G,hydro}$ (nm)	10.67 ± 0.03	10.75 ± 0.02	10.70 ± 0.01
s (10^{-13} s)	62.8 ± 0.2	64.0 ± 0.1[b]	63.8 ± 0.1[b]
D (10^{-7} cm^2/s)	1.64 ± 0.01	1.67 ± 0.00	1.66 ± 0.00
Consensus models: ab initio			
Number of models included	27	11	11
$R_{G,hydro}$ (nm)	10.71 ± 0.04	10.71 ± 0.15	10.70 ± 0.04
s (10^{-13} s)	61.1 ± 0.5	61.3 ± 1.1[b]	61.6 ± 0.8[b]
D (10^{-7} cm^2/s)	1.59 ± 0.01	1.60 ± 0.03	1.61 ± 0.02
Consensus models: electron microscopy biased			
Beads on lattice points:			
Number of models included	8 (r_b = 0.8 nm)	8 (r_b = 0.8 nm)	8 (r_b = 0.8 nm)
$R_{G,hydro}$ (nm)	10.67 ± 0.07	10.73 ± 0.05	10.70 ± 0.05
s (10^{-13} s)	62.3 ± 0.5	62.6 ± 0.6[b]	62.2 ± 0.8[b]
D (10^{-7} cm^2/s)	1.62 ± 0.01	1.63 ± 0.02	1.62 ± 0.02
Number of models included	8 (r_b = 0.6 nm)	6 (r_b = 0.6 nm)	6 (r_b = 0.6 nm)
$R_{G,hydro}$ (nm)	10.68 ± 0.09	10.74 ± 0.05	10.70 ± 0.07
s (10^{-13} s)	63.3 ± 0.4	63.3 ± 0.5[b]	63.4 ± 0.7[b]
D (10^{-7} cm^2/s)	1.65 ± 0.01	1.65 ± 0.01	1.65 ± 0.02
Beads on local centers of gravity:			
Number of models included	4 (r_b = 0.8 nm)	4 (r_b = 0.8 nm)	4 (r_b = 0.8 nm)
$R_{G,hydro}$ (nm)	10.67 ± 0.14	10.76 ± 0.03	10.73 ± 0.03
s (10^{-13} s)	62.6 ± 0.5	64.1 ± 0.2[b]	63.6 ± 0.3[b]
D (10^{-7} cm^2/s)	1.63 ± 0.01	1.67 ± 0.01	1.66 ± 0.01

[a]Experimental details are given in the references cited
[b]Assuming the same molar mass (3,500 kg/mol [1, 20]) and partial specific volume (0.733 cm^3/g [22]) as for the protein from *L. terrestris*

struction methods available apply different strategies to obtain a fit to the experimental scattering profiles: use of a genetic algorithm (program DAL.LINUX [12, 13]), simulated annealing (program DAMMIN [14, 15]), or a Monte Carlo type reconstruction algorithm (program SAXS3D [16]). Application of these ab initio approaches to selected test proteins, including *L. terrestris* hemoglobin, disclosed that in the absence of constraints similar, but sometimes unrealistic, models were obtained, in all cases the uniqueness of the models remaining a crucial problem [9, 11]. To some extent the programs DAL.LINUX and particularly DAMMIN can exploit additional information (e.g., size, shape, symmetry), if available, to improve the structural

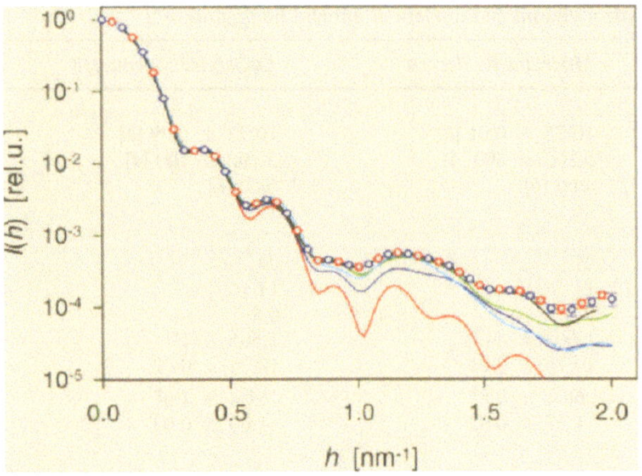

Fig. 1 Comparison of the experimental scattering curve of the hemoglobin from *Lumbricus terrestris* (*black line*) with the scattering curves of various models. Average of the scattering curves of 27 DAMMIN ab initio models (*red circles*); average of the scattering curves of 20 DAMMIN models biased by 3D reconstructions from cryogenic electron microscopy (*EM*) (*blue circles*). The *colored lines* represent the scattering curves calculated for averaged structures (consensus models, visualized in Figs. 3, 5). Average of 27 ab initio models [bead radius $r_b = 0.8$ nm, threshold (minimum frequency of beads) of 5] (*red line*); average of 20 EM-biased models of different types ($r_b = 0.8$ or 0.6 nm, beads arranged on hexagonal lattice points or on local centers of gravity, threshold of 1) (*blue line*); average of eight EM-biased models ($r_b = 0.6$ nm, beads on lattice points, threshold of 3) (*cyan line*); average of four EM-biased models ($r_b = 0.8$ nm, beads on local centers of gravity, threshold of 1) (*green line*)

reconstruction of the resultant models. It turned out to be an obvious advantage of the DAMMIN procedure to allow the use of symmetry constraints and other input modifications (e.g., use of templates derived from SAXS or EM studies) for predicting more realistic models [11].

The present paper is concerned with giant annelid hemoglobins from different sources (*L. terrestris*, *Macrobdella decora*, *Eudistylia vancouverii*) and includes a comparison of their structure and of the modeling approaches used. In particular, the following modeling strategies were applied:

1. The low-resolution solution structure of the proteins was restored from experimental SAXS data by the ab initio modeling program DAMMIN [14, 15].
2. In a few pilot tests the consensus models derived from previous SAXS studies [4] were used as additional a priori information (templates) for calculations by DAMMIN.
3. Based on these tests, models derived from earlier EM studies [5] were used as templates for the analysis of SAXS data by DAMMIN.
4. Consensus models were created by averaging the structures of the best-fitting models.
5. For all the models obtained, structural and hydrodynamic data (R_G, s, D) were computed; critical error assessments were performed and the predicted values were compared to literature data. For predicting the scattering and hydrodynamic behavior of the models, well-established procedures and programs were used (Debye modeling [17], program HYDRO [18, 19]).

Annelid hemoglobins

Most annelids have extracellular and/or intracellular hemoglobins [1, 6]. The giant annelid hemoglobins consist of two types of polypeptide chains: globin chains and nonglobin linker chains; the role of the linker subunits is believed to be rather structural. Early EM studies indicated that the giant extracellular hemoglobins of annelids have an HBL appearance: two superimposed layers of six subassemblies surround a central cavity (bracelet model). These findings were supported by SAXS studies [4, 5, 20], and the results

Fig. 2a–c Top and side views of the averaged structure of 49 DAMMIN ab initio models obtained for the hemoglobin from *L. terrestris*. The underlying models consist of equal-sized spheres (bead radius $r_b = 0.8$ nm). The frequency of occurrence of the spheres in the models is expressed by different shades of *red* (low frequency, *light red*; high frequency, *dark red*), in **a** and **b** additionally by the size of the spheres (the bead volume is proportional to the frequency). In **a** and **b** all beads are shown ($N_b = 4,461$), while in **c** only beads occurring in at least five models are visible (threshold of 5, $N_b = 3,371$). The scattering curve corresponding to **c** is very similar to the red line in Fig. 1

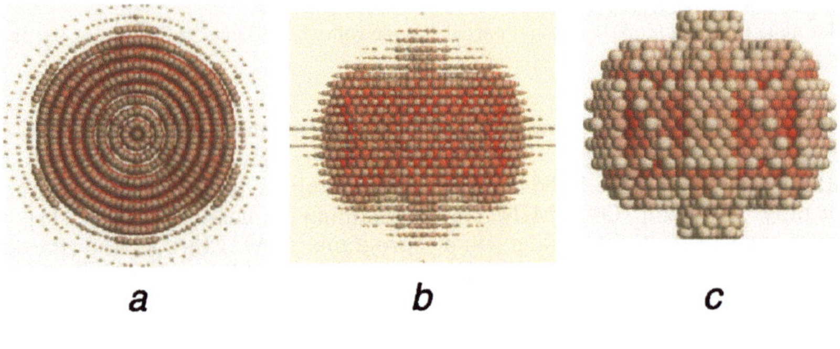

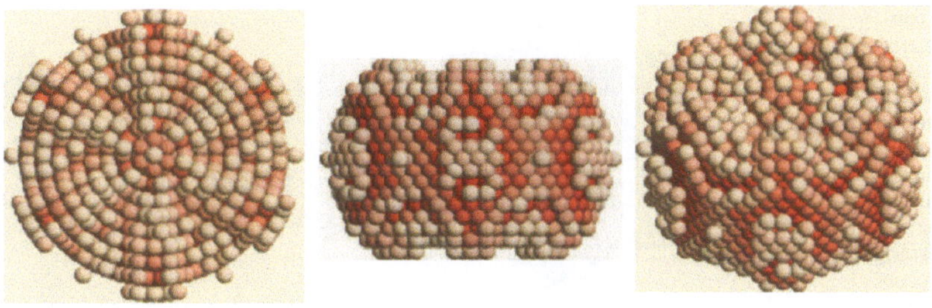

Fig. 3 Different views of the consensus model for *L. terrestris* hemoglobin, obtained by averaging the structures of 27 selected DAMMIN ab initio models and omitting all spheres which are not present in at least five of the underlying models. The 2,997 spheres are of equal size ($r_b = 0.8$ nm). The frequency of occurrence in the models is expressed by different shades of *red*

were critically compared with the patterns derived from EM, including also recent 3D reconstructions from cryo-EM [1, 2, 3, 21]. Very similar quaternary structures for the hemoglobins from *L. terrestris*, *M. decora*, *E. vancouverii* ("chlorocruorin") and some others have been established, at least at low resolution [6]. Among the HBL hemoglobins, that from the common earthworm (*L. terrestris*) is the most extensively investigated hemoglobin.

SAXS and hydrodynamic reference data for the various annelid hemoglobins under investigation were taken from the literature: SAXS data from Refs. [4, 5, 20], and hydrodynamic properties from Refs. [6, 22, 23, 24]. In particular, the following data were used: scattering intensity, $I(h)$, distance distribution function, $p(r)$, molar mass, M, partial specific volume, $\bar{v}$, hydrated volume, V, radius of gyration, R_G, sedimentation coefficient, s, and translational diffusion coefficient, D.

3D reconstructions of the *L. terrestris* hemoglobin structure from cryo-EM were taken from Ref. [5] and those of the other hemoglobin species were provided by F. de Haas and J.-C. Taveau.

Methods

General line

To be able to compare the information provided by SAXS, 3D reconstructions from EM, and hydrodynamics, we used the SAXS curves [4] of the three annelid hemoglobins under analysis as starting points. Recently, comparative studies have revealed the advantage of the DAMMIN procedure [14, 15] for the anticipation of biophysically relevant protein solution structures [11]. By this advanced modeling approach low-resolution shape and internal structure of the hemoglobins were restored, exploiting the experimental SAXS curves without additional structural information (ab initio modeling) or utilizing models from previous SAXS or EM studies as templates. On the basis of the results created by DAMMIN, hydrodynamic parameters (s and D) were predicted by means of HYDRO [18, 19], in addition to the calculation of R_G values. For this purpose some intermediate steps were performed: superimposing all models, selection of appropriate models, generation of consensus models by averaging. The validity of the results was checked by comparison with experimental reference data.

Advanced modeling by simulated annealing

The program DAMMIN [14, 15] starts with a search space filled with densely packed dummy atoms and assigns the atoms to the protein or to the solvent. By simulated annealing a protein configuration is obtained that fits the experimental SAXS curve almost perfectly. Scattering curves are calculated by means of spherical harmonics. The DAMMIN program requires a GNOM file [25] for the primary data input. Running the program in the "expert mode" allows the input of additional structural information, such as symmetry constraints, expected anisometry, or a template to replace the default spherical search space.

Fig. 4a–c Views of the templates used for the DAMMIN analysis of the experimental small-angle X-ray scattering curve of *L. terrestris* hemoglobin. The templates were derived from 3D EM reconstructions and consist of equal-sized spheres (**a**, **b** $r_b = 0.8$ nm; **c** $r_b = 0.6$ nm), arranged in a hexagonal lattice (**a**, **c**) or at local centers of gravity (**b**)

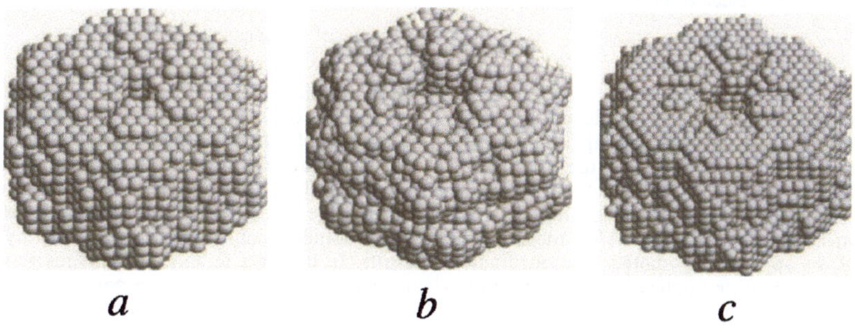

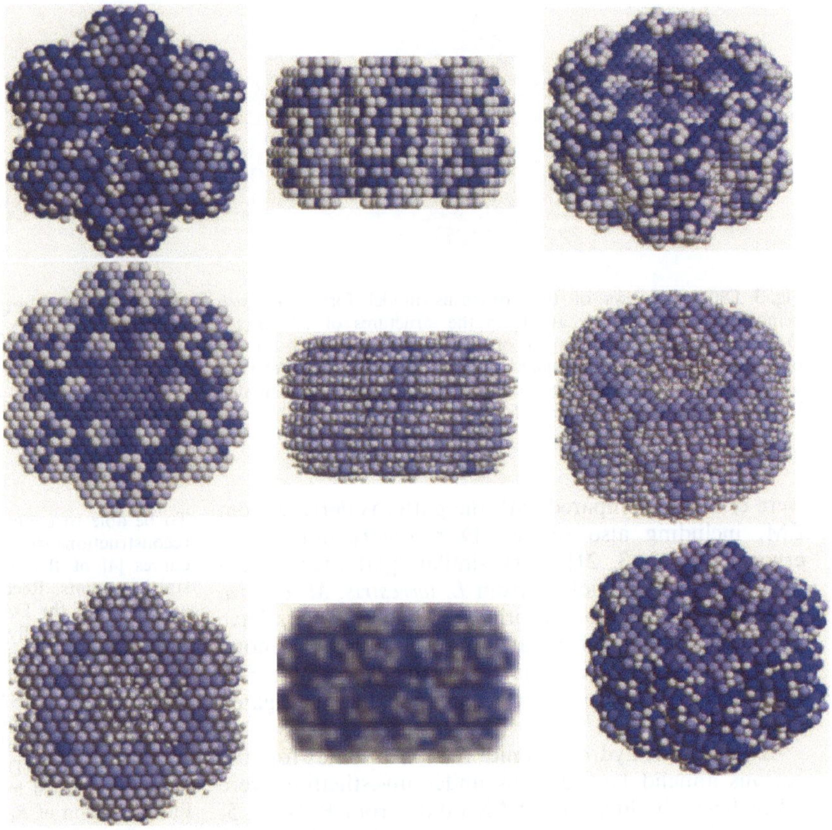

Fig. 5 Different views of consensus models for *L. terrestris* hemoglobin, obtained by averaging the structures of DAMMIN models biased by templates (3D EM reconstructions). *Top:* Consensus model based on 20 models consisting of equal-sized spheres ($r_b = 0.6$ and 0.8 nm, respectively) in different arrangements; the frequency of occurrence of the spheres in the models is expressed by different shades of *blue* (low frequency, *light blue*; high frequency, *dark blue*) and additionally by the size of the spheres. *Middle:* Consensus model based on eight models consisting of spheres ($r_b = 0.6$ nm) on hexagonal lattice positions; the frequency of occurrence of the spheres is only expressed by different shades; only spheres present in at least three models are shown. *Bottom:* Consensus model based on four models consisting of spheres ($r_b = 0.8$ nm) on local centers of gravity

Ab initio approach

DAMMIN calculations without any constraints did not lead to reasonable models for the hemoglobins, in accordance with previous results [11]. By contrast, application of symmetry constraints turned out to be essential for predicting realistic models, whereas use of shape information was helpful but by no means imperative. All ab initio models which apply the symmetry constraint consist of slightly overlapping spheres, owing to the packing mode inherent in the DAMMIN program.

A large number of models was created, most of them, however, in disagreement from the conclusive EM images. In our approach unrealistic individual models were discarded, while the remaining models were selected for the further analysis which included the calculation and averaging of scattering curves and molecular parameters (R_G, s, D) as well as the creation of consensus models by averaging the structures of the selected models.

Use of templates

A more realistic analysis by DAMMIN is possible by including additional structural information in terms of a template replacing the original spherical search space and reducing the number of possible configurations. Any kind of realistic model of the structure under analysis can be used as a template, for example, a 3D reconstruction from cryo-EM or a tentative SAXS model. The template must consist of an appropriate number of coordinates and, if symmetry constraints are to be applied, the template must fulfill the symmetry conditions strictly.

To establish appropriate templates from 3D EM reconstructions or previous SAXS consensus models, the following procedure was applied. The selected model was rescaled and centered if required, and superimposed in six rotation states. A data reduction step [8] using a definite voxel size (1.2 or 1.6 nm) led to an intermediate template. Finally 1/12 of the reduced structure was used to create the final symmetric template.

For data reduction, the spheres were placed either on a hexagonal lattice or at local mass centers of gravity based on this lattice. The latter variant yielded templates of more realistic appearance, however, it resulted in considerable numbers of overlapping spheres, whereas the approach using hexagonal lattice points is devoid of overlapping. In an alternative approach, the templates were scaled up by a factor of 1.05 in the data reduction step, in order to reduce effects of the peripheral penalty in the DAMMIN procedure.

Creation of consensus models

Averaging the models obtained as described previously led to new consensus models for the various hemoglobins. The frequency of occurrence of spheres in the underlying models was expressed in the consensus models by the size of the spheres (their volume was assumed to be proportional to their frequency) or by their electron density. The first approach resulted in models consisting of unequal spheres, influencing both SAXS and hydrodynamic modeling results. In contrast to this, the spheres resulting from the second approach are unequal only in context with scattering and can be treated as equal when hydrodynamic aspects are considered (the

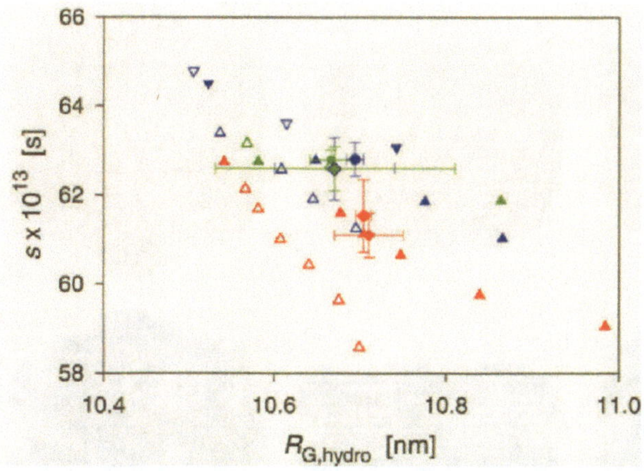

Fig. 6 The sedimentation coefficient, *s*, versus the hydrodynamic radius of gyration, $R_{G,hydro}$, of various models of *L. terrestris* hemoglobin. The *circles* with *error bars* represent the mean values and standard deviations for DAMMIN ab initio models (*red*) and DAMMIN models biased by different EM templates (beads on lattice points, *blue*; beads on local centers of gravity, *green*). The *triangles* show the parameters calculated for models obtained by averaging the structures of DAMMIN ab initio (*red*) or EM-biased models (*blue* and *green*; $r_b = 0.8$ nm, *up triangles*; $r_b = 0.6$ nm, *down triangles*) under variation of the threshold of lowest acceptable frequency of the beads. The threshold increases from the right to the left. *Empty triangles* characterize models expressing the different frequency of spheres by their size, *filled triangles* correspond to models consisting of equal-sized spheres of differing electron density (for the calculation of *s* and $R_{G,hydro}$, however, only the size but not the density of the beads is of importance). The *diamonds* (with *error bars*) indicate the values for *s* and $R_{G,hydro}$ of the best-fitting consensus models

program HYDRO uses the radius of the spheres but not the density). In some cases the creation of the consensus models required only spheres of sufficiently high frequency ("threshold") to be chosen.

Visualization of models

All the models were visualized by the program RASMOL [26], but can also be viewed by any other 3D molecular graphics program.

Parameter predictions

SAXS curves of the bead models were calculated by the Debye formula [17]; distance distribution functions and molecular parameters such as R_G were derived as described in Ref. [27]. To conform to the scattering curves created by DAMMIN, the calculation of scattering curves by means of the Debye formula requires a rescaling of the spheres (beads) by a factor of 0.33 (for a bead radius $r_b = 0.6$ nm) or 0.25 (for $r_b = 0.8$ nm). The reason for this requirement is presumably concerned with some peculiarities of the DAMMIN program.

Hydrodynamic parameters (*s, D*) were obtained by the García de la Torre approach [18, 19] using version 13 of the program HYDRO. For our calculations the program was adapted to allow calculations on models consisting of a huge number of beads ($N_b \leq 5{,}500$), and the input routine was modified to ease the handling of our input data. Our ad hoc expression for the interaction tensor of overlapping spheres of unequal size [28, 29, 30] is already implemented in this version of the program [31].

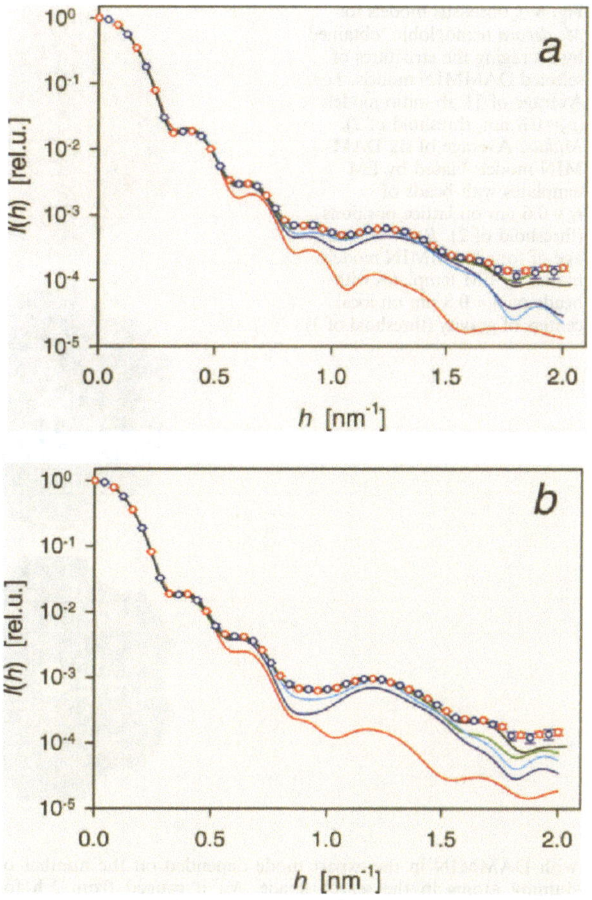

Fig. 7a, b Comparison of the experimental scattering curves (*black lines*) of **a** the hemoglobin from *Macrobdella decora* and **b** the chlorocruorin from *Eudistylia vancouverii* with the scattering curves of various models. Average of the scattering curves of 11 DAMMIN ab initio models (*red circles*); average of the scattering curves of 18 DAMMIN models biased by 3D EM reconstructions (*blue circles*). The *colored lines* represent the scattering curves calculated for averaged structures (consensus models, visualized in Figs. 8, 9). Scattering curve corresponding to the averaged structure of 11 ab initio models ($r_b = 0.8$ nm, threshold of 2) (*red line*); average of 18 EM-biased models of different types ($r_b = 0.8$ or 0.6 nm, beads on hexagonal lattice points or on local centers of gravity, threshold of 1) (*blue line*); average of six EM-biased models ($r_b = 0.6$ nm, beads on lattice points, threshold of 2) (*cyan line*); average of four EM-biased models ($r_b = 0.8$ nm, beads on local centers of gravity, threshold of 1) (*green line*)

The calculated values for the radius of gyration ($R_{G,hydro}$, $R_{G,SAXS}$) differ with respect to the procedures and input parameters applied for their determination. While the hydrodynamic calculations by means of HYDRO only use r_b, the calculations of SAXS data evaluate both r_b and the electron density of the beads [27].

Computational requirements

For our calculations personal computers with Intel Pentium III or AMD Athlon processors with 1 GHz and 512 MB random access memory were used. The computation time required for calculations

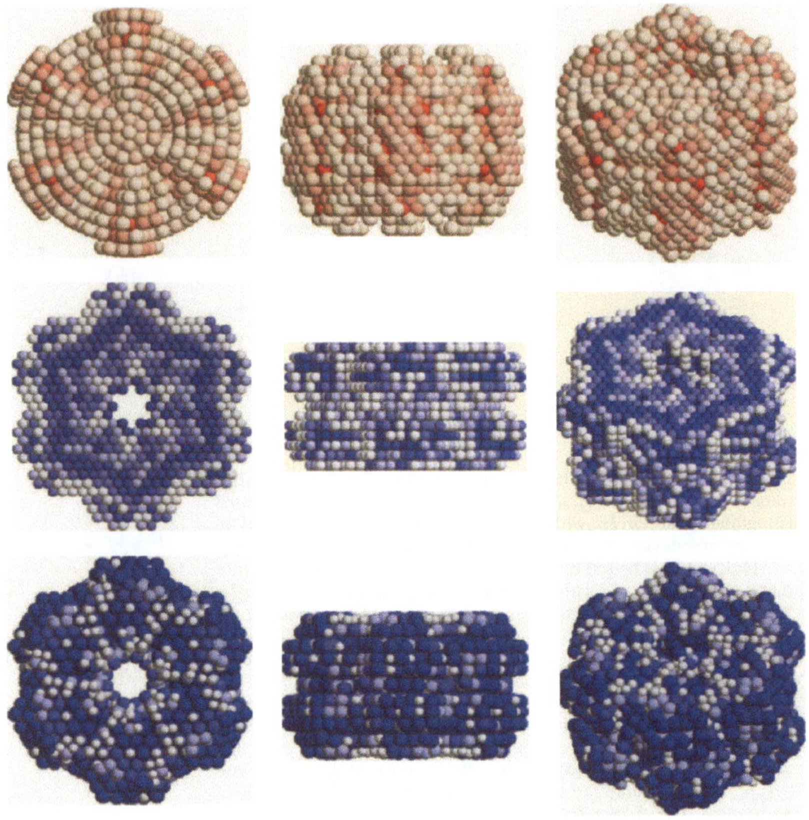

Fig. 8 Consensus models for *M. decora* hemoglobin, obtained by averaging the structures of selected DAMMIN models. *Top:* Average of 11 ab initio models ($r_b = 0.8$ nm, threshold of 2). *Middle:* Average of six DAMMIN models biased by EM templates with beads of $r_b = 0.6$ nm on lattice positions (threshold of 2). *Bottom:* Average of four DAMMIN models biased by EM templates with beads of $r_b = 0.8$ nm on local centers of gravity (threshold of 1)

with DAMMIN in the expert mode depended on the number of dummy atoms in the search space, N_s; it ranged from 2 h for $N_s = 2{,}000$ up to nearly 15 h for $N_s = 6{,}500$. The computation time for calculations with HYDRO amounted to about 35 min for models with $N_b = 2{,}000$, and about 10 h for $N_b = 5{,}100$.

Results and discussion

Starting from the experimental SAXS curves of the hemoglobins from *L. terrestris*, *M. decora* and *E. vancouverii*, a variety of modeling approaches by means of the program DAMMIN were performed, varying the manner of acting and several input variables. Parameter predictions for the radius of gyration, R_G, and the hydrodynamic parameters, s and D, are summarized in Table 1, based either on the mean values for ab initio and EM-biased models or the respective consensus models. Procedures were tested systematically for *L. terrestris* hemoglobin, and were then applied to the other two hemoglobin species in a similar manner. The values for D are more relevant than those for s, since D is obtained more directly than s (it does not require information regarding M and $\bar{v}$).

All models generated by using EM templates yielded larger values for s and D than the ab initio models. Compared to the experimental data, the ab initio models seem to fit better than the EM-biased models. The comparison of the various models shows that usage of EM templates with beads positioned on local centers of gravity may result in larger values for s and D than with beads on lattice points. Similarly, the consensus models obtained for models with beads of $r_b = 0.6$ nm on lattice points or beads of $r_b = 0.8$ nm on centers of gravity yield larger values for s and D than consensus models with beads of $r_b = 0.8$ nm arranged on lattice points.

Hemoglobin from *L. terrestris*

All models generated by the program DAMMIN were able to simulate almost perfectly the experimental scattering curve. The scattering curves provided by DAMMIN are not shown, since they are identical to the experimental curve within the accuracy of drawing. The scattering curves of the models calculated by means of the Debye formula also agree with the experimental profile (Fig. 1), except for the angular range $h > 1.7$ 1/nm (where $h = 4\pi \sin\theta/\lambda$, θ is half of the scattering angle, and λ the wavelength).

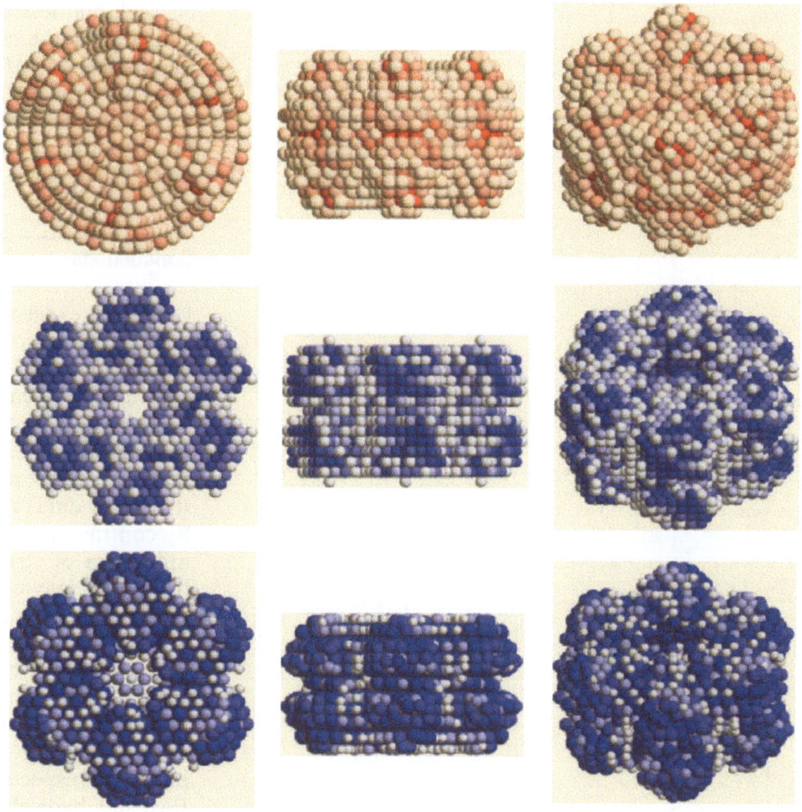

Fig. 9 Consensus models for *E. vancouverii* chlorocruorin, obtained by averaging the structures of selected DAMMIN models. *Top:* Average of 11 ab initio models ($r_b = 0.8$ nm, threshold of 2). *Middle:* Average of six DAMMIN models biased by EM templates with beads of $r_b = 0.6$ nm on lattice positions (threshold of 2). *Bottom:* Average of four DAMMIN models biased by EM templates with beads of $r_b = 0.8$ nm on local centers of gravity (threshold of 1)

As shown in Fig. 2, even models created without templates and differing considerably yield after averaging some relevant structural characteristics (hard core). It is plausible, however, that individual models with a highly unrealistic arrangement of spheres or with protuberances should be discarded; therefore only those models were eventually selected for averaging which were not in clear contradiction to the structures observed in the EM images. Models based on templates exhibit a priori much less scatter in structural details than the ab initio models.

The further evaluation was performed only with the selected models. The models obtained after averaging the selected models are shown in Figs. 3, 4 and 5. The plausible consensus model of the selected structures of ab initio models are represented in Fig. 3. The templates used for creating EM-biased DAMMIN models are depicted in Fig. 4. Averaging of these models yielded the consensus models illustrated in Fig. 5.

The radius of gyration and the hydrodynamic parameters s and D of averaged structures were computed by applying different thresholds (different N_b). In Fig. 6 the values for s and $R_{G,hydro}$ are shown for thresholds varying between 1 (on the right-hand side) and 6 (on the left-hand side), both for the averaged structures of the ab initio and EM-biased models. In addition the mean values for s and $R_{G,hydro}$ of the original models are compared to the expectation values for consensus models. These consensus models were adopted using the agreement of the s values rather than agreement of $R_{G,hydro}$ as a selection criterion.

The models shown in Figs. 3 and 5 are the best realization of the consensus models. Their scattering curves are depicted in Fig. 1. A perfect fitting to the experimental curve or the averaged scattering curves cannot be expected, since, as a consequence of averaging the structures, slightly different distance distributions result. All consensus models based on templates yield better fits of the experimental scattering curve than the ab initio consensus model. Even for this model the distance distribution function deviates only marginally from the function derived from the averaged scattering curves (not shown).

Hemoglobins from *M. decora* and *E. vancouverii*

The models for the *M. decora* and *E. vancouverii* hemoglobins were treated in analogy to the procedures outlined previously for *L. terrestris* hemoglobin, and, of

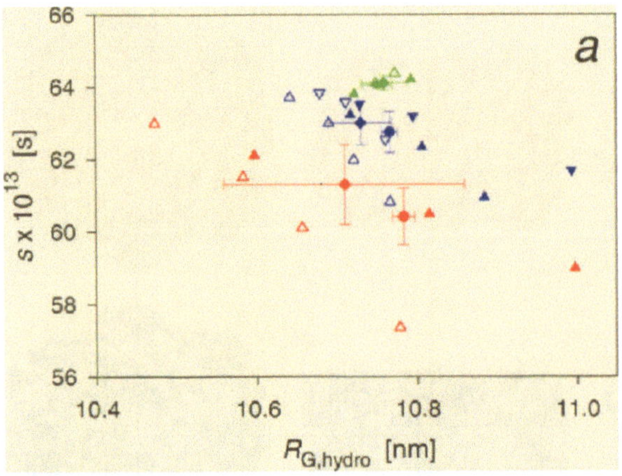

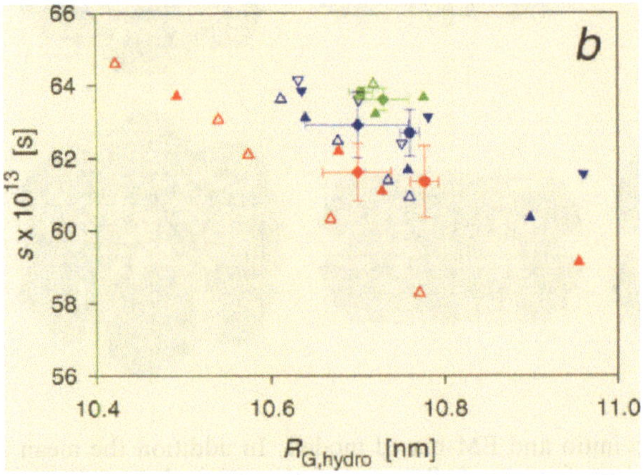

Fig. 10a, b The sedimentation coefficient versus the hydrodynamic radius of gyration of various models of **a** *M. decora* hemoglobin and **b** *E. vancouverii* chlorocruorin. The *circles* with *error bars* represent the mean values and the standard deviations for DAMMIN ab initio models (*red*) and DAMMIN models biased by different EM templates (beads on lattice points, *blue*; beads on local centers of gravity, *green*). The *triangles* show the parameters calculated for models obtained by averaging the structures of DAMMIN ab initio (*red*) or EM-biased models (*blue* and *green*; $r_b = 0.8$ nm, *up triangles*; $r_b = 0.6$ nm, *down triangles*) under variation of the threshold of lowest acceptable frequency of the beads. The threshold increases from the right to the left. *Empty* triangles characterize models expressing the different frequency of spheres by their size, *filled triangles* correspond to models consisting of equal-sized spheres of differing electron density. The *diamonds* (with *error bars*) indicate the values for s and $R_{G,hydro}$ of the best-fitting consensus models

course, by adopting the experience made with this example. A comparison of the experimental scattering curves with those of various models is shown in Fig. 7.

An inspection of the EM-biased consensus models (Figs. 8, 9) discloses significant differences of these models from those of *L. terrestris* hemoglobin (Figs. 3, 4, 5), particularly with regard to the central cavity. This cavity appears to be most pronounced for the *M. decora* hemoglobin, less expressed for *E. vancouverii*, and only moderate in the case of *L. terrestris*. The evaluation of the models for *M. decora* and *E. vancouverii* hemoglobins in terms of s and $R_{G,hydro}$ values (Fig. 10) leads to consensus models which are in accord with the features of the scattering curves (Fig. 7). The results are similar to those found for *L. terrestris* hemoglobin.

Conclusions

Using the experimental SAXS curves as starting material, the ab initio modeling program DAMMIN allows detailed models of giant annelid hemoglobins to be generated. Statements on the validity of models under consideration necessitate computation of a huge variety of models; calculation of a single model is by no means sufficient. Strict ab initio conditions (neglect of existing structural data) may lead to rather ambiguous models. By contrast, use of additional information, for example, utilization of structural templates based on the results from EM, may considerably improve the uniqueness of the models to be derived. Several procedures have to be applied (selection and averaging of individual models), in order to derive consensus models which fit the experimental SAXS curves and the hydrodynamic data (s and D) as far as possible.

The current calculations combine a variety of important aspects: (1) combination of data from SAXS, EM and hydrodynamics, (2) combined use of various highly efficient modeling approaches for the data from these techniques, (3) systematic application of the ab initio modeling technique DAMMIN to such a problem, (4) use of structural templates (e.g. based on EM) as auxiliary means for advanced modeling approaches, (5) derivation of consensus models to test the validity of the models obtained, (6) comparison of the results when applying a variety of models and different modeling input variables, (7) usage of the modeling approaches mentioned to extreme demands (e.g. application to giant molecules, huge number of beads), (8) comparison of the structure of three (very similar) annelid hemoglobins, etc.

The predictions for the hydrodynamic parameters s and D seem, in general, to be satisfactory, with some tendency to overestimate these quantities. This may be partially influenced by the 3D reconstructions used for the EM templates, a conclusion also suggested by the comparison with the results of earlier model calculations [9]. The neglect of the packing density of the spheres in the models may additionally contribute to the mentioned overestimation of s and D values: The volume of all models, calculated as the total volume of spheres, is always smaller than the experimental hydrated volume of the various proteins; corrections for the packing density (thereby rescaling the bead radii) would enhance the volume and, simultaneously, reduce the values for s and D by about 1%. These corrections and some other fine tuning of input parameters and procedures as well as the application of the approaches to other types of molecules remain to be performed in the future.

Acknowledgements The authors are much obliged to F. de Haas and J.-C. Taveau for providing the 3D reconstructions of the annelid hemoglobins analyzed, and to D.I. Svergun, J. García de la Torre, and R.A Sayle for use of the programs DAMMIN and GNOM, HYDRO, and RASMOL, respectively. A.K. is thankful to the Austrian Academy of Sciences for supporting the project.

References

1. Lamy JN, Green BN, Toulmond A, Wall JS, Weber RE, Vinogradov SN (1996) Chem Rev 96:3113–3124
2. de Haas F, Boisset N, Taveau J-C, Lambert O, Vinogradov SN, Lamy JN (1996) Biophys J 70:1973–1984
3. de Haas F, Taveau J-C, Boisset N, Lambert O, Vinogradov SN, Lamy JN (1996) J Mol Biol 255:140–153
4. Krebs A (1996) Thesis. University of Graz, Austria
5. Krebs A, Lamy J, Vinogradov SN, Zipper P (1998) Biopolymers 45:289–298
6. Weber RE, Vinogradov SN (2001) Physiol Rev 81:569–628
7. Royer WE, Strand K, van Heel M, Hendrickson WA (2000) Proc Natl Acad Sci USA 97:7107–7111
8. Zipper P, Durchschlag H (2000) J Appl Crystallogr 33:788–792
9. Zipper P, Krebs A, Durchschlag H (2002) Prog Colloid Polym Sci 119:141–148
10. Zipper P, Durchschlag H (2002) Physica A 314:613–622
11. Zipper P, Durchschlag H (2003) J Appl Crystallogr 36:509–514
12. Chacón P, Morán F, Díaz JF, Pantos E, Andreu JM (1998) Biophys J 74:2760–2775
13. Chacón P, Díaz JF, Morán F, Andreu JM (2000) J Mol Biol 299:1289–1302
14. Svergun DI (1999) Biophys J 76:2879–2886
15. Svergun DI (2000) J Appl Crystallogr 33:530–534
16. Walther D, Cohen FE, Doniach S (2000) J Appl Crystallogr 33:350–363
17. Glatter O, Kratky O (eds) (1982) Small angle X-ray scattering. Academic, London
18. García de la Torre J, Navarro S, López Martínez MC, Díaz FG, López Cascales JJ (1994) Biophys J 67:530–531
19. García de la Torre J, Huertas ML, Carrasco B (2000) Biophys J 78:719–730
20. Krebs A, Zipper P, Vinogradov SN (1996) Biochim Biophys Acta 1297:115–118
21. de Haas F, Kuchumov A, Taveau J-C, Boisset N, Vinogradov SN, Lamy JN (1997) Biochemistry 36:7330–7338
22. Shlom JM, Vinogradov SN (1973) J Biol Chem 248:7904–7912
23. David MM, Daniel E (1974) J Mol Biol 87:89–101
24. Svedberg T, Eriksson I-B (1933) J Am Chem Soc 55:2834–2841
25. Svergun DI (1992) J Appl Crystallogr 25:495–503
26. Sayle RA, Milner-White EJ (1995) Trends Biochem Sci 20:374–376
27. Durchschlag H, Zipper P (2002) Prog Colloid Polym Sci 119:131–140
28. Zipper P, Durchschlag H (1997) Prog Colloid Polym Sci 107:58–71
29. Zipper P, Durchschlag H (1998) Biochem Soc Trans 26:726–731
30. Zipper P, Durchschlag H (1999) Prog Colloid Polym Sci 113:106–113
31. Carrasco B, García de la Torre J, Zipper P (1999) Eur Biophys J 28:510–515

Progr Colloid Polym Sci (2004) 127: 136–147
DOI 10.1007/b98023

K.-J. Tiefenbach
H. Durchschlag
R. Jaenicke

Hydrodynamic and spectroscopic analysis of the denaturation of serum albumin induced by guanidinium chloride and sodium dodecyl sulfate

Abstract The combination of high-speed sedimentation equilibria and specific volume estimates under isomolal or isopotential conditions allows the molar mass of proteins, denatured by chaotropic reagents such as guanidinium chloride (GdmCl) or ionic detergents such as sodium dodecyl sulfate (SDS), to be analyzed. The results depend strictly on the procedures applied: dialyzed samples allow the mass of the unliganded protein to be determined, whereas undialyzed samples yield the masses of the protein–ligand complexes. The first alternative is the only way to obtain thermodynamically sound data; it was applied in the case of GdmCl-denatured bovine serum albumin (BSA). In the case of detergents which do not allow equilibrium dialysis to be applied, mixtures of stock solutions of BSA and SDS were used to determine the molar masses of the BSA–SDS complexes and their binding isotherms. Spectroscopic techniques (UV absorption, fluorescence and circular dichroism) were used to monitor protein unfolding in terms of transition curves and free energies of denaturation. The free-energy change obtained for the SDS-induced denaturation of BSA is found to be considerably smaller than that for the denaturation by GdmCl, in agreement with the lower denaturation capacity of SDS and other detergents compared with highly potent chaotropic denaturants such as GdmCl.

Keywords Protein denaturation · Guanidinium chloride · Sodium dodecyl sulfate · Analytical ultracentrifugation · Spectroscopy

K.-J. Tiefenbach · H. Durchschlag (✉)
R. Jaenicke
Institute of Biophysics and Physical Biochemistry, University of Regensburg, Universitätsstrasse 31, 93040 Regensburg, Germany
e-mail: helmut.durchschlag@biologie.uni-regensburg.de
Tel.: +49-941-9433041
Fax: +49-941-9432813

Introduction

Protein denaturation has been defined in several ways, considering either changes in physical or chemical or biological properties [1, 2, 3, 4]. Relevant solvent parameters involved in the reaction are extremes of temperature, pH and pressure, presence of high concentrations of organic solvents, detergents or chaotropic agents. The alterations observed are manifestations of conformational changes occurring in the native folded structure of a given protein. They lead to a more or less pronounced unfolding of the protein, a process accompanied by the exposure of buried, preferably nonpolar, amino acid residues. Despite the different definitions of protein denaturation, the structural changes may be operationally described by any measurable quantity capable of monitoring noncovalent alterations of the secondary, tertiary or quaternary structure of proteins. Transitions may take place either by an all-or-none "two-state" mechanism involving exclusively the native folded (F) and denatured unfolded (U) states, or by multistate transitions accumulating substantial numbers of intermediates [5, 6, 7]. Their recognition depends on the method applied:

usually spectroscopic techniques or techniques such as analytical ultracentrifugation (AUC), which use spectrometers as monitoring devices, are superior to other approaches in detecting small conformational changes or small numbers of intermediates.

In the case of "chemical denaturation" by chaotropic additives such as guanidinium chloride (GdmCl) or urea, hydrogen bonds stabilizing both the polypeptide chain and protein subunits are disrupted, owing to the high hydrogen-bond capacity of the denaturants. Given their high concentrations in the F↔U transition, their level is sufficiently high to compete with the intramolecular interactions stabilizing the protein in its native state. To achieve this goal, high molar concentrations are required (e.g., 6 M GdmCl, 8 M urea). By contrast, millimolar concentrations of synthetic ionic detergents such as sodium dodecyl sulfate (SDS) are sufficient to initiate a significant perturbation of the native protein structure. Because of their amphiphilic nature, these denaturants act on both hydrogen and hydrophobic bonds, but the absolute denaturing power of SDS at low additive concentrations is by far more effective than that of GdmCl.

In principle, both chaotropic chemical denaturants and detergents may bind to proteins in a nonlinear fashion; they do so to a quite different extent, depending on the agents and proteins applied. Moderate amounts only bind to the protein surface and do not influence the structure significantly, while excess amounts generally cause protein denaturation. Exact binding constants and stoichiometries of binary GdmCl or urea complexes with standard proteins are not available, presumably owing to the low affinities. Indirect determinations made use of precise densimetric measurements and sophisticated thermodynamic considerations [8, 9, 10, 11, 12, 13, 14, 15]. By contrast, the amount of detergent binding is well established. In the case of binding of SDS to proteins, the following binding levels have been elaborated: (1) at moderate detergent concentrations about 0.4 g detergent per gram of protein is bound to both nonreduced and reduced protein; (2) excess detergent leads to about 0.9 g/g for the nonreduced protein and about 1.4 g/g for the reduced protein; (3) at large excess of detergent even larger amounts can be achieved [16, 17, 18, 19, 20, 21].

AUC is the method of choice to provide thermodynamic information, here the extent of ligand binding, preferably by determining the molar masses of proteins and their complexes with the respective ligands. As follows from the thermodynamics of multicomponent systems [22, 23, 24], the interpretation of the data with respect to ligand-free protein or protein plus ligands, critically depends on the experimental procedure applied. AUC experiments and density measurements of samples dialyzed to equilibrium generally yield the masses of the proteins without bound ligands, because interactions (binding and/or exclusion of all ligands and water [11, 24]) are canceled out by use of the "isopotential specific volume" [25, 26]. This procedure is commonly applied if the influence of chaotropic chemical agents such as GdmCl or urea on proteins is studied. In using these additives, owing to the high density of their concentrated solutions, also high rotor speeds are required. In contrast to these experimental conditions, studies of protein–detergent complexes make use of undialyzed samples, for the simple reason that protein–detergent complexes often cannot be dialyzed to thermodynamic equilibrium. In this case, isopotential specific volumes and partial specific volumes of the protein–detergent complexes cannot be unambiguously determined; instead, they have to be calculated on the basis of the volumes of the respective components (protein and detergent) and the putative level of complex formation. AUC experiments then yield the masses of the protein–ligand complexes rather than the masses of the protein alone [25, 26, 27, 28, 29, 30]; in this approach, as a first approximation, ligand-induced changes in hydration are neglected.

Spectral data have been widely used to monitor protein unfolding. Though the spectral properties of proteins do not represent thermodynamic quantities, a refined analysis of observed spectral data, based on thermodynamic principles, allows us to obtain thermodynamic quantities such as the equilibrium constant, K, and the Gibbs free-energy change, ΔG [31, 32]. This thermodynamic evaluation of spectral data concerning ligand-induced protein unfolding can be used to derive the free energies of stabilization of standard proteins and to compare the results obtained under varying denaturation conditions, for example, GdmCl, urea or heat [33, 34, 35, 36, 37]. Recently we succeeded in applying this approach also to the effect of ionic detergents [38, 39].

Numerous studies have been performed with bovine serum albumin (BSA) as a model protein, since the molecular properties of this protein are well known [40, 41]. In the native state the monomeric protein has a molar mass of 66.4 kg/mol; it contains 583 amino acid residues, among them one or fewer cysteines (SH groups), 17 cystins (disulfides), and several aromatic residues (2 tryptophan, 20 tyrosine, 27 phenylalanine). In continuation of our previous work [19, 20, 21, 38, 39, 42, 43, 44, 45, 46, 47], we also used BSA to be able to compare the results from different studies.

In the present study, an attempt was made to apply different principles of evaluating AUC-based masses of GdmCl- and SDS-treated BSA complexes, using either dialyzed or undialyzed samples to obtain the masses of the protein and its ligand complexes, respectively. To achieve this goal, isopotential or partial specific volumes of the respective protein–ligand complexes were applied. Also, an attempt was made to correlate hydrodynamic and spectral data, aiming for thermodynamic information from both sorts of experiments. This was accomplished by evaluating the spectral data from absorption, fluorescence and circular dichroism (CD) experiments in terms of transition curves and free-energy changes, and by comparing the results with the molar mass and specific volume estimates based on AUC studies and experimental and predicted volume quantities.

Materials and methods

Samples and solutions

BSA and SDS were obtained from Serva, Heidelberg, and GdmCl from Merck, Darmstadt; all the other reagents were of analytical grade. The protein and the additives were dissolved in 0.05 M sodium phosphate buffer pH 7.0 (ionic strength $I \cong 0.1$ mol/l). To investigate the binding and denaturation behavior of the additives, their concentrations were varied over a wide range: 0–6 M GdmCl, 0–3 g/l SDS. Stock solutions of BSA and additives were used to prepare the desired protein–additive solutions; the concentrations of GdmCl solutions were controlled refractometrically. In order to make sure that the binding and denaturation processes had reached equilibrium, the solutions were incubated for at least 12 h at 25 °C. All experiments were performed at 25 °C.

Experimental techniques

For density measurements a Paar digital density meter was applied. Experimental and calculated specific volumes were determined as described in Refs. [25, 26, 48, 49, 50]. Concentration determinations of the GdmCl solutions made use of refractive index measurements in a Zeiss Optron refractometer.

AUC was performed using a Beckman model E ultracentrifuge equipped with a high-sensitivity photoelectric scanner, a multiplexer system, and a 10-in. recorder. Runs were performed in a six-hole rotor (AnG), using 12-mm double-sector cells (charcoal-filled Epon) and sapphire windows. Molar masses, M, and molar mass weight averages, M_w, of the macromolecular compounds (proteins, protein–ligand complexes) were determined by means of high-speed sedimentation equilibria (HSSE), using the meniscus-depletion technique [51]. Ultracentrifuge data were evaluated as described elsewhere [52].

Absorption, fluorescence and CD spectra were recorded by means of a Perkin-Elmer Lambda 5 spectrophotometer, a Jobin Yvon-Spex Fluoromax-2 spectrofluorometer and an Aviv 62A-DS spectropolarimeter, respectively. Spectroscopic data were interpreted as described elsewhere [38, 53, 54].

The investigation of proteins in the presence of denaturants and detergents requires consideration of several adaptations and precautions.

1. The dialysis of protein solutions containing high molar cosolutes (such as 6 M GdmCl) requires very long dialysis times. The dialysis of proteins in the presence of ionic detergents (such as SDS) does not lead to complete equilibrium [44].
2. In context with AUC experiments, the following pitfalls had to be considered. The investigation of GdmCl-treated solutions requires the same amount of GdmCl to be present in the reference cell, preferably achieved by equilibrium dialysis. In the case of protein–detergent complexes, detergent molecules have to be absent in the reference (to avoid formation of micelles near the cell bottom) [38]. To avoid misinterpretations, AUC experiments were performed under varying experimental conditions (in particular at several rotor speeds). The magnitude of the monitored signals was modulated by suitable registration conditions, for example, by choosing optimal wavelengths and signal amplifications.

Evaluation of data

Molar masses of proteins and protein–ligand complexes

In principle, the molar masses of proteins in the absence or presence of ligands such as chaotropic chemical agents and detergents can be determined using HSSE [8, 9, 24, 25, 26, 55, 56].

In the case where the impact of chaotropic agents is the subject of the AUC analysis, proteins have usually been dialyzed to thermodynamic equilibrium against the complex solvent (e.g., 6 M GdmCl in water or buffer). Then the mass of the protein without the bound ligands can be obtained, provided the proper volume quantity is applied:

$$M = \frac{2RT}{(1 - v'_s\rho_s)\omega^2} \frac{\mathrm{d}\ln c}{\mathrm{d}r^2}, \tag{1}$$

where M is the molar mass of the ligand-free protein in the complex solvent, v'_s the isopotential specific volume of the protein of concentration c in the complex solvent, ρ_s the density of the complex solvent, ω the angular velocity, r the distance of a selected mass point in the rotor cell from the center of rotation, R the gas constant, and T the absolute temperature. In the case of polydisperse systems, the weight-average molar mass, M_w, is obtained instead of M.

If solutions cannot be dialyzed to thermodynamic equilibrium, as in the case of proteins solubilized by detergents such as SDS, then the mass of the liganded protein is obtained:

$$M_c = \frac{2RT}{(1 - \bar{v}_c\rho)\omega^2} \frac{\mathrm{d}\ln c}{\mathrm{d}r^2}. \tag{2}$$

Here, M_c and $\bar{v}_c$ represent the molar mass and partial specific volume of the corresponding protein–ligand complex, and ρ the density of the dilute solvent. Thus, the molar mass of the complex considers the amount of bound ligand (e.g., detergent), δ_l, in grams per gram of protein:

$$M_c = M(1 + \delta_l). \tag{3}$$

The partial specific volume of the complex, $\bar{v}_c$, required for this procedure, can be derived from the partial specific volume of the ligand-free protein, $\bar{v}$, and the assumed amount and partial specific volume of the bound ligand, δ_l and $\bar{v}_l$:

$$\bar{v}_c = \frac{\bar{v} + \delta_l\bar{v}_l}{1 + \delta_l}. \tag{4}$$

Again, in the case of heterogeneous protein–ligand complexes, the masses obtained represent weight averages.

Accuracy of molar mass estimates

The interpretation of the AUC results in terms of molar masses and number of bound ligands requires a critical assessment of the errors involved. The investigation of the particle weights of the BSA–SDS complexes is affected by several aspects [19, 21, 38]: (1) the unavoidable presence of appreciable numbers of dimers, trimers and higher oligomers in native samples of BSA [40, 41]; (2) the formation of SDS micelles upon excess of free detergent in the solvent; (3) the validity of the assumptions required for predicting the values of the partial specific volume of the complex, $\bar{v}_c$, required for the mass estimates of the complex under analysis (Eqs. 2, 3, 4). The first two aspects may be influenced by variation of the experimental conditions. At low rotor speed, overestimations of the masses are found, owing to the simultaneous presence of monomeric protein ($M = 66.4$ kg/mol) and oligomeric protein cross-links (primarily 133 and 199 kg/mol) in the absence of detergent, or of the corresponding complexes between protein monomers and oligomers and the detergent. On the other hand, studies at very high rotor speed are disturbed by the formation of SDS micelles ($M = 20$–30 kg/mol [19, 21]) if the free detergent concentration exceeds the critical micelle concentration (cmc). However, because of the low absorbance of SDS at the wavelengths used for monitoring the protein, absorption optics can only detect high levels of the detergent.

In principle, both types of experimental problems (presence of oligomers and detergent micelles) can be avoided by working at intermediate rotor speeds, since the masses of the previously mentioned disturbing molecular species differ drastically from those of the native protein and its complex, thus allowing adequate separation of the particles. In addition, perturbations may be minimized by using not too high detergent concentrations, preferably below the cmc. However, owing to the accumulation of material near the cell bottom, the formation of detergent micelles cannot be avoided completely in the case where significant detergent binding is intended. If small proteins with masses close to those of the detergent micelles are to be analyzed [19], specific methods involving distribution functions have to be applied.

Previous HSSE experiments have shown that at intermediate rotor speeds the accuracy of the masses of the unliganded proteins obtained amounts to a maximum of $\pm 5\%$, while the masses of protein–detergent complexes may be afflicted with errors up to 10–20% or even deviations beyond this range; the errors increase with increasing detergent concentration [19, 21]. Thus, the error in δ_l (Eq. 3) is of the same size, if no additional sources of error are included. Owing to the propagation of errors, however, it can be appreciably higher, if further variables need to be included in the error calculations. The most important problem remaining is the validity of the values for $\bar{v}_c$ to be applied.

The present estimate of detergent binding rests on the assumptions that for the mass of the native unliganded protein, the value deduced from the sequence (66.4 kg/mol) can be used. The validity of this assumption can be proved by the experimental determination of the mass of the ligand-free protein: within the accuracy of the determination applied, the measured mass must be identical to the value deduced from the sequence. If this is what is observed, perturbations caused by oligomers or detergent micelles can be neglected under the experimental conditions applied.

Thermodynamic analysis of unfolding curves

The spectral investigation of protein unfolding follows the directions outlined recently [38]. The analysis exploits wavelengths which provide maximum effects and marginal noise of signals [33, 34]: 287 nm in the case of UV absorption, 320 nm for fluorescence emission and 220 nm for far-UV CD experiments. Finally, the spectroscopic measurements allow the determination of the free energy, ΔG, for the transition from the native folded (F) to the denatured unfolded (U) state. ΔG at 25 °C extrapolated to zero denaturant concentration was taken as a measure for the stability of a protein.

The results depend on various assumptions regarding the folding/unfolding mechanism (two-state or multistate transitions), reaching equilibrium conditions and reversibility of the F$\leftrightarrow$U transition, quality of pretransition and posttransition regions, evaluation methods (e.g., extrapolation methods for fixing the reference states) and experimental conditions, etc. Owing to these assumptions, the calculated free energies are prone to large errors; however, the results may be useful in comparing stability differences after preceding denaturation steps.

Results and discussion

The hydrodynamic analysis of GdmCl- and SDS-treated BSA at various denaturant concentrations by AUC yields information on the particle weight of the protein moiety or the weights of the respective protein–additive complexes. The masses obtained depend on the approaches applied for the analysis of the HSSE data and the type of specific volume used for the evaluation. The results for the protein–additive complexes can be used to derive the amount of ligand binding; this property may be compared with other representative thermodynamic parameters derived spectroscopically using absorption, fluorescence and CD.

Hydrodynamics

Native (nonreduced) BSA and samples incubated with varying amounts of GdmCl (0–6 M) or SDS (0–3 g/l) were investigated by means of HSSE, varying the experimental conditions over a wide range: rotor speed (10,000–26,000 rpm), scanning wavelength (230, 280, 295 nm), amplification factor of monitored electronic signals (1–4), and evaluation methods for determining the masses (direct evaluation of data points, DE, or nonlinear regression analysis, RA). In this context, it has to be mentioned that values for the masses of BSA may be affected by the unavoidable presence of small numbers of dimers and higher oligomers [38, 40, 41], particularly when investigating the native protein at low rotor speed. Examples at appropriate rotor speeds are outlined in Table 1. Within the accuracy of the measurements ($\pm 5\%$), the masses obtained for the unliganded protein at 16,000 or 26,000 rpm (65–69 kg/mol; evaluation by RA) are in agreement with the value derived from the amino acid sequence (66.4 kg/mol). This proves that at these rotor speeds oligomers are either not populated to a significant extent or they sediment to the cell bottom without perturbating the HSSE of the monomer.

Representative profiles for native BSA and for GdmCl- and SDS-denatured protein, all obtained at 26,000 rpm, are illustrated in Fig. 1. The comparative investigation of these samples requires relatively high rotor speeds, since the high density of 6 M GdmCl

Table 1 Molar masses, M, of native bovine serum albumin (*BSA*) and of BSA treated with 6 M guanidinium chloride (*GdmCl*) or varying amounts of sodium dodecyl sulfate (*SDS*) (0–3 g/l), obtained by high-speed sedimentation equilibria (*HSSE*) under varying experimental conditions. The amount of detergent, δ, bound to the protein was derived from the molar masses of the BSA–SDS complexes. Analytical ultracentrifugation conditions: c_0 of the protein 0.66 g/l; 16,000 or 26,000 rpm; scanning at $\lambda = 230$, 280 or 295 nm; amplification factor 1 or 4; 25 °C; the reference cell was filled with buffer (in the case of native BSA and BSA–SDS complexes) or 6 M GdmCl in buffer (in the case of GdmCl-treated BSA)

Additive c_3 (M or g/l)[a]	Protein $\bar{v}_2$ or ϕ'_2 (cm^3/g)[b]	Solvent ρ (g/cm^3)[c]	Ultracentrifugation				
			Rotor speed (rpm×10^{-3})	M_w (kg/mol)[d]		δ (g/g)[e]	
				DE[f]	RA[f]	DE[f]	RA[f]
None	0.735 $\bar{v}_2, \phi'_2$	1.0074	26	62	69		
6 M GdmCl	0.720 ϕ'_2	1.1411	26	65	68		
None	0.735 $\bar{v}_2, \phi'_2$	1.0074	16	71	65		
0.100 g/l SDS	0.753 $\bar{v}_{2,calc}$	1.0074	16	75	72	0.13	0.08
0.268 g/l SDS	0.774 $\bar{v}_{2,calc}$	1.0074	16	83	81	0.25	0.21
0.300 g/l SDS	0.777 $\bar{v}_{2,calc}$	1.0075	16	87	82	0.31	0.24
0.500 g/l SDS	0.793 $\bar{v}_{2,calc}$	1.0075	16	99	94	0.50	0.42
0.618 g/l SDS	0.800 $\bar{v}_{2,calc}$	1.0076	16	101	99	0.52	0.49
0.800 g/l SDS	0.809 $\bar{v}_{2,calc}$	1.0077	16	118	115	0.77	0.74
1.000 g/l SDS	0.816 $\bar{v}_{2,calc}$	1.0078	16	125	119	0.88	0.80
2.000 g/l SDS	0.836 $\bar{v}_{2,calc}$	1.0079	16	129	134	0.95	1.01
3.000 g/l SDS	0.846 $\bar{v}_{2,calc}$	1.0081	16	145	135	1.19	1.04

[a] The concentrations, c_3, given for SDS are initial concentrations
[b] Values for the specific volumes of native and GdmCl-treated BSA, ϕ'_2, $\bar{v}_2$, are taken from Refs. [25, 26]; values for the BSA–SDS complexes, $\bar{v}_{2,calc}$, are calculated for the predicted amounts of detergent binding, assuming partial specific volumes of 0.735 cm^3/g for BSA, $\bar{v}_2$, and of 0.870 cm^3/g for bound (micellar) SDS, $\bar{v}_3$ [25, 26]
[c] Values for the density, ρ, of buffer in the absence or presence of SDS were determined by densimetry; the value for 6 M GdmCl was derived from the data in Ref. [63]
[d] The molar mass estimates listed are molar mass weight averages, M_w, and refer to the native (ligand-free) or GdmCl-denatured (ligand-free) protein or to the protein–detergent complexes. Accuracy of M_w values ±5 % for the ligand-free protein and errors up to ±20% for the protein–detergent complexes (see Evaluation of data)
[e] Detergent binding estimates, δ, are based on the molar masses of the protein–detergent complex, M_c, and the native protein, $M_N = 66.4$ kg/mol [41] derived from the amino acid composition: $\delta = (M_c - M_N)/M_N$. Accuracy of δ values ±5–20% for the protein–detergent complexes; if further uncertain parameters are involved, even bigger errors may occur (see Evaluation of data)
[f] Molar mass determinations from HSSE experiments are based on a direct evaluation, DE, of data points [51] or nonlinear regression analysis, RA, using iterative fitting by the Marquardt–Levenberg algorithm [64]

solutions results in a small value for the buoyancy term; on the other hand, the influence of 3 g/l SDS on the density of the principal solvent (buffer) is marginal. In the figure the data for native protein are compared with those for the samples denatured by elevated concentrations of the applied denaturants. Though the initially derived A versus r profiles look very similar for these protein samples (Fig. 1a), the calculated molar masses differ considerably (Fig. 2) after applying the proper values for specific volume and density in the buoyancy term. This may be taken from the slopes (proportional to molar masses and buoyancy) in Fig. 1b, and the numerical values listed in Table 1.

In the case of 6 M GdmCl, using the partial or isopotential specific volume, $\bar{v}$ or v' (or the corresponding apparent quantity, ϕ or ϕ'), yields similar results, since the volumes are similar, and also similar to the value for the native protein. However, this does not mean that the values for the specific volumes over the whole concentration range between 0 and 6 M GdmCl would differ only slightly from the value of the native protein; as was shown earlier [25, 26, 42, 43], the contrary is true (cf. the v' values in Fig. 3a). Therefore, to be on the safe side, applying dialyzed samples both for the AUC experiment and for the determination of the specific volume is recommended. Canceling all interactions by using the isopotential specific volume v' (or ϕ') means that the mass of the GdmCl-free protein is obtained, though the experiment is performed in the presence of 6 M GdmCl. The detailed investigation of GdmCl binding to BSA was not tried in this study, since several previous experiments indicated only low binding capacities [8, 9, 10, 57, 58, 59, 60, 61, 62].

HSSE studies on proteins in the presence of the detergent SDS have to be performed on solutions which have not been dialyzed. They require usage of $\bar{v}$ (or ϕ), preferably obtained from calculations based on the values of the constituent protein and detergent moieties in the BSA–SDS complexes. The resulting volumes vary nonlinearly between the value of the native protein and that of protein saturated with detergent (Table 1, Fig. 3b). At high SDS-binding

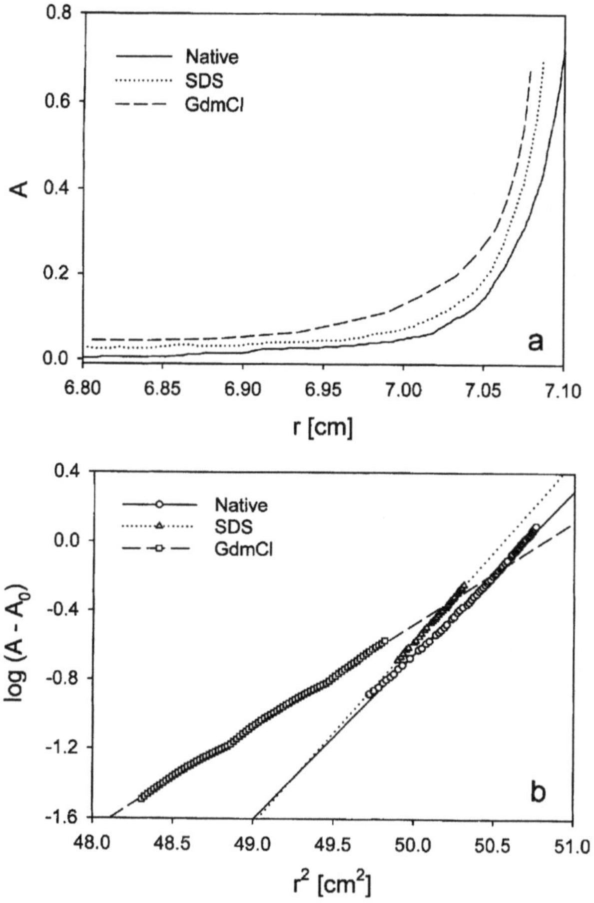

Fig. 1a,b High-speed sedimentation equilibria (*HSSE*) of native (nonreduced) bovine serum albumin (*BSA*) ($c_0 = 0.66$ g/l) and of BSA treated with 6 M guanidinium chloride (*GdmCl*) or 3.0 g/l sodium dodecyl sulfate (*SDS*) at 26,000 rpm: **a** original data in A versus r format; **b** $\log(A-A_0)$ versus r^2 with slopes proportional to the molar mass, M, and the buoyancy term, $(1-v'\rho)$ in the case of GdmCl and $(1-\bar{v}\rho)$ in the case of SDS. In these patterns A and A_0 represent the absorbances of the sample and the baseline, respectively, and r is the radial distance of selected mass points from the center of rotation. Scanning conditions: recording at $\lambda = 280$ nm and amplification factor 1

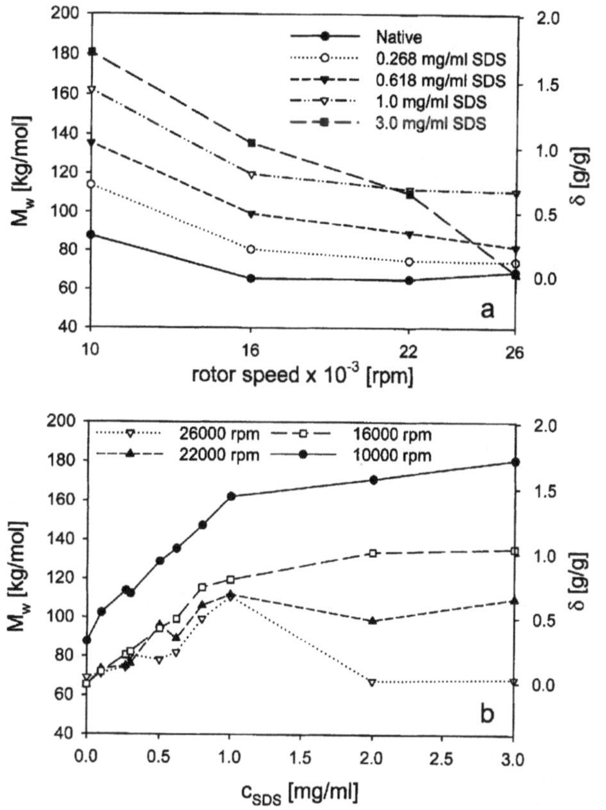

Fig. 2a,b Molar mass weight averages, M_w, of representative BSA–SDS complexes (0.66 g/l nonreduced BSA), obtained by HSSE at different rotor speeds (10,000–26,000 rpm) and varying amounts of SDS (0–3.0 g/l). The amount of detergent binding, δ, to the protein may be derived from the given M_w values. **a** M_w versus rotor speed. **b** M_w versus c_{SDS}. The M_w values represent averages of different scanning conditions (λ, amplification factor), obtained by nonlinear regression analyses; accuracy of M_w values $\pm 5\%$ for the detergent-free protein, and errors up to $\pm 20\%$ for the protein–detergent complexes

levels, HSSE experiments may be perturbed by the presence of detergent micelles (with masses significantly lower than the native protein) [19, 21], preferably occurring at high rotor speed. Therefore, for a detailed analysis of the binding isotherms, HSSE experiments were performed at several rotor speeds and varying SDS concentrations. The resulting molar masses differ over a wide range (Fig. 2). Owing to the presence of oligomers, the masses observed at very low rotor speed (10,000 rpm) were unreasonably high (detergent binding to monomers and oligomers). At elevated detergent concentrations (greater than 1 g/l, i.e. above the cmc), the masses obtained at very high rotor speed (26,000 rpm) were too low, obviously caused by the presence of detergent micelles at concentrations beyond the cmc of the detergent. Thus, experiments at medium rotor speeds (16,000–22,000 rpm) were preferred, in order to avoid artifacts. They exhibit only a marginal effect of the rotor speed on the molar masses; within the accuracy of the data, they are unperturbed by the presence of either oligomers or SDS micelles. Only the value at 3 g/l SDS and 22,000 rpm is perturbed by micelle formation. Averaging the results over all rotor speeds only yields very rough estimates [38].

Figure 3 depicts the volumetric and densimetric properties required for the determination of the buoyancy terms and the molar masses of the respective protein–additive samples, plotted as a function of the

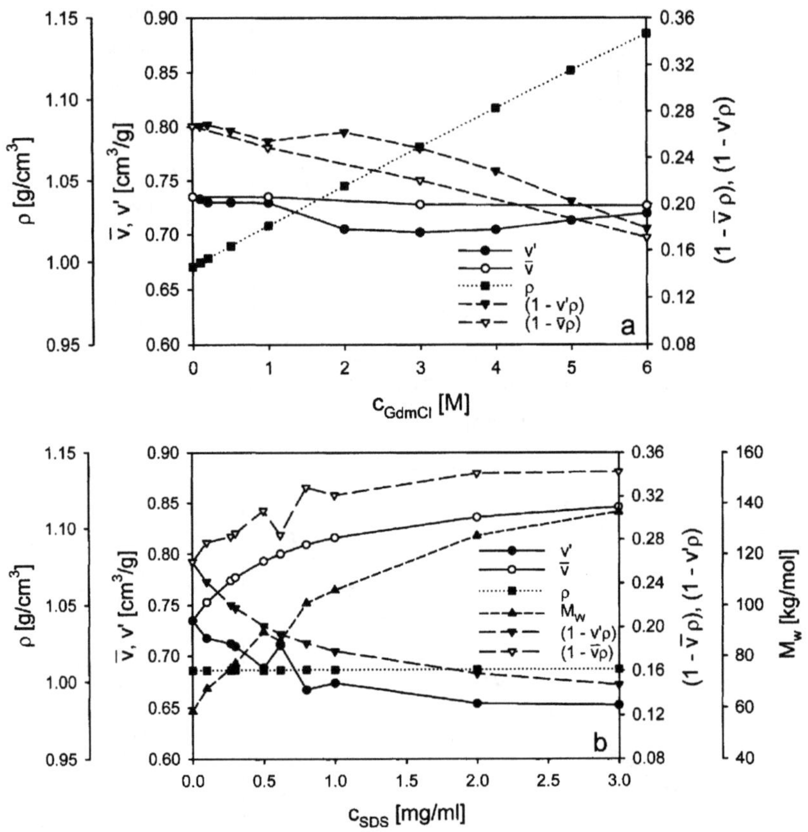

Fig. 3a,b Hydrodynamic properties of nonreduced BSA treated with varying amounts of GdmCl (0–6 M) or SDS (0–3 g/l), together with the densities, ρ, of the respective complex solvents. The hydrodynamic parameters (isopotential specific volume, v', partial specific volume, $\bar{v}$, solvent density, ρ, and the respective buoyancy terms, $(1-v'\rho)$ and $(1-\bar{v}\rho)$, were derived from or predicted for densimetric and ultracentrifugal experiments (experimental v' and $\bar{v}$ values were taken from Refs. [25, 26, 59, 62], temperature corrections from Ref. [42]; ρ values were calculated according to the method in Ref. [63]); the molar mass weight averages, M_w, were obtained from HSSE runs in the analytical ultracentrifuge at 16,000 rpm. In the case of the BSA–SDS complexes, the given values for $\bar{v}$ and $(1-\bar{v}\rho)$ represent calculated values based on the $\bar{v}$ values of the components (BSA, SDS) and the amount of binding, δ, while the values for v' and $(1-v'\rho)$ are derived from the molar mass estimates in the centrifuge. **a** Hydrodynamic properties versus GdmCl concentration. **b** Hydrodynamic properties versus SDS concentration

additive concentrations. In the case of GdmCl (Fig. 3a), the density, ρ, of the solvent increases strongly in a close-to-linear fashion; the experimental values of the partial specific volume, $\bar{v}$, and isopotential specific volume, v', of the protein, and the buoyancy terms $(1-\bar{v}\rho)$ and $(1-v'\rho)$, decrease in a nonlinear fashion. For SDS solutions (Fig. 3b), ρ of the solvent changes only insignificantly, while $\bar{v}$ of the BSA–SDS complexes exhibits marked nonlinear increases, owing to the progressive binding of the detergent. The M_w values derived from experiments at 16,000 rpm show the remarkable increase outlined earlier. On the basis of these M_w values, an rearrangement of Eq. (1) allows the prediction of v' values, which otherwise can hardly be obtained. In spite of the fact that the experiments are afflicted with large errors, pilot tests have shown that experimental ϕ' values decrease. Consequently, the resulting buoyancy terms, $(1-\bar{v}\rho)$ and $(1-v'\rho)$, must differ considerably.

Spectroscopy

Recent thermodynamic analyses of spectroscopic data obtained for the binding of SDS to BSA have enabled us to determine the free energy of stabilization of this protein [38]. To compare these results with data obtained for GdmCl denaturation, in the following the spectroscopic analysis of the BSA–GdmCl system is presented. The results concerning the limiting states of the native and the additive-treated protein are summarized in Fig. 4: UV absorption, fluorescence emission and CD are presented for native BSA and for the protein saturated with 6 M GdmCl or 3 g/l SDS. The UV absorption of native BSA and its SDS or GdmCl complex (Fig. 4a) induces only moderate blueshifts of the maximum at about 280 nm, owing to alterations in the π/π^* transitions. The wavelength shifts are accompanied by slight decreases in absorbance. For a better resolution of the effects, difference spectra were recorded; they reveal maximum effects at 287 nm. Monitoring the intrinsic fluorescence emission, Fig. 4b and c shows that the interaction between GdmCl and BSA leads to redshifts, while BSA–SDS complex formation is accompanied by blueshifts. The reason for this is a loss of the emission energy, caused by enhanced solvent interactions of the BSA–GdmCl system compared with the BSA–SDS system. The BSA–SDS complex obviously gains even more energy than native BSA. Measurements at excitation wavelengths of 280 or 295 nm reveal significant differences for SDS. There is a more pronounced

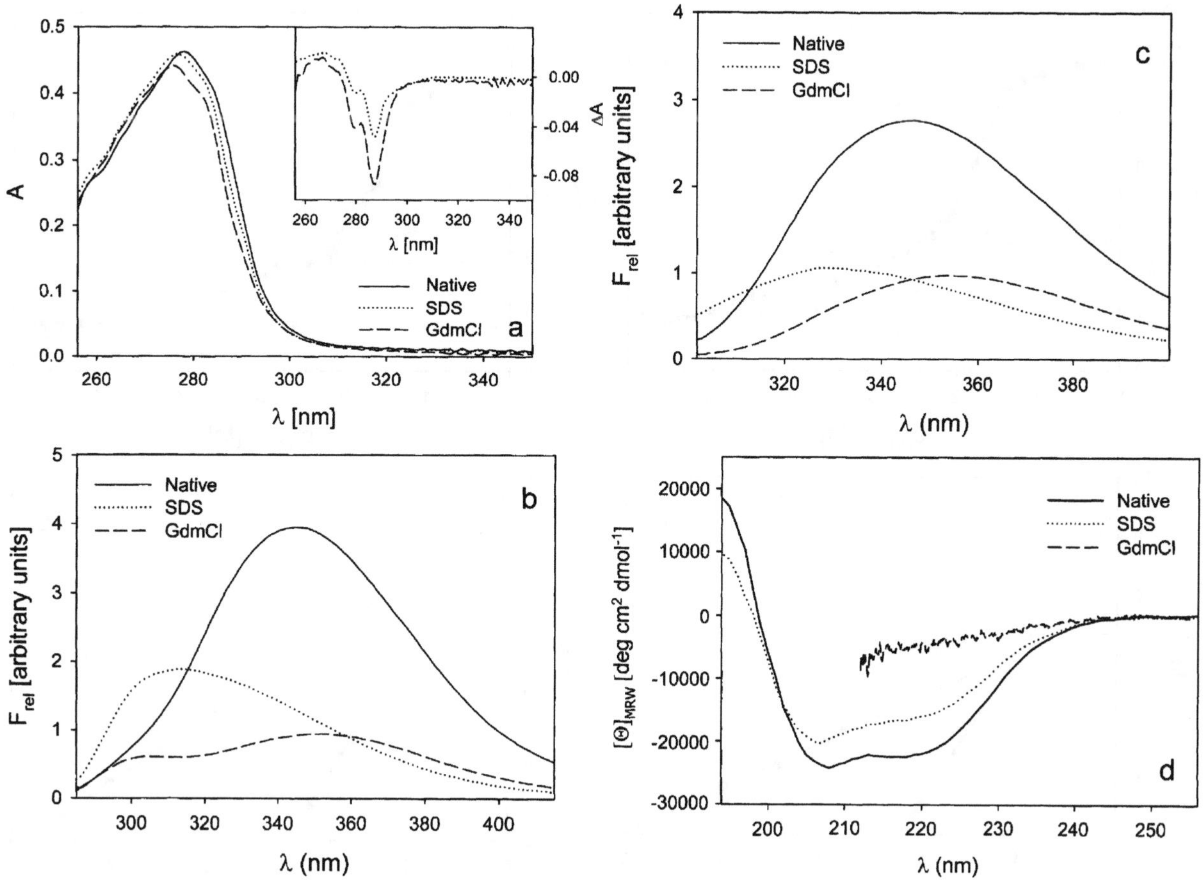

Fig. 4a–d Absorption, fluorescence and circular dichroism (*CD*) spectra of native (nonreduced) BSA ($c_0 = 0.66$ g/l, *solid line*) and of BSA treated with 6 M GdmCl (*dashed line*) or 3.0 g/l SDS (*dotted line*). **a** UV absorption; *inset*: difference spectra, maximum effect at 287 nm. **b** Fluorescence emission: $\lambda_{exc} = 280$ nm. **c** Fluorescence emission: $\lambda_{exc} = 295$ nm. **d** Far-UV CD

wavelength shift at $\lambda_{exc} = 280$ nm; in the case of GdmCl, the fluorescence intensity at $\lambda_{exc} = 280$ nm shows a drastic decrease and two separate peaks. Obviously, the quenching for the two denaturants at $\lambda_{exc} = 280$ nm differs significantly, in contrast to the effects at $\lambda_{exc} = 295$ nm. While the excitation at 280 nm reflects effects in the vicinity of tryptophan and tyrosine residues, excitation at 295 nm exclusively reflects changes in the neighborhood of tryptophan residues. From this one can conclude that SDS and GdmCl induce similar denaturation effects on both tryptophan residues present in BSA, whereas, in the case of the 20 tyrosine residues, GdmCl has a more pronounced effect than SDS.

As may be expected, the CD spectra in the far-UV regime (Fig. 4d) show distinct differences for native BSA, on the one hand, and the complexes formed with GdmCl and SDS, on the other. Owing to the different extents of unfolding, the changes in helicity differ widely: maximum helicity for the native protein, complete loss after incubation with 6 M GdmCl, and only moderate changes after SDS binding, in full accord with previous results [21, 44, 46].

Based on measurements at varying additive concentration (0–6 M GdmCl), transition curves have been measured and evaluated for all three spectroscopic techniques (Figs. 5, 6, 7, 8). All the transition curves could be fitted in terms of a three-state model; as in the case of the SDS-induced denaturation [38], fitting by two-state models failed. The transition energies, ΔG, were calculated from the logarithm of the equilibrium constant, *K*, extrapolating the pretransition and post-transition regions and taking the ordinate intercepts (corresponding to zero denaturant concentration) as being proportional to ΔG.

According to the three-state model, calculated Gibbs free-energy changes, ΔG, were obtained as the sum of two transitions. The values found for GdmCl denaturation may now be compared with the data obtained for SDS or other denaturants. While for GdmCl denaturation ΔG values of –29, –25 to –28, and –27 kJ/mol (standard

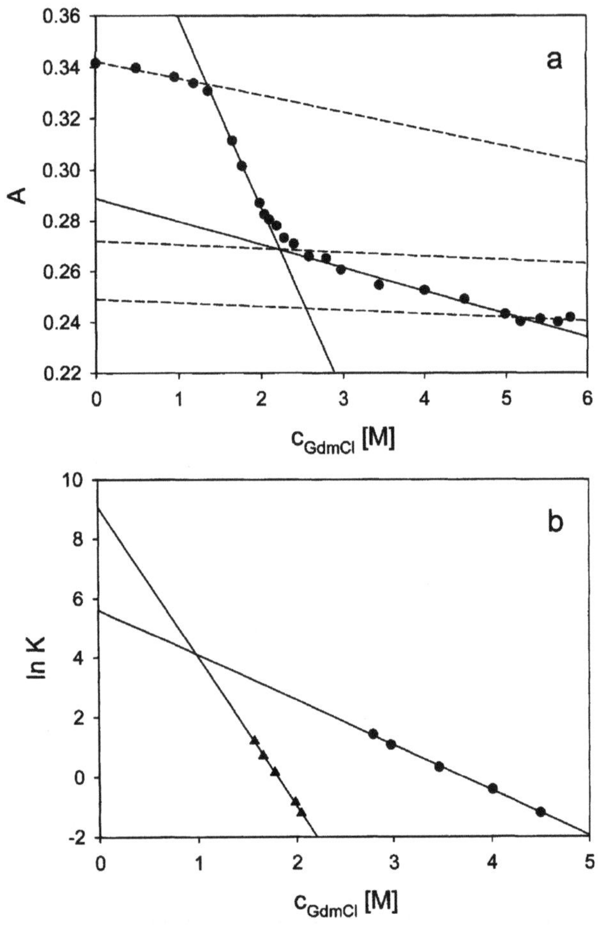

Fig. 5a,b Conformational stability of BSA in the presence of GdmCl, as monitored by UV absorption spectroscopy (0.66 g/l BSA, 0–6 M GdmCl). **a** Transition curve at 287 nm. **b** Calculation of ΔG. The lines illustrate the three-state evaluation (*solid lines*) and the pretransition and posttransition regions (*dashed lines*)

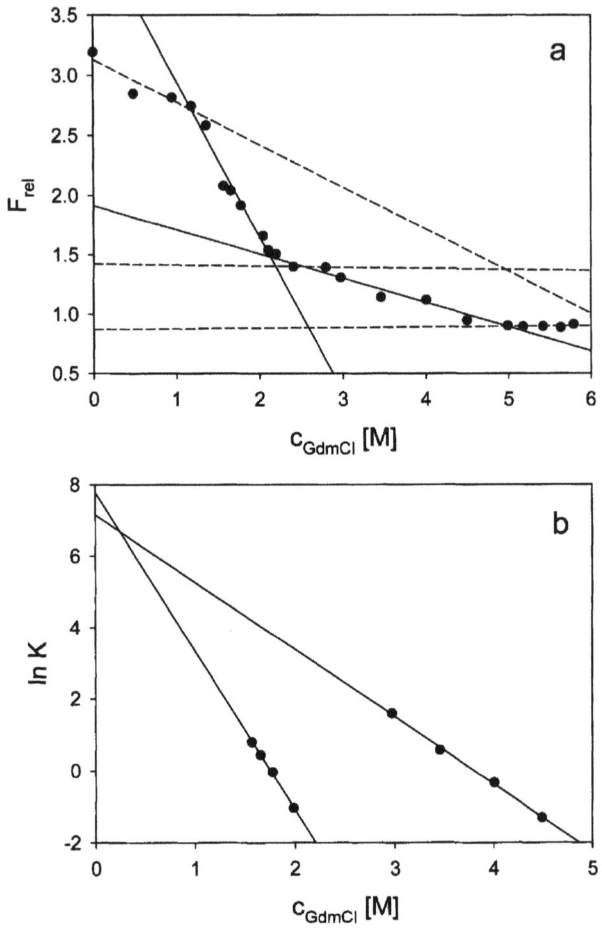

Fig. 6a,b Conformational stability of BSA in the presence of GdmCl, as monitored by fluorescence emission spectroscopy (0.66 g/l BSA, 0–6 M GdmCl). **a** Transition curve at 320 nm for $\lambda_{exc} = 280$ nm. **b** Calculation of ΔG. The lines illustrate the three-state evaluation (*solid lines*) and the pretransition and posttransition regions (*dashed lines*)

deviation ± 2 kJ/mol) were obtained for UV absorption (A at 287 nm), fluorescence ($\lambda_{em} = 320$ nm, $\lambda_{exc} = 280$ or 295 nm), and CD ($[\Theta]_{MRW}$ at 220 nm), the corresponding values for SDS denaturation were reported to be −18, −14 to −18, and −12 kJ/mol [38], exhibiting slightly larger deviations (standard deviation ± 3 kJ/mol). This proves significantly higher ΔG values for the GdmCl-induced denaturation, as compared to the energy changes accompanying the BSA–SDS complex formation.

Correlation between hydrodynamic and spectroscopic data

The transition curves derived from the optical properties at varying denaturant concentration are summarized in Fig. 9. All the transition curves reveal significant changes of their signals at about 1.7 M GdmCl or 0.3 g/l SDS, in addition to smaller transitions at higher additive concentrations (about 4.0 M GdmCl and 1.4 g/l SDS). In the case of SDS denaturation, also a correlation between the results derived from spectroscopy and AUC can be given: The changes in M_w obtained from HSSE runs at 16,000 rpm (Fig. 9b), and the detergent binding derived from Fig. 2, correlate quantitatively with alterations of the optical properties (at about 0.3 g/l SDS); the correlation between M_w, on the other hand, and the specific volumes and buoyancy terms has already been given in Fig. 3.

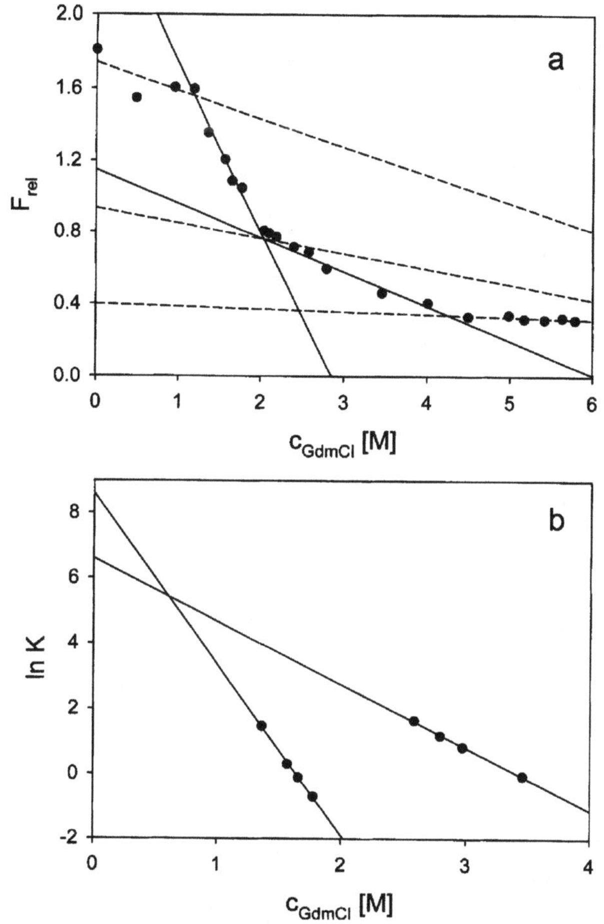

Fig. 7a, b Conformational stability of BSA in the presence of GdmCl, as monitored by fluorescence emission spectroscopy (0.66 g/l BSA, 0–6 M GdmCl). **a** Transition curve at 320 nm for $\lambda_{exc} = 295$ nm. **b** Calculation of ΔG. The lines illustrate the three-state evaluation (*solid lines*) and the pretransition and posttransition regions (*dashed lines*)

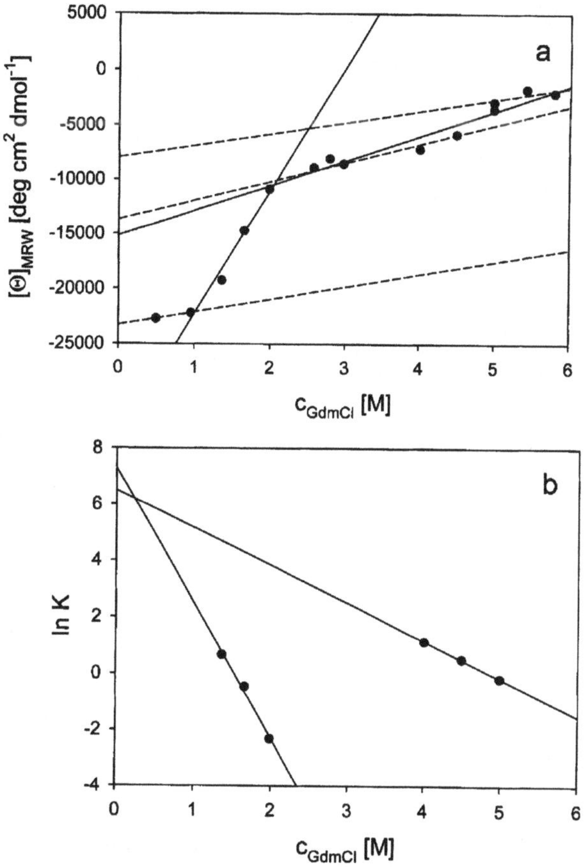

Fig. 8a,b Conformational stability of BSA in the presence of GdmCl, as monitored by far-UV CD spectroscopy (0.66 g/l BSA, 0–6 M GdmCl). **a** Transition curve for $[\Theta]_{MRW}$ at 220 nm. **b** Calculation of ΔG. The lines illustrate the three-state evaluation (*solid lines*) and the pretransition and posttransition regions (*dashed lines*)

Conclusions

AUC in combination with specific volume estimates was used to quantify the particle weights of BSA and BSA–ligand complexes. The results of mass determinations critically depend on the experimental procedures applied. Using AUC to study protein–ligand complexes requires several precautions, particularly in the case of the BSA–SDS complexes: the experimental conditions, especially the rotor speeds, have to be varied over a wide range in order to avoid perturbations caused by protein oligomers, on the one hand, and formation of detergent micelles, on the other.

For the study of the GdmCl-denatured protein, dialyzed samples were used. The resultant mass corresponds to the mass of the unliganded protein, i.e. within the accuracy of the measurement, it is identical to the mass of the native protein. If both the AUC and density measurements are performed with the dialyzed material, all interactions with the solvent (binding or exclusion of water and ligand) cancel out.

In the case of the SDS-denatured protein, the samples cannot be dialyzed to reach thermodynamic equilibrium. The specific volumes of the various protein–detergent complexes have to be deduced from the volume increments of the constituents and the approximate stoichiometry of detergent binding. The masses obtained from AUC experiments are those of the protein–detergent complexes. In applying this approach, changes in protein hydration are neglected. With this caveat in mind, on the basis of the resultant HSSE masses of the complexes, the isopotential quantities, v' and $(1-v'\rho)$, may be predicted;

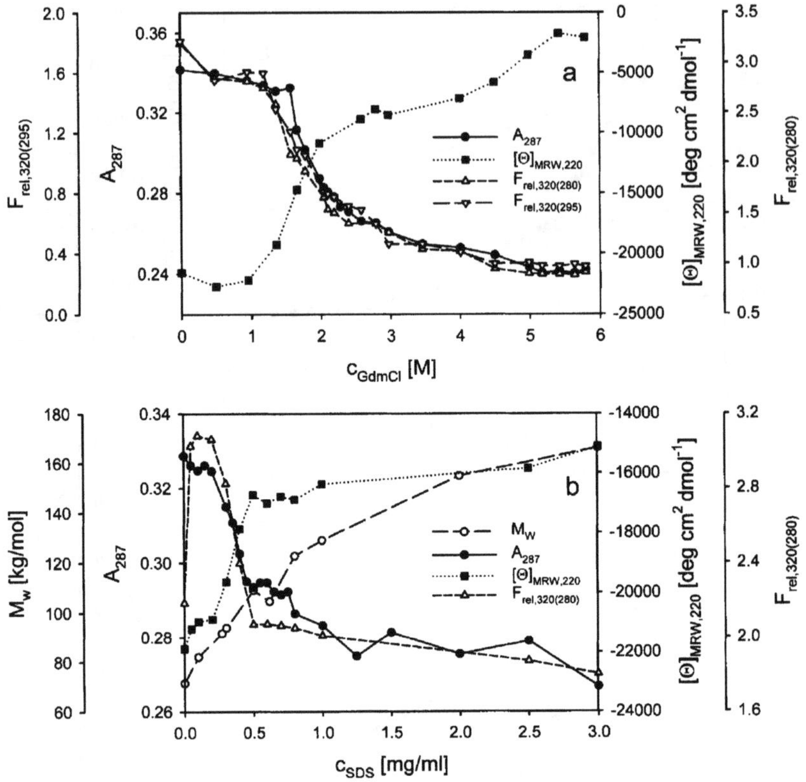

Fig. 9a, b Spectral properties and molar masses of nonreduced BSA (0.66 g/l) treated with varying amounts of GdmCl (0–6 M) or SDS (0–3 g/l). The spectral signals were derived at the wavelengths applied for monitoring the transition curves: A_{287}, F_{rel} at $\lambda_{em} = 320$ nm ($\lambda_{exc} = 280$ or 295 nm), and $[\Theta]_{MRW}$ at 220 nm; the masses of the BSA–SDS complexes were obtained from HSSE runs in the analytical ultracentrifuge at 16,000 rpm. **a** Spectral properties versus GdmCl concentration. **b** Spectral properties versus SDS concentration

no direct measurement of these quantities was possible (Fig. 3b).

As taken from the resulting specific volumes, ν' and $\bar{\upsilon}$, and buoyancy terms, $(1-\nu'\rho)$ and $(1-\bar{\upsilon}\rho)$, the differences seem to be small in the case of GdmCl (Fig. 3a), but differ substantially for SDS binding to the protein. Though this result can only be regarded as a first hint, it may tentatively be interpreted as being caused by considerable detergent binding to the protein, compared to moderate binding of the chaotropic reagent GdmCl.

UV absorption, fluorescence and CD spectra have been applied to quantify the denaturing effect of SDS on BSA [38]. Complementary studies with GdmCl now allow a comparison of the denaturation transitions and the Gibbs free-energy changes, ΔG, induced by both denaturants. Assuming one intermediate on the folding pathway to be populated, use of a three-state model allows an interpretation of the results of the denaturants applied. The ΔG values obtained for the SDS-induced denaturation of BSA are lower by a factor of about 2, compared with the values observed for the protein denatured by 6 M GdmCl. This behavior is in accordance with the partial denaturation of proteins in the presence of SDS, compared with the strong denaturing effect of the "chaotropic" reagent GdmCl, which practically randomizes the polypeptide chain.

Inspecting the hydrodynamic and spectral data summarized in Fig. 9 clearly indicates that the transitions occur at definite denaturant concentrations, irrespective of the experimental technique applied: a substantial transition at 1.7 M GdmCl and at 0.3 g/l SDS, and a further transition at elevated denaturant concentrations. The behavior of SDS is in agreement with the occurrence of two binding levels of SDS, confirming earlier studies [16, 17, 18, 19, 20, 21]. The one at moderate detergent concentrations is caused by SDS binding to the protein surface, the one at elevated concentrations by the partial unfolding of the protein. All data confirm the clear correlation between the hydrodynamic and spectroscopic effects.

References

1. Timasheff SN, Gibbs RJ (1957) Arch Biochem Biophys 70:547–560
2. Tanford C (1968) Adv Protein Chem 23:121–282
3. Lapanje S (1978) Physicochemical aspects of protein denaturation, Wiley, New York
4. Creighton TE (1993) Proteins. Structures and molecular properties, 2nd edn. Freeman, New York
5. Creighton TE (ed) (1992) Protein folding. Freeman, New York
6. Creighton TE (1994) In: Pain RH (ed) Mechanism of protein folding. IRL, Oxford, pp 1–25
7. Jaenicke R (1999) Prog Biophys Mol Biol 71:155–241
8. Eisenberg H (1994) Biophys Chem 53:57–68
9. Eisenberg H (2000) Biophys Chem 88:1–9
10. Timasheff SN (1998) Adv Protein Chem 51:355–432
11. Timasheff SN (2003) In: Hinz H-J (ed) Landolt–Börnstein new series biophysics – proteins, vol VII/2A. Springer, Berlin Heidelberg New York, pp 5/1–23
12. Schellman JA (1987) Biopolymers 26:549–559
13. Schellman JA (1990) Biophys Chem 37:121–140
14. Schellman JA (2002) Biophys Chem 96:91–101
15. Makhatadze GI, Privalov PL (1992) J Mol Biol 226:491–505
16. Reynolds JA, Tanford C (1970) Proc Natl Acad Sci USA 66:1002–1007
17. Tanford C (1980) The hydrophobic effect: formation of micelles and biological membranes, 2nd edn. Wiley, New York
18. Takagi T, Tsujii K, Shirahama K (1975) J Biochem 77:939–947
19. Tiefenbach K-J, Durchschlag H, Jaenicke R (1997) Prog Colloid Polym Sci 107:102–114
20. Durchschlag H, Kuchenmüller B, Tiefenbach K-J, Jaenicke R (1998) Jorn Com Esp Deterg 28:353–365
21. Durchschlag H, Tiefenbach K-J, Weber R, Kuchenmüller B, Jaenicke R (2000) Colloid Polym Sci 278:312–320
22. Casassa EF, Eisenberg H (1964) Adv Protein Chem 19:287–395
23. Eisenberg H (1976) Biological macromolecules and polyelectrolytes in solution. Clarendon, Oxford
24. Eisenberg H (1981) Q Rev Biophys 14:141–172
25. Durchschlag H (1986) In: Hinz H-J (ed) Thermodynamic data for biochemistry and biotechnology. Springer, Berlin Heidelberg New York, pp 45–128
26. Durchschlag H (2003) In: Hinz H-J (ed) Landolt-Börnstein new series biophysics – proteins, vol VII/2A. Springer, Berlin Heidelberg New York, pp 4/1–157
27. Tanford C, Nozaki Y, Reynolds JA, Makino S (1974) Biochemistry 13:2369–2376
28. Steele JCH Jr, Tanford C, Reynolds JA (1978) Methods Enzymol 48:11–23
29. Reynolds JA, McCaslin DR (1985) Methods Enzymol 117:41–53
30. Schubert D, Schuck P (1991) Prog Colloid Polym Sci 86:12–22
31. Schellman JA (1987) Annu Rev Biophys Biophys Chem 16:115–137
32. Jones MN (ed) (1988) Biochemical thermodynamics, 2nd edn. Elsevier, Amsterdam
33. Pace CN, Shirley BA, Thomson JA (1989) In: Creighton TE (ed) Protein structure: a practical approach. IRL, Oxford, pp 311–330
34. Shirley BA (1992) In: Ahern TJ, Manning MC (eds) Stability of protein pharmaceuticals, part A: Chemical and physical pathways of protein degradation. Plenum, New York, pp 167–194
35. Eftink MR (1995) Methods Enzymol 259:487–512
36. Pfeil W (1998) Protein stability and folding: a collection of thermodynamic data. Springer, Berlin Heidelberg New York
37. Pfeil W (2001) Protein stability and folding, supplement 1: a collection of thermodynamic data. Springer, Berlin Heidelberg New York
38. Tiefenbach K-J, Durchschlag H, Schneider G, Jaenicke R (2003) Prog Colloid Polym Sci 122:130–140
39. Durchschlag H, Tiefenbach K-J, Bayersdorfer F, Pachmann G, Jaenicke R (2003) Jorn Com Esp Deterg 33:281–293
40. Peters T Jr (1985) Adv Protein Chem 37:161–245
41. Peters T Jr (1996) All about albumin: biochemistry, genetics, and medical applications. Academic, San Diego
42. Durchschlag H, Jaenicke R (1982) Biochem Biophys Res Commun 108:1074–1079
43. Durchschlag H, Jaenicke R (1983) Int J Biol Macromol 5:143–148
44. Durchschlag H, Weber R, Jaenicke R (1996) In: Proceedings of the 4th world surfactants congress, vol 1. CESIO, Barcelona, pp 519–534
45. Tiefenbach K-J, Durchschlag H, Jaenicke R (1999) Prog Colloid Polym Sci 113:135–141
46. Durchschlag H Tiefenbach K-J, Fischer A, Weber, R, Jaenicke R (2000) In: Proceedings of the 5th world surfactants congress, vol 1. CESIO, Florence, pp 634–646
47. Durchschlag H, Tiefenbach K-J, Gebauer S, Jaenicke R (2001) J Mol Struct 563–564:449–455
48. Kratky O, Leopold H, Stabinger H (1973) Methods Enzymol 27:98–110
49. Durchschlag H, Zipper P (1994) Prog Colloid Polym Sci 94:20–39
50. Durchschlag H, Zipper P (1995) Jorn Com Esp Deterg 26:275–292
51. Yphantis DA (1964) Biochemistry 3:297–317
52. Chervenka CH (1973) A manual of methods for the analytical ultracentrifuge. Spinco Division of Beckman Instruments, Palo Alto
53. Schmid FX (1997) In: Creighton TE (ed) Protein structure: a practical approach, 2nd edn. IRL, Oxford, pp 261–297
54. Shirley BA (ed) (1995) Methods in molecular biology, vol 40. Protein stability and folding: theory and practice. Humana, Totowa, NJ
55. Durchschlag H (1989) Colloid Polym Sci 267:1139–1150
56. Eisenberg H (1990) In: Saenger W (ed) Landolt–Börnstein New series biophysics – nucleic acids, vol VII/1C. Springer, Berlin Heidelberg New York, pp 257–276
57. Hade EPK, Tanford C (1967) J Am Chem Soc 89:5034–5040
58. Reisler E, Eisenberg H (1969) Biochemistry 8:4572–4578
59. Reisler E, Haik Y, Eisenberg H (1977) Biochemistry 16:197–203
60. Noelken ME, Timasheff SN (1967) J Biol Chem 242:5080–5085
61. Lee JC, Timasheff SN (1974) Biochemistry 13:257–265
62. Arakawa T, Timasheff SN (1984) Biochemistry 23:5924–5929
63. Kawahara K, Tanford C (1966) J Biol Chem 241:3228–3232
64. Marquardt DW (1963) J Soc Ind Appl Math 11:431–441

Progr Colloid Polym Sci (2004) 127: 148–160
DOI 10.1007/b94227

GUEST COLUMN

Michael Bommes

The shrinking inclusive capacity of the national welfare state: international migration and the deregulation of identity formation*

Michael Bommes
Institut für Sozialwissenschaften,
Pädagogische Hochschule Freiburg,
Kunzenweg 21, 79117 Freiburg
e-mail: bommes@ph-freiburg.de

* For Janaka Quell, a source of inspiration not just because of her name

Samuel Huntington's vision in the early 1990s of a "clash of civilisations" struck a chord to such an extent that his core theme was reignited after the attacks of September 11th. International migration seems to be viewed as an issue that signifies this so-called clash (Bade and Bommes 1996). Migration, culture, ethnicity and conflict have become linked. The result is that conflicts arising from migration are more likely to be seen as an outcome of the multiplication of different cultures within one country. Germany, the UK, the Netherlands and the Switzerland have all been described as multicultural societies and advised to pay attention to this "fact". This has been combined with the view that even if there was no road to multiculturalism without social conflicts, there was also no viable alternative to tolerance as a device for the interaction of cultures (Leggewie 1990).

The formula of the "multicultural society" has been criticised in many ways (Bade 1996a, Hoffmann-Nowotny 1996, Joppke 2001, Nassehi 1997, Radtke 1990). Yet, simultaneously, the multicultural frame of reference for describing the consequences of migration has lost much of its mobilising potential for both science and politics. The Netherlands or Sweden used to be heralded as examples of multicultural societies, but have become far more cautious about basing their policies on the diagnosis of the emergence of a multicultural society (Entzinger 1998, Ring 1998). It also became obvious that the term "multicultural" was used in very different ways in different countries. The same word came to mean different things in Canada, the USA, Australia, the UK and Germany. The different usages were closely related to the different political traditions and ways of dealing with immigration in these countries. Meanwhile, policy oriented or so-called applied sciences like political science, education, but also political philosophy (Taylor 1993, Kymlicka 1995) became fascinated, at least for a while (although some still are) by the political debates about the "multicultural society" and began to participate in these deliberations. Sociology, however, has remained distant and appeared to be unconvinced by the appropriateness of the concept.

If one interprets Huntington's thesis of "the clash of civilisations" as a – certainly problematic – proposal to understand recent globalisation processes then it is also worth noting that most of the contributions of migration research to the political and scientific debates about globalisation are rarely linked to discussions of the multicultural society. Instead "transnational migrations" have been identified as a central part of ongoing globalisation processes.[1] "Multiculturalism" had been observed in the frame of nation state societies. It was viewed as a challenge to the efforts by nation states to seek cultural homogenisation (Gellner 1988). It would appear that "transnationalism" has become the key term for migration research when participating in the globali-

[1] See, for example, Faist 2000, Glick Schiller/ Basch/ Blanc-Szanton 1995, Hannerz 1996, Pries 1997, 2001, Portes 1995, 1996, Vertovec/ Cohen 1999.

sation debate. The idea of transnationalism stresses that the various forms of transnational migration and the emergent social networks that come with it need to be understood as both a part and as an expression of the nation state's loss of both centrality and sovereignty. It could be argued that many recent proposals that deny the continuing importance of nation states are premature and unconvincing (Bommes 1999). As far as research on transnational migrations is based on these assumptions it inherits some of the problems linked with them (Bommes 2002).

That said, this article[2] has a different target. It explores the distinction between the empirical relations referred to by multiculturalists and transnationalists on the one hand and the conceptual and theoretical tools used by them, on the other. By concentrating on empirical phenomena described as multicultural or transnational it will be argued that these can actually be interpreted as indicative of the reduced inclusive capacity of national welfare states without implying 1) any prognosis about the end of the welfare state or 2) any assumption that multi-culturalism and transnationalism indicate crises or at the least phenomena that threaten nation states. To develop these points, I discuss some of the implications of Joppke's argument that European nation states have given up – certainly to different degrees – the strong programme of national homogenisation.[3] This occurs as a result of post-war-immigration history and because of ongoing and more general processes of social differentiation. The result is that a "deregulation" of collective forms of social identity can be observed. National or ethnic belonging have lost much of their political and legal relevance. I will then develop this point by arguing that these political deregulations of identity formation come as a relief for (nation) states because it signifies the reduced emphasis on communal relations, the downgrading of claims based on the idea of national solidarity and equality, and is also part of the changing relations between states and their constituencies. My argument is that immigration and immigrant group formations are thus an integral element of ongoing differentiation and reconstruction processes of welfare state constituencies.

I.

Observers of the multicultural society and of an emerging transnationalism tend to share, albeit in different ways, the proposition that central structural elements of the nation state are being eroded. The multiplicity of cultures arising from migration indicates growing cultural heterogeneity within the state territory. In turn, this questions the old program of the nation state, i.e., the creation and/ or the safeguard of the cultural homogeneity of the population. In a different mode transnationalism refers to the emergence of social structures which result from enduring migration streams and links, i.e., networks and organisations that stabilise such links and render them transnational in the sense that they become links that transcend the borders of nation states in a permanent manner. At the same time, transnationalism underlines the shifting orientations of migrants who orient themselves to these transnational structures of social opportunities. As a consequence of this the nation state and its classical aim of "social integration" loses some of its centrality as a frame of social orientation.

We are, though, racing ahead of ourselves. If a theoretical interpretation of some of the social developments indicated by multiculturalism and transnationalism is to be developed then first of all some historical contextualisation is useful. Since the Second World War until roughly the end of the 1980s, large scale migration movements in the European immigration countries, as well as in the USA, Canada and Australia – such as refugees and expellees, labour migrants, postcolonial migrants and family migration – were met by a remarkably high level of provision by national welfare states and relatively successful and effective programmes of social integration (see among others Hollifield 1992, Soysal 1994). The social integration of migrants can initially be seen as politically unintended, but this did not greatly affect the common outcome best summarised by Hammar (1989, 1994) when he argued that the majority of migrants were included in civil, social and to some extent political rights to an extent which was previously seen as the privilege of citizens. Moreover, this inclusion process took place without major efforts to secure cultural assimilation aiming at homogenisation. In this way migrants became what Hammar termed "denizens". In this context, the multiplication of cultures in an empirical sense, i.e. the immigration of individuals from different cultural backgrounds and the related multiplication of forms of life was not necessarily combined with multiculturalism in the social sense that cultural differences and the ethnic and national meanings ascribed to them became a dominant mechanism for the allocation of socio-structural positions and rights. To put it bluntly: culture and ethnicity were not socially dominant mechanisms for the creation of minorities.[4] The expanded welfare state after the Second World War possessed enough capacity – at least in Europe, Canada and

[2]A revised version of a paper prepared for the conference "Power and Democracy in Multicultural Societies" at Rosendal Norway in May 2001.

[3]See Joppke 2001 and Therborn (1995) on the emergent post-nationalism of European states after the second world war.

[4]For the USA and Germany see Santel/ Hollifield 1998, for Australia Castels 1996, 1998, in a comparative perspective for the USA, Germany and France Hollifield 1992 and for Germany Bommes 1995.

Australia – to provide chances for the social inclusion of migrants in a way that devalued to a certain extent ethnic and national forms of social closure.

It appeared that this constellation of the 1970s and 1980s came to an end as a result of post-1989 political changes. These changes centre on the immigration of refugees and asylum seekers, the arrival of new labour migrants and ethnic migrants like Jews and Ethnic Germans from Eastern Europe, the growth of ethnic and national conflicts in the former socialist countries and in former Yugoslavia. During this time, we can also see the diminished interventionist capacity of national welfare states in the context of political, economic, legal and scientific globalisation and European integration. In addition, countries such as Germany, France, Great Britain, Canada and the USA have begun competing for highly qualified migrants especially in sectors such as IT and health care.

This leaves the question of what precisely in the field of migration has changed and in which ways have welfare states reacted to this? Is it justified to assume that fundamental shifts have occurred as indicated by the formulas provided by transnationalism and multiculturalism? Before discussing this point, it is useful to remind ourselves about the specific role of national welfare states in modern society for the political regulation of migration and its social consequences. This will provide some background for a theoretical interpretation of recent social changes.

II.

The modern nation state can be defined by the execution of sovereignty over a territory and a population. As Swaan (1988) has shown this has been linked to the emergence of the welfare state. By providing chances for participation in the social realms of the economy, law, health or education – that is within a welfare state – the nation state created the social preconditions for a process in which former subjects became political citizens. This process saw the inclusion of the population into the political system as individual citizens while legitimising and rendering valid the claim of political sovereignty over these people. The welfare state left behind its former role as a "polizey" and it became the central event in modern society that moderated relations between the principle of universal access to and inclusion in the social realms of the economy, the law, education, health or politics and the empirical reality of social exclusion.

This result of this universalism of inclusion in modern society is that nobody should be excluded from claiming economic, legal or educational provisions if they can fulfil the social preconditions for these claims (Luhmann 1989). For example, one can participate in education if one is perceived as educable, one can participate in the economy if one finds access to monetary means, one can participate in law if one knows how to act on behalf of one's rights. Of course, none of these necessarily imply that inclusion always empirically succeeds.[5]

A central condition for success or failure is access to membership roles in organisations and to the income earned by fulfilling these roles through labour. Finding employment in modern society means offering competence and specific services to an organisation in exchange for payment (and other rewards). The opportunities for individuals to gain access to the functional realms of the economy, law, health or education are mediated in many ways by organisations (Bommes and Tacke 2001). Organisations (such as firms, administrations, hospitals, courts and universities) engage individuals as members by specifying formally the conditions of participation. As a particular type of social system, organisations rely on the formal definition of membership that distinguishes between members and non-members. By engaging individuals as members then these *organisations* distribute the central conditions and chances for inclusion and exclusion in other social spheres. Recruitment of personnel, the allocation of positions and careers in organisations then become linked with the distribution of money, reputation and influence or – in the words of Bourdieu – economic, cultural and social capital. Organisational memberships based on careers and the corresponding conditions of engaging individuals create different chances for individuals to receive services and resources as consumers, patients, clients, pupils, electors and so on. (Bommes 2001).

The organisational means of allocating positions and incomes is a central frame of reference for *welfare state* policies that aim to increase and secure the transferability of these chances for inclusion. That is, they seek to ensure that organisational membership equips individuals with the necessary means and rights that provide them with options for participation in the social realms of the economy, education, law, health, politics and the family.[6] Historically the welfare state

[5]It may be worth stressing that this is a very technical concept of inclusion and exclusion implying no normative assumptions: Inclusion formally speaking is the way in which individuals are addressed by social systems as relevant for actually ongoing social processes; exclusion correspondingly refers to the irrelevance of individuals-and this does not always indicate a "social problem": e.g. the irrelevance of an individual for the health systems means the individual is healthy and therefore not a case for a doctor or hospital.

[6]Residual welfare states perceive the safeguard of transferabilities only in a limited sense as a political duty. The "conservative" type of welfare state however aims at the maintenance of the standard of living, whereas the "social democratic" type is based on the concept of "social citizenship" as interpreted by Marshall (s. Esping-Anderson 1990). For "the southern model of welfare in Europe" see Ferrera 1996.

became a mediator that tried to reduce barriers of participation and access to the provisions of the different social realms as well as providing the resources to transcend these barriers through education, money, legal advice, health care and so on. We can then understand such measures as efforts to secure the empirical openness of social systems for individuals.

From its beginnings welfare provided by the nation state was linked to territory. The provisions of welfare states initially addressed only the citizens of that state. Nation building transformed a population into a unitary nation, a people ("Volk") on a politically defined territory.[7] The nation may have been defined in either ethno-cultural (Germany) or political terms (France), but in the historical context of competitive state building processes in Europe the common welfare of the people of the nation – of *the community of national citizens* – evolved as the general frame of reference for the state (Bommes 1999).

III.

A principal result of this European tradition of nation state building was the tense relationship between immigration and the national welfare state. Welfare states sought to mediate social participation for their citizens, but organisations and the differentiated social realms of society found it difficult to permit social inclusion on the basis of the distinctions and criteria preferred by nation states. By this I mean that inclusion and exclusion in the economy, education, law or health did not necessarily correspond to the political distinction of belonging or non-belonging to the nation. In addition to this migrations were not accidental or unwanted. They were often politically or legally initiated or induced, as was the case with the labour migrations of the 1960s and 1970s, the different forms of ethnic migration or contract labour migration, refugee migration and migration of highly qualified individuals.

Within the national welfare state then the experiences of social exclusion provokes the question of why exclusion also hits those who 'belong' to the nation and not just those who do not belong since the welfare state institutionalises the expectation that it cares primarily for its people on the territory. Why are foreigners or new immigrants recruited by organisations if major parts of the indigenous population are unemployed? Why can immigrants access health and education when the general capacity of provision of welfare states is shrinking? This occurs because border-crossing international migrations induce a tendency to copy the distinction between those included and those excluded into the distinction between the nation/people and the residual population on a state territory. This articulates the political expectation that both distinctions should classify the same group of individuals.

This tendency increases when it becomes apparent that national welfare states with a diminished capacity to offer provision can for several reasons only regulate but not control migration. For example, labour migration in the 1960s and 1970s was initiated on the premise that states could control these migration processes and their consequences – that the 'guests' would return to their countries of origin at some point in the future. But political efforts to reverse these immigrations encountered the legal barriers that arose when "guestworkers" were included within the social insurance systems of the welfare states (Hollifield 1992, Bommes 1997). In the case of family migration then international and national laws limit the control capacities of states. Refugees and asylum seekers can also seek the protection of international and national laws. To this litany of lost control we can add the development of internal migration within the EU even when such intra-EU migrations undermine national welfare regimes, as was the case with 'posted workers' in Germany. (Eichenhofer 1996, 1997, Santel and Hunger 1997).

To put it more systematically: the tendency to copy the distinction between inclusion and exclusion into the distinction between the nation (people) on the one hand and the immigrants (foreigners) on the other hand is structurally based on the historical claim of national welfare states to regulate access to the social realms of the economy, law, education and health. These states do so by providing welfare for their citizens and securing the chances for social inclusion within these realms. This tendency was only mildly relevant from the 1960s to 1980s, when the expansion of welfare was combined with the political restriction of international mobility based on the East-West-confrontation. Welfare states seemed to be able to fulfil their promises, and, as a result of a progressive differentiation between politics and law, migrants in Western Europe had become 'denizens'. Until the end of the 1980s, European migration policies were largely the social consequences of the immigrations of the 1960s and 1970s. This led to programmes of social integration for those already settled, coupled with efforts to limit the numbers of new migrants such as refugees and asylum seekers. In these terms, policies that included immigrants into the expansive programmes of welfare provision were part of a singular

[7]See Koselleck 1992, Brubaker 1992. Quite contrary to the development in Europe was the emergence of "the people" in the US a result of immigration. Consequently immigration was never used in the same way for the distinction between those belonging to the nation and those who did not. Historically the existence of the immigration country USA itself has been a chance for social inclusion and this made the emergence of a welfare state of the European type unlikely.

phase due to a specific post-war constellation of economic prosperity and welfare state expansion.

The texture of debate has changed since this time in ways that bring into much sharper focus the relations between migration and the inclusive capacity of these national welfare states. Globalisation since 1989 combined with the development of supranational EU governance erode the constitutive principle of territoriality of the welfare state.[8] But by doing so it actually makes this constitutive principle even more visible. This has been accompanied by a reduction in the level of provision in a number of welfare states. It matters little whether migrations are politically initiated, allowed, tolerated or forbidden, they are an attempt to realise either legally or illegally the chances for social inclusion that are opened up to individuals by these processes of globalisation. These attempts affect all states even if they declare that they are not an immigration country and do not want any further immigration. Germany was a telling example for this and has embarked on an attempt to rethink this relationship.

IV.

Germany, the Netherlands, Britain and France have all reacted to this loss of control over access and residence on the state territory by diversifying and making more specific their immigration regulations. They introduce new forms of border control and highly differentiated distinctions between migrant groups. They also restrict access to resident permits and social entitlements while constructing different paths of access to citizenship. One result of this diversified treatment of migrants through national welfare states is the constitution of *differentiated migrant populations*. It is worth recalling that 'immigration' comes in many and diverse forms: postcolonial migrants, ethnic migrants, labour migrants, family migrants, refugees, asylum seekers, contract labour, illegal migrants and so on. There are widely different implications of these different status positions. This will be illustrated by an analysis of six different migrant categories in Germany:

- *Refugees, expellees* and *Ethnic Germans* could move to the Federal Republic because for historical reasons they counted as Germans. Reunification in 1990 ended the status of an 'incomplete nation' and symbolised the link between the population and territory. Yet, the result was that the right of access for Ethnic Germans to the German territory was reduced and finally limited to those born before 1993. The migrant category of Ethnic Germans had been a rather privileged one for reasons of national belonging but had been closed. This change can be seen as part of the readjustment of the German concept of citizenship. In 1990, 1993 and 1999 amendments of the foreigner and citizenship laws had allowed guestworkers and their children to access German citizenship. In such terms, restrictions on the legal status of Ethnic Germans can be interpreted as an aspect of the strengthening of territoriality as criterion for the granting of citizenship coupled with a weakening of the *ius sanguinis* tradition (Bommes 2000).
- *Labour migrants* and their families. Guestworkers migrants became a normal client group of the welfare state and its programmes of social integration even if their position on the labour market exposed them to far higher risks of poverty and exclusion (Geißler 1996). One consequence of the irreversible settlement process and the inclusion into welfare has been improved access to naturalisation since 1990.
- *Family migrations* can be politically regulated but not avoided because national and international laws guarantee the right of the family. These migrations utilise a right established in modern society as a result of the institutionalisation of the nuclear family. That is, the right of each individual to found a family and not be separated from it. As a result of this right, individual migrants can, in a sense, transfer the option of geographical mobility to family members. This then creates access to better chances of social inclusion for their family members. This holds true independently from the question of whether migrant families themselves form nuclear families because they can rely on the legal protection of the family. Migration policies trying to limit and control the foundation of families react to these implications[9] but encounter legal limits in countries like Germany because of the constitutional protection of the family.
- Since the end of the 1970s the arrival of *asylum seekers* and *refugees* has been seen as a problem of maintaining territorial sovereignty. It is also a challenge to the institutionalised difference between the people and the population. In reaction to the world wide growth of refugee and asylum migration since the early 1980s European welfare states have restricted access to asylum and refugee status and weakened the residence

[8] For a comparative discussion of the perspectives of different welfare states see the volume Esping-Anderson 1996. This does not imply that there is a direct relation between globalisation and the shrinking capacity of welfare states (Fligstein 2000): Different types of welfare states have different adaptive capacities to new constellations and it seems that the model of Esping-Andersen (1990) does not adequately account for these different capacities (Ganßmann/Haas 1999, Goodin et al. 1999, Alber 2000).

[9] The recently introduced immigration law in Germany reduces the age of children allowed to follow their immigrated parents to 12, allowing at the same time a number of exemptions from this general rule.

status of asylum applicants (Opitz 1997). Attempts to harmonise at EU level asylum policies via the Schengen agreement and the Dublin convention have led to the creation of 'safe' third countries, strict visa requirements for migrants from most non-EU countries and 'carrier sanctions' on forms of international transport that wittingly or unwittingly transport 'unwanted' migrants. These measures have allowed EU member states to regain control over access to their territory (Joppke 1997). The numbers of refugees have fallen since the beginning of the 1990s without breaking national or international law. This has been done by making it extremely difficult for migrants to reach their territory independently from the question of whether or not they are legitimate refugees/asylum seekers or economic migrants . The ways in which rich welfare states like Germany subvert national and international refugee law corresponds to the political regulations concerning the options of social inclusion and entitlements for immigrated refugees and asylum seekers. By this is meant that the dispersal schemes and restrictions on movement mean that asylum applicants have either no or extremely limited labour market access. Meanwhile, health protection and public assistance are reduced to minimum standards. The welfare state thus reduces the social participation of refugees and asylum seekers to a level that allows the administrative process to function but which makes very visible the provisional nature of the asylum-seekers residence by limiting the possibilities for an asylum seeker to lead a normal life (Bank 2000).

- National welfare states do not have the capacity to completely control migration. Some forms of migration continue, such as *seasonal* or *contract labour* (Hönekopp 1997). For demographic reasons a number of European states have started to compete for certain types of migrants. One striking element of these forms of migration is their highly organised and regulated character. The residence status of seasonal and contract migrants is strictly limited and they are not included in the welfare provisions of the state. These types of regulation aim to maintain labour migration as a flexible instrument of labour market policy hindering the chance for these seasonal and contract labourers to develop into irreversible migration and to build up legal and social entitlements as did former "guestworkers". The welfare state can be seen to operate in way that maintains what can be called a threshold of inequality. By this is meant that barriers are established that limit access for various categories of migrants to the social security system and the resources and entitlements linked with them. Simultaneously, the highly qualified migrants are actively sought to ensure the success of sectors such as IT and healthcare.
- *Undocumented* or *illegal migrants* are, of course, defined by their effort to avoid a confrontation with the state and its attempts to ascribe a legal status.[10] Portugal, Spain or Italy are newer immigration countries that have reacted to illegal immigration through the regularisation of the status of 'illegal' immigrants. These efforts to regularise have, to a large extent failed because the acquisition of legal status is often linked with the loss of competitive advantages on labour markets (Santel 1995). In Germany these regularisations have never been seriously tried. Instead a stronger focus has been placed on rigorous border controls and stronger efforts to force migrants to return to their countries of origin.

This overview of different migrant categories shows the efforts of national welfare states to maintain their sovereign authority and to govern access to their territory with the effect that various forms of migration are created. These differences between forms of migration can then help explain differences in legal status and entitlements. From this we can say that belonging to a category of migrants is not a quality of migrants themselves[11] but a result of the ways in which states distinguish between them and ascribe a status to them.

V.

The ways in which states differentiate migrants and define their options of social participation has the effect of structuring in a peculiar way the empirical results of post-war migration . Since the 1990s there has been a creeping erosion of the core programme of the national welfare state. This core programme centred on the "integration" of the state's citizens and legally resident migrants "into society". This erosion also undermines to a certain extent those perspectives from migration research that focuses on the integration or assimilation of migrants and continues to implicitly presuppose that

[10]It may be worth noting that the different terminologies used in this field-"undocumented" (usually preferred for reasons of political correctness) versus "illegal migrants"-refer to different social aspects linked with this type of migration: The reference to the undocumented status of migrants indicates the extent to which migration research and its capability to observe and analyse migration still rely methodologically on states and their capacity to control and to document migration (via statistical accounts). Illegality refers to the fact that these migrants are more or less forced to avoid situations that provoke the relevance of their legal status. For this reason these migrants can make only a limited use of rights and this implies a restricted capacity to risk social conflicts-most obvious in the case of international prostitution.

[11] Migrants need to take these categories into account. They have to decide how they will present themselves-e.g. as asylum seekers or family members-or if they prefer illegality. Many migrants move through several of these different statuses.

the welfare state and its programme will remain a central frame of reference (Favell 2002).

This is not to say that political programmes aimed at integration have been abandoned. Attempts to manage the "integration" of legally residing foreigners remain on the agenda in parallel with attempts to limit the possibility for migrant groups such as refugees, asylum seekers, contract and seasonal labourers, foreign students and illegal migrants to acquire a more permanent status. The result is that the enduring validity of the programme of integration is politically emphasised even in the context of the reduced welfare capacity of the state. The point is that it acquires a different meaning.

There is another side to these political efforts to differentiate migrant groups and include parts of the legally residing immigrants into welfare provisions. This is that a growing number of social groups lead their lives in ways that include recurrent transnational migrations or become embedded in social contexts structured by these migrations, the related organisational forms and the corresponding forms of state intervention. This means that it is less useful to describe migration as "an enduring move to another [national] society or region of a single or a number of individuals" (Treibel 1990: 21), as a standard German text book on the sociology of migration puts it.

It would seem that the national state is losing some aspects of its central position as a frame of reference for the activities and orientations of a substantial number of migrants. To illustrate this point we can return to the migrant categories outlined above.

a) Because of their supposed national belonging Ethnic Germans were treated by welfare state programmes until 1990 as though they had spent their whole lives on the territory of the state. They were treated as though they were part of "the German people". This and rendered them invisible as a migrant group through extensive welfare programmes (Bommes 2000). Since the early 1990s these programmes have been heavily reduced and the "Ethnic German" migration route has been successively closed. But this has not changed the fact that once in Germany Ethnic Germans are seen as part of the state population and their "social integration" is still seen as a political duty. The nation state remains the frame that defines this type of immigration – but their status has been altered in such a way that it becomes similar to that of other legal immigrant groups. With the introduction of freedom of movement in the former Soviet Union and its successor states migration networks have evolved that organise constant back and forth migrations and maintain transnational resources. A trade in documents that testify German descent and belonging has emerged, as too have schools that teach the German language. Both descent and German language proficiency are required if "Germanness" is to be officially acknowledged. Those developments could be interpreted as simply transitional phenomena, but they also help build more enduring transnational social structures.
b) *Labour migrants* of the 1960s and 1970s moved under state direction. As a result of their employment they acquired welfare and residence entitlements that were followed by the granting of naturalisation entitlements and the expectation that they would be "fully integrated into society (the community of citizens)". From the perspective of the welfare state this can be interpreted as a normalisation with the denizens (finally) becoming full citizens.[12]
c) *The children of these labour migrants, the second and third generation and the Ethnic Germans arriving after 1990* encounter high barriers for their inclusion and corresponding face a greater risk of exclusion, especially on the labour market. The specific risk for these generations is that they depend on educational qualifications and thus career chances if they are to realise their chances of inclusion in other social realms. This reliance arises because their parents are less likely to have established the social networks that can help their offspring get good jobs. The position of the second and third generation is further weakened because many migrant children are unable to compete effectively in the education system.
d) *Family migrations* have become the most important migration stream to most immigration countries. They establish durable relations of migration between the regions of origin and destination. These family migrations and corresponding chain migrations are a major part of the crystallisation of transnational networks and communities.[13] These migrations, the networks, requirements and opportunities linked with them then prompt the development of organisations which provide the necessary political, legal, religious, health, cultural and economic resources. If both are placed in the context of origin and destination these structures acquire a self-relying character. "Transnationalism" of this kind clearly does not imply that the nation state becomes irrelevant. Rather migrants - but not only migrants – orient themselves and their lives to structures of social inclusion offered by these

[12]In the Netherlands the labour migrants of the 1960s and 1970s have been quite effectively excluded from the labour market-even under conditions of nearly full employment during the 1990s (Thränhardt 2000).

[13]See Glick Schiller/Basch/Blanc-Szanton 1995, Levitt 1998a, 1998b, Müller-Mahn 2000, Pries 2000. The brilliant case study of Müller-Mahn demonstrates the emergent career structures for migrants between an Egyptian village and Paris as a consequence of transnational migration processes during the 1980s and 1990s.

organisations and networks that are centred around transnational migration.

The result is that family migrations and the connected social structures have major effects on the difference between "the people" and "the population". The transnational forms of inclusion resulting from these migrations open up paths of mobility for undocumented or illegal migrants, but they also provide multiple new opportunities and career structures for migrants that transcend those of the immigration context (Müller-Mahn 2000). In this way transnational organisations and networks also provide some compensation for migrants equipped with only weak welfare entitlements. For those migrants participating in transnational relations, the frame of the national welfare state looses some relevance. The result is that the distinction between citizens and foreigners, the people and the population is de-valorised. Transnational organisations, networks and bonds of loyalty along ethnic lines open up chances of inclusion that are alternative to or 'beyond' the welfare state. The state may be instrumentalised, but it is neither the origin nor the central frame of reference for such social relations that crystallise around migration.

e) Asylum-seeking has been politically restricted in the ways outlined above. As a result of civil wars, ethnic cleansing or political repression, this type of migration depends on the availability of support networks organised around family, ethnic or regional bonds (as was the case with Bosnians, Croatians, Serbians, Albanians and Kurds). Restrictions combined with the dubious legal status direct these migrants to transnational organizations for access to income, finance, employment, transport, law, health and politics.

f) If welfare states allow foreign workers to compete on the labour market in a highly organised form, as is the case for *contractual and seasonal labour* then this can lead to the creation of organisations that exploit the constructed difference between contract labour and free labour. This type of contract labour migration forms the context for new types of short-term, cross-border pendel migration networks for labour and trade (Morawska 1998).

g) The number of *illegal or undocumented migrants* is difficult to assess, but this type of migration can be seen as the other side of immigration control attempts made by the national welfare state. Illegal migration is a result of restrictions on immigration from non-EU countries. Illegal migration assumes different guises (e.g. overstayers or clandestine immigration), but makes visible the enormous extent to which national welfare states intervene politically in the opportunities for inclusion in different social realms. It also shows how state organisations attempt to avoid the inclusion of non-citizens even where citizens do not take up opportunities such as low-skilled jobs.[14] A central aim of the control and exclusion of migrants is the attempt to maintain the standards of social inclusion protected by welfare states and to ensure that these standards are not undermined by unregulated immigration. However, increased chances of mobility because of developments in communications and transport technologies provide opportunities for migratory networks that sustain world-wide people smuggling operations (Kyle 1998). The result is that welfare states create chances for social inclusion based on illegality.

Illegal migrants do not necessarily lead socially separated lives. They do not live entirely beyond the organisations regulated by the state. However, their social participation is more or less hidden from the state and its efforts to observe and control. In various areas of production and services all kinds of illegal employment evolve. It is not just various organisations or private households that are interested in preserving this employment because of the low wages for intensive and effective labour performances. It is also the case that a large proportion of the migrants themselves are interested in keeping up these relations since illegality itself provides their chances of social inclusion. Their competitive labour market advantage is their illegality and the by-passing of welfare state regulations. The threshold of inequality maintained by national welfare states is successfully ignored by organisations and private households that employ 'illegal immigrants' and, in turn, can provide a chance for the social inclusion of migrants. Attempts by illegal immigrants to gain an income are not determined by the standards of equality and distribution of welfare states. In this way the differentiated efforts of welfare states to maintain the difference between "the people" and "the population" and to privilege their citizens produces a paradoxical effect. By this I mean the emergence of a type of immigration that acquires its dynamic precisely from these political efforts to hinder further immigration. These efforts are the basis for the reciprocal interest of organisations and private households on the one hand and illegal migrants on the other. The maintenance of the welfare threshold by national welfare states in relation to "outsiders" leads to a growth of distribution differences and inequality

[14] Partly because citizens can afford to do this, partly because they are not allowed to use opportunities because they imply the ignorance of welfare regulations concerning working hours, minimum wages or social insurance. At the same time we see however that states are confronted with pressure to legalise some of these types of migration because of the expanding demand. For instance for the care of elderly people.

"inside" and feeds the growth of an illegal population and related social forms of organisation and networks.

To sum up, examples c) to g) provide evidence that the national welfare state is loosing some of its centrality as a frame in the double sense that either the efforts of various migrant groups to realise chances of social inclusion are decreasingly oriented to the programme of the welfare state – "integration into society" – or that migrants are largely excluded from this programme.

VI

For a quite substantial number of migrants the welfare state offers limited options. Family migrations and the other forms of migration including illegal migration not only limit the sovereignty of national welfare states over access to their territory. The resultant social structures may have the effect that a growing part of the population on a state territory does not aim to become part of the collectivity of citizens. This does not preclude naturalisation, although this cannot be simply taken as a sign of "integration into society" but rather as the utilisation of an important social option which enhances mobility. It is important to stress here that this does not imply that the state is of no use for these migrants. It remains important in several respects but not in the classical holistic sense as the institution mediating membership in society by making individuals members of the people.

What are the effects on national welfare states that have historically regulated the relation to the territorial population by transforming that population into their people ? We may understand some of the meaning of these developments better if we do not just observe them as "external" migration phenomena with specific effects on welfare states but as one element of an ongoing reconstruction process of the internal relation between states and their populations. States obviously try to preserve some sovereignty over access to their territory. Yet, the social consequences of international migration may articulate an ongoing process in which the distinction of national welfare states between the people and the population is dissolved and reconstructed in a different way.

Modern states are presently faced with a double challenge: On the one hand they have to accommodate their territorially defined sovereignty in relation to supra- and international governance. On the other, international migration articulates a constellation in which a growing number of social processes in the realms of the economy, law, education, science, health, sports or the family retain a territorial index but are not territorially limited or restricted. This is shown by the erosion of the classification of the world population into nation state populations through processes of international migration that subvert the internal differentiation between the people and the population traditionally made by states. This distinction always presupposed a threshold of inequality maintained by welfare states. Writing about the relation between migration and the welfare states in "the world society", Stichweh argues "in several respects it is more informative to know if somebody is from Switzerland than from Morocco (...) than to know his class position" (Stichweh 1998: 54) and this still applies. But international migration does not just refer to the successful overcoming of politically institutionalised thresholds of inequality by migrants but is also part of an internal process of restructuration of the relations of distribution and inequality inside these welfare states. This restructuration process affects the internal differentiation of the addressees of both welfare state politics and the political semantics of identifying and addressing these welfare state beneficiaries. These are now considered in turn.

a) Welfare State Politics. An important reaction to the reduced capacity of provision of European welfare states has been increasing differentiation within welfare provisions of different groups of receivers.[15] On the one hand welfare states cover illness, invalidity or old age for the more competitive segments of the population while seeking to safeguard or improve their standards of living by providing social insurance schemes and/or fiscal relief. On the other hand, welfare states restrict access of insufficiently competitive segments of the population to comprehensive social insurance schemes by means of increasing individual responsibilities to cover social risks requiring growing excess payments for the costs of health and other forms of care. The result is that these groups are less equipped to cover these costs so that welfare provisions for them are redesigned as various types of basic provision concerning income and healthcare. These types of basic provision are combined with expanded means testing and "active employment" measures orienting individuals to the demands of the labour market and to less qualified jobs with reduced payments. The provision of welfare is therefore combined with the effort to motivate these groups to offer the labour market services that are currently supplied by (legal or illegal) migrants.
   Migrants are not the only weaker group on the labour market. One consequence of the internal differentiation of welfare provision and the implied revision of the classical promise of national welfare states – full membership in the national community as described by Marshall (1950) – is what can be called a parallel differentiation of the addressees of welfare state politics. Competitive and successful groups on the

[15]See footnotes 6 and 8; the organisational structure and the modes of welfare provision of European welfare states differ enormously but they share this general trend referred to in this text.

labour market are in command of effective political voice- and/or exit-options (Hirschman 1970) and are the main addressees of the welfare state. They become a kind of core population-the political centre-whose loyalty defines one central frame of reference for state policies and politics. This core population is decreasingly defined in national terms. It includes successful international migrants and is open to successful migrants. The main criterion for access to this profane community of legitimate receivers of welfare and social security is the individual's capacity to succeed and pay for social security.

Complementary to this core population emerges a population with a rather weak position in the welfare state equipped with basic provisions and security in exchange for social discipline and (passive) political loyalty. The composition of this population is also nationally heterogeneous and internally socially differentiated. It includes those migrants with limited access to social provisions. This depends to a large extent on their history of immigration and/or their legal status. This is not an argument that necessarily points to increasing exclusion processes. European welfare states emphasise "social integration" mediated by markets. They support individuals and expect them to strengthen their capacity to cope with these forms of integration.[16] But these states maintain basic welfare provisions and provide even those migrants with weak legal status with political, legal and social security.[17] The common criterion for belonging to one of the two types of population is legal residence.

b) This restructuration of the addressees of welfare state politics corresponds to a change of the political semantics used to address the welfare state beneficiaries. The modification of the classical promise of national welfare states as described by Thomas H. Marshall – to offer each individual full membership in the national community – is linked with a political redesign of semantics:

The *core population* of states is no longer built by an emphatic collectivity of "the people" but is increasingly defined by individuals and their competitiveness and success. They form a rather profane community expecting a policy that supports individual and collective efforts to handle social risks and to secure a decent standard of living. This community is not based on national closure and its borders are rather porous allowing for the entrance of those who meet the expectations of competitiveness and success. This part of the population forms the political centre and it is addressed and mobilised in political terms that signal support for competitiveness and success. The semantics of national solidarity has lost much of its meaning and value for this constituency. Its relation to the state however remains based on the classical exchange relation of loyalty for provision.

The complementary *residual population* is not built primarily by non-nationals but by those who fail to become part of the core population, i.e., those that do not succeed in gaining a secure standard of living and who have limited exit- or voice-capacities . This segment of the population plays a more marginal political role. It is addressed in the language of social problems-security, education, health, poverty – and the promise of resources for basic needs is still provided. The old programme of the national welfare state – equality in a substantial sense for members of the national community – is substituted by the message: the state still cares.

This promise is made to a varying degree to all those that reside legally on the state territory, but with different implications for transnational migrants or autochthonous parts of the population. The basic provisions of welfare states become an inbuilt part of the transnational social relations with which different migrant populations are involved. Autochthonous parts of the population opt instead for national closure in order to restrict competition for scarce welfare resources. The idea of national equality retains much of its meaning for the effort to justify closure and international inequality – with limited chances of political success that vary from country to country.

So far, it has been argued that international migration, transnationalism and multiculturalism make the historical improbability of the welfare state's political project visible. By this is meant the effort to distinguish the people and the population living on its territory, to supervise this difference and to limit it. It appears that not only has a rather large difference between the classical "core population", the community of the citizens, and the "residual population" of migrants of all kinds become apparent but that the difference itself has become precarious. It also seems that we must understand this as one element of an ongoing reconstruction process of the constituency of modern states changing the meaning of the difference between the "core population" and "the residual population". Migrants will be found in both parts of the population.

The differentiation of international migrants into various categories thus indicates two things: a shrinking capacity of control and provision of the welfare state as well as a general shift of the political inclusion of

[16]We can see that recent changes of the "integration programs" in countries like the Netherlands and Sweden (Entzinger 1998, Thränhardt 2000, Ring 1998) are linked with a shift of direction, pointing to individual migrants and expecting them to acquire skills to cope with the demands of labour markets. The recent implementation of an immigration law in Germany is linked with similar programmatic intentions to strengthen the capacity of "self-responsibility" of individual migrants through language and social training programs.

[17]These provisions of welfare states are in many cases an internal element of the building and reproduction of transnational social structures.

individuals into the nation state. Questions of political inclusion are successively separated from questions of emphatically "belonging" to a national community legitimating related claims of "integration into the society" by means of the welfare state. This certainly refers to the internal side of a loss of sovereignty of the nation state. In the future European welfare states may no longer be following the universalistic programme of full civic, political and social integration of all individuals subjected to its authority. The classical programme of the welfare state-"social integration", – as addressed to the migrants of the 1960s and 1970s- is no longer maintained in the same kind of way. Welfare states differentiate the treatment of individuals and allow for larger internal social inequalities. Parallel to this, the options for the social integration of migrants have been differentiated stressing individual responsibility and competitiveness. At the same time social integration has been disconnected from the former political teleology of making migrants national citizens in the thick meaning of this term.[18]

Cultural pluralisation processes, the emergence of transnational social structures and the development of ethnically differentiated populations have been interpreted by multiculturalism and transnationalism as challenges for national welfare states in Europe. They have, however, tended to overlook that the political semantics of the nation and corresponding programs of cultural homogenisation – certainly in each country to a different extent – have lost much of their relevance. This is linked to international migration, but is also connected with the more general expansion of the modern welfare state in Europe after the Second World War as well as the resulting social differentiation and individualisation (Bommes 1995, Joppke 2001). In this context collective forms of identity lost much of their social value and at the same time individual and collective forms of identity became, so to speak, politically deregulated.[19] This devaluation has now acquired a peculiar relevance in the recent processes of the restructuration of the population of welfare states described above. The administrative operations of welfare states and their internally differentiated forms of provision are both sober and profane. Access to provisions and services is made dependent on entitlements that individuals have to accumulate in their working careers. Claims on welfare are decreasingly based on the emphatic idea of solidarity between the members of a national community of citizens.

Seen from this perspective another meaning of the emergence and pluralisation of collective forms of identity becomes visible. This is that the relevance of the population for the state is reconstructed in political, legal, administrative and welfare terms. The political devaluation of the semantics of the national community and of the related welfare implications on the level of the state is, however, accompanied by the emergence of so-called new identities below or beyond the state and its borders. Such identities may be articulated ethnically, regionally or locally. It is left to the different parts of the population to create their own identities. This may be one of the meanings of those processes that have been discussed under the headings of "multicultural society" and "transnationalism". The emergence of ethnic identities is no longer structured by either orientation to or refusal of belonging to national welfare states. These identities organise loyalties and chances of access that are less centred on states. The ethnic, multicultural or transnational *milieus* signal and to some extent also allow the retreat of the state from the political project to integrate the population on its territory in a national mode. These *milieus* articulate the shrinking capacity of the welfare state. If multiculturalism claimed that different cultures and identities should be accepted in a "multicultural society" this seems to be the ironic mode of acceptance for this in European welfare states reconstructing the relation to their populations.

Conclusion

The formula "integration into society" has been identified as part of the political semantics of a national welfare state. This national welfare state was viewed in the post-war period as sovereign to the extent that the identification of society with the nation state could be socially persuasive. The debate about the "multicultural society" shared this identification of society and the nation state. The nation state was confronted with the assumption that the ongoing processes of cultural pluralisation would force a correction of the presupposition that national homogeneity could be maintained as a political objective. "Transnationalism" certainly reacts to the recent trends of migration but even here the concept of national societies remains central. An alternative concept of the modern society which conceptualises society as functionally differentiated world society (Luhmann 1997) allows us to interpret the transnational networks and organisations resulting from

[18] At first glance it may appear that recent efforts in Scandinavia or Great Britain point in an opposite direction since political demands have been made that migrants have to accept the values of "western" or "occidental culture". A closer look reveals however that usually no more is expected than the acceptance of so-called civic rules, i.e. an implication of the institutionalisation of social differentiation – which is very different from the classical assimilation programmes of European nation states constructing national communities. In that sense they have become "postnational" (s. also Therborn 1995).

[19] Looked at it from this perspective some of the political hopes put forward by multiculturalists that nation states will have to give up their classical program of cultural homogeneity appear somewhat belated.

migration as part of globalisation processes and as structural elements of the world society. The modern national welfare state – world-wide the established organisational form of politics in modern society – turns out to be an institution with specific conditions of functioning and specific limits on its interventionist capacity. Its shrinking capacity to provide social inclusion makes this diminished interventionist capacity apparent. The consequence of this however is not that the (post-) national welfare state has become dispensable. Despite its reduced capacity to provide social inclusion, the welfare state remains the central instance for the production of security and collectively binding decisions. Where the internal enforcement of the state and the political inclusion of the population fail, vital social preconditions for the functioning of the economy, law, science, health, education and the family are eroded. Yet, it appears that political inclusion and the provision of welfare can decreasingly be based on the construction of national communities of citizens. International migration will certainly remain one important element of the ongoing processes of the reconstruction of the relation between states and their populations in Europe.

References

Alber, Jens (2000) Sozialstaat und Arbeitsmarkt: Produzieren kontinentaleuropäische Wohlfahrtsstaaten typische Beschäftigungsmuster?-Gleichzeitig eine Abhandlung über einige Probleme komparativer statistischer Analyse. In: Leviathan 28 (4): 535–569

Bade, Klaus J. (ed) (1996a) Die multikulturelle Herausforderung: Menschen über Grenzen – Grenzen über Menschen. München

Bade, Klaus J (ed) (1996b) Migration – Ethnizität – Konflikt. Systemfragen und Fallstudien. Osnabrück

Bade, Klaus J/Bommes, Michael (1996) Migration-Ethnizität-Konflikt. Erkenntnisprobleme und Beschreibungsnotstände: eine Einführung, in: Bade, 11–40

Bank, Roland (1998) Europeanisation of the Reception of Asylum Seekers: the Opposite of Welfare State Politics. In: Bommes/Geddes (2000), 148–169

Bommes, Michael (1995) Migration and Ethnicity in the National Welfare State. In: Martiniello, M (ed) Migration, Citizenship and National Identities in the European Union. Aldershot, Avebury, 120–143

Bommes, Michael (1997) Von »Gastarbeitern« zu Einwanderern: Arbeitsmigration in Niedersachsen, in: Bade, Klaus J (Hg.): Fremde im Land: Zuwanderung und Eingliederung im Raum Niedersachsen seit dem Zweiten Weltkrieg. Osnabrück, 249–322

Bommes, Michael (1999) Migration und nationaler Wohlfahrtsstaat. Ein systemtheoretischer Entwurf. Wiesbaden

Bommes, Michael (2000) National Welfare State, Biography and the Impact of Migration – Labour Migrants, Ethnic Germans and the Re-Ascription of Welfare State Membership. In: Bommes/ Geddes, 90–108

Bommes, Michael (2001) Organisation, Inklusion und Verteilung. Soziale Ungleichheit in der funktional differenzierten Gesellschaft. In: Tacke, Veronika (Hg.): Organisation und moderne Gesellschaft. Wiesbaden, 236–258

Bommes, Michael (2002) Migration, Raum und Netzwerke. Über den Bedarf einer gesellschaftstheoretischen Einbettung der transnationalen Migrationsforschung. In: Oltmer, Jochen (ed) Migrationsforschung und Interkulturelle Studien. Osnabrück, 91–105

Bommes, Michael/Geddes, Andrew (eds) (2000) Welfare and Immigration: Challenging the Borders of the Welfare State. London

Bommes, Michael/Halfmann, Jost (eds) (1998) Migration in nationalen Wohlfahrtsstaaten. Theoretische und vergleichende Untersuchungen. Osnabrück

Bommes, Michael/Tacke, Veronika (2001) Arbeit als Inklusionsmedium moderner Organisationen. Eine differenzierungstheoretische Perspektive. In: Tacke, Veronika (Hg.): Organisation und moderne Gesellschaft. Wiesbaden, 61–83

Brubaker, Rogers (1992) Citizenship and Nationhood in France and Germany, Cambridge, Mass

Castles, Stephen (1996a) Immigration and Multiculturalism in Australia, in: Bade, 249–270

Castles, Stephen (1998) Einwanderung und Sozialpolitik in Australien, in: Bommes/Halfmann, 171–197

Eichenhofer, Eberhard (1996) Arbeitsbedingungen bei Entsendung von Arbeitnehmern, In: Zeitschrift für ausländisches und internationales Sozialrecht, 10: 55–82

Eichenhofer, Eberhard (1997) (ed) Social Security of Migrants in the European Union of Tomorrow. Osnabrück

Entzinger, Han (1998) Zu einem Modell der Inkorporation von Einwanderern: das Beispiel der Niederlande, In: Bommes/Halfmann, 105–122

Esping-Andersen, Gøsta (1990) The Three Worlds of Welfare Capitalism. Cambridge

Esping-Andersen, Gøsta (ed) (1996a) Welfare States in Transition. National Adaptations in Global Economies. London, Thousand Oaks, New Delhi

Faist, Thomas (2000) The Volume and Dynamics of International Migration and Transnational Social Spaces. Oxford

Favell, Adrian (2002) The 'Integration' of Immigrants in Western Europe: Contours and Constraints of a Research Paradigm, in: Bommes, Michael/ Morawska Ewa (eds): Reflections on Migration Research: Constructions, Omissions, and Promises of Interdisciplinarity. London (forthcoming)

Ferrera, Maurizio (1996) The "Southern Model" of Welfare in Social Europe, In: Journal of European Social Policy 6: 17–37

Fligstein, Neil (2000) Verursacht Globalisierung die Krise des Wohlfahrtsstaates? In: Berliner Journal für Soziologie 10(3): 349–378

Ganßmann, Heiner/Haas, Martin (1999) Arbeitsmärkte im Vergleich. Rigidität und Flexibilität auf den Arbeitsmärkten der USA, Japans und der Bundesrepublik Deutschland. Marburg: Schüren

Geißler, Rainer (1996) Die Sozialstruktur Deutschlands. Opladen, (2. edition)

Gellner, Ernest (1988) Nations and Nationalism. Oxford

Glick Schiller, Nina/Basch, Linda/Blanc-Szanton, Christina (1995) From Immigrant to Transmigrant: Theorizing Transnational Migration, In: Anthropological Quarterly, 68: S48–63

Goodin, Robert E/Headey, Bruce/Muffels, Ruud/Dirven, Henrik-Jan (1999) The Real Worlds of Welfare Capitalism, Cambridge
Hammar, Tomas (1989) State, Nation and Dual Citizenship, in: Brubaker, Rogers (ed), Immigration and the Politics of Citizenship in Europe and North America. Lanham/London, 81–95
Hammar, Tomas (1994) Democracy and the Nation State. Aliens, Denizens and Citizens in a World of International Migration. Aldershot
Hannerz, Ulf (1996) Transnational Connections: Culture, People, Places, New York
Hirschman, Albert O (1970) Exit, Voice and Loyalty. Responses to Decline in Firms, Organizations and States. Cambridge, Mass
Hönekopp, Elmar (1997) The New Labour Migration as an Instrument of German Foreign Policy, In: Münz, Reiner/ Weiner, Myron (Hg.): Migrants, Refugees, and Foreign Policy. Providence, Oxford, 165–181
Hoffmann-Nowotny, Hans-Joachim (1996) Soziologische Aspekte der Multikulturalität, In: Bade, 103–126
Hollifield, James F. (1992): Immigrants, Markets and States: The Political Economy of Postwar Europe. Cambridge, Mass
Joppke, Christian (1997): Asylum and State Sovereignty. A Comparison of the United States, Germany, and Britain, in: Comparative Political Studies, 30. Jg., S259–298
Joppke, Christian (2001) Multicultural Citizenship: A Critique, in: Archives européennes de sociologie, 42: 431–447
Koselleck, Reinhart (1992) Volk, Nation. Einleitung, in: Brunner, Otto/Conze, Werner/Koselleck, Reinhart (ed) Geschichtliche Grundbegriffe. Historisches Lexikon zur politisch-sozialen Sprache in Deutschland. Bd. 7., Stuttgart, 142–149
Kyle, David (1998) Transparent Economies and Invisible Workers: Human Smuggling under Global Economic Liberalism. Europäisches Hochschulinstitut, Conference Paper MIG/60, Florenz
Kymlicka, Will (1995) Multicultural Citizenship. A Liberal Theory of Minority Rights. Oxford
Leggewie, Claus (1990) Multi Kulti-Spielregeln für die Vielvölkerrepublik. Berlin
Levitt, Peggy (1998a) Forms of Transnational Community and their Implications for Immigrant Incorporation: Preliminary Findings. Europäisches Hochschulinstitut, Conference Paper MIG/35, Florenz
Levitt, Peggy (1998b) Local Level Global Religion: The Case of U.S.-Dominican Migration, In: Journal for the Scientific Study of Relegion, 37: S74–89
Luhmann, Niklas (1989) Individuum, Individualität, Individualismus. In: Niklas Luhmann: Gesellschaftsstruktur und Semantik. Studien zur Wissenssoziologie der modernen Gesellschaft. 3: 149–258
Luhmann, Niklas (1997): Die Gesellschaft der Gesellschaft. Frankfurt a.M
Marshall, Thomas H. (1950) Citizenship and Social Class – and Other Essays. Cambridge
Morawska, Ewa (1998) Structuring Migration in a Historical Perspective: The Case of Travelling East Europeans. Europäisches Hochschulinstitut, Working Papers EUF No. 98/ 3, Florenz
Müller-Mahn, Detlef (2000) Ein ägyptisches Dorf in Paris. Eine empirische Studie zur Süd-Nord-Migration am Beispiel ägyptischer "Sans-papiers" in Frankreich, in: IMIS-Beiträge, H. 15: 79–110
Nassehi, Arnim (1997) Das stahlharte Gehäuse der Zugehörigkeit. Unschärfen im Diskurs um die "multikulturelle Gesellschaft, In: Nassehie, Arnim (ed) Nation, Ethnie, Minderheit. Beiträge zur Aktualität ethnischer Konflikte. Köln, Weimar, Wien, 177–208
Opitz, Peter J. (ed) (1997) Der globale Marsch. Flucht und Migration als Weltproblem. München
Portes, Alejandro (ed) (1995): The Economic Sociology of Immigration: Essays on Networks, Ethnicity and Entrepreneurship. New York
Portes, Alejandro 1996: Transnational Communities: Their Emergence and Significance in the Contemporary World-System, In: Korzniewidcz, Roberto Patricio and Smith, William C. (eds.): Latin America in the World Economy. Westport, Connecticut: Greenwood
Pries, Ludger (ed) (1997) Transnationale Migration. Soziale Welt, Sonderband 12. Baden Baden
Pries, Ludger (2000) Transnationalisierung der Migrationsforschung und Entnationalisierung der Migrationspolitik. Das Entstehen transnationaler Sozialräume durch Arbeitswanderung am Beispiel Mexiko – USA, in: Michael Bommes (Hg.), Transnationalismus und Kulturvergleich. IMIS-Beiträge 15: 55–77
Pries, Ludger 2001: The Disruption of Social and Geographic Space. Mexican-US Migration and the Emergence of Transnational Social Spaces, In: International Sociology, 16: S55–74
Radtke, Frank-Olaf (1990) Multikulti: Das Gesellschaftsdesign der 90er Jahre?, in: Informationsdienst zur Ausländerarbeit, H.4: 27–34
Ring, Hans (1998) Einwanderungspolitik im schwedischen Wohlfahrtsstaat, In: Bommes/Halfmann, 239–249
Santel, Bernhard (1995): Migration in und nach Europa. Erfahrungen, Strukturen, Politik. Opladen
Santel, Bernhard/Hollifield, James F (1998) Erfolgreiche Integrationsmodelle? Zur wirtschaftlichen Situation von Einwanderern in Deutschland und den USA, in: Bommes/Halfmann, 123–145
Santel, Bernhard/ Hunger, Uwe (1997) Gespaltener Sozialstaat, gespaltener Arbeitsmarkt. Die Etablierung postwohlfahrtsstaatlicher Einwanderungspolitiken in Deutschland und den Vereinigten Staaten, in: Soziale Welt, 48: 379–396
Soysal, Yasemin (1994) Limits of Citizenship. Migrants and Postnational Membership in Europe. Chicago, Illinois
Stichweh, Rudolf (1998) Migration, nationale Wohlfahrtsstaaten und die Entstehung der Weltgesellschaft. In: Bommes/Halfmann, 49–61
Swaan, Abram de (1988) In Care of the State. Cambridge
Taylor, Charles (1993) Multikulturalismus und die Politik der Anerkennung. Frankfurt a.M
Therborn, Göran (1995) European Modernity and Beyond: the Trajectory of European Societies, 1945–2000. London etc.
Thränhardt, Dietrich (2000) Multiculturalism and Integration in Germany and the Netherlands, in: Bulletin of the Royal Institute for Inter-Faith Studies 2(1): 43–69
Treibel, Annette (1990) Migration in modernen Gesellschaften. Weinheim/ München
Vertovec, Steven/Cohen, Robin (Hrsg.) (1999) Migration, Diasporas and Transnationalism. Cheltenham/ Northhampton, MA

AUTHOR TITLE INDEX

Progr Colloid Polym Sci (2004) 127: 162

KEY WORD INDEX

MIX
Papier aus verantwortungsvollen Quellen
Paper from responsible sources
FSC® C105338

If you have any concerns about our products,
you can contact us on
ProductSafety@springernature.com

In case Publisher is established outside the EU,
the EU authorized representative is:
Springer Nature Customer Service Center GmbH
Europaplatz 3, 69115 Heidelberg, Germany

Printed by Libri Plureos GmbH
in Hamburg, Germany